Human Factors in Driving, Vehicle Seating, and Rear Vision

SP-1358

GLOBAL MOBILITY DATABASE
All SAE papers, standards, and selected books are abstracted and indexed in the Global Mobility Database

Published by:
Society of Automotive Engineers, Inc.
400 Commonwealth Drive
Warrendale, PA 15096-0001
USA
Phone: (724) 776-4841
Fax: (724) 776-5760
February 1998

Permission to photocopy for internal or personal use of specific clients, is granted by SAE for libraries and other users registered with the Copyright Clearance Center (CCC), provided that the base fee of $7.00 per article is paid directly to CCC, 222 Rosewood Drive, Danvers, MA 01923. Special requests should be addressed to the SAE Publications Group. 0-7680-0178-1/98$7.00.

Any part of this publication authored solely by one or more U.S. Government employees in the course of their employment is considered to be in the public domain, and is not subject to this copyright.

No part of this publication may be reproduced in any form, in an electronic retrieval system or otherwise, without the prior written permission of the publisher.

ISBN 0-7680-0178-1
SAE/SP-98/1358
Library of Congress Catalog Card Number: 97-81275
Copyright © 1998 Society of Automotive Engineers, Inc.

Positions and opinions advanced in this paper are those of the author(s) and not necessarily those of SAE. The author is solely responsible for the content of the paper. A process is available by which the discussions will be printed with the paper if is is published in SAE Transactions. For permission to publish this paper in full or in part, contact the SAE Publications Group.

Persons wishing to submit papers to be considered for presentation or publication through SAE should send the manuscript or a 300 word abstract to: Secretary, Engineering Meetings Board, SAE.

Printed in USA

PREFACE

This Special Publication, <u>Human Factors Issues in Driving, Vehicle Seating, and Rear Vision</u> (SP-1358), includes papers from three sessions of the 1998 SAE International Congress and Exposition: "Human Factors Issues in Vehicle Seating," "Human Factors in Driving," and "Rear Vision."

This publication continues to feature our effort in informing seat and automotive interior designers and engineers of the recent activities in seating comfort issues. The papers have a strong emphasis on industrial practice in seating comfort enhancement. In line with the general approach to seating and interior comfort presented in previous special publications, papers in this publication cover broad areas from seating accommodation, seat component characteristics, static and dynamic seat comfort, and environmental comfort. In particular, a number of presentations address dynamic comfort issues in primary supporting materials, namely, cushion foam, trim cover, and lumbar support mechanism. Topics on seat noise absorption and variable temperature seats are interesting additions. Furthermore, a few papers from Korea reflect the differences in comfort investigation between Asia and North America.

The field of rear vision broadly encompasses systems which allow drivers to gain visual information about the scene behind and adjacent to their vehicle. Such systems can include conventional mirrors, optic mirror devices, and cameras. The majority of current research in rear vision involves non-planar rearview mirrors and their effects on drivers. Specific studies include distance perception and driver acceptance of various nonplanar rearview mirrors, as well as binocular disparity in aspherical mirrors. An additional study presents an image analysis method for measuring curvature of non-planar mirrors.

Another topic of research interest is optic mirror devices designed for automotive exterior mirror applications. Such systems use optical lens systems to minimize the protrusion of conventional rearview mirrors from the vehicle body. These systems are intended to decrease aerodynamic drag and wind noise, and to place the mirror closer to the driver's line of forward vision.

Research has also focused on the trend toward adding features to automotive mirrors. This highlights the fact that rearview mirrors pose an ideal location for various functions in addition to rear vision. Such features include lighting, turn signals, information centers, security systems, and car area networks. A final area of research involves school bus mirror systems and their effect on driver visibility of objects around the vehicle.

Today's rear vision research reflects an exciting variety of interests within the field which culminate in achieving the common goals of improved safety, convenience, and performance for tomorrow's vehicles.

This Special Publication also contains six papers involving human factors in driving. Three different models of driver behavior or positioning are included. Renski discusses a vehicle-driver-environment model with data collected from roads and a drum type test stand driving simulator. His model can be used to better match vehicle dynamics properties to driver characteristics. A driver eye position model developed by the University of Michigan Transportation Research Institute provides additional data to the current SAE eyellipse position. This model is beneficial for occupant packaging in vehicle development. Majjad et al. describe a hybrid driver model developed at the University of Karlsruhe, Germany to model driver behavior. They use a queuing network and GPC controllers to simulate different driver types.

A driver's internal state is measured through finger blood pulse in a paper by Hashimoto et al. The paper identifies the ability to use finger blood pulse to monitor the arousal level and emotional state of a driver, which has useful implications for improving driving safety by reducing accidents caused by fatigue and inattention.

Potential improvements in driving safety are also addressed in Mr. Morita's paper on dusk-related traffic accidents. This paper contains statistical analyses to determine if accidents are more likely to occur during the dusk viewing period.

Improvements to the interior driving environment may be possible based on the research of Kosaka et al., who examine electronic sounds in automobiles. This paper addresses how people respond to different sounds and sound qualities. The paper links physical characteristics of sounds to a potential customer's perception and feelings. The papers on human factors in driving should be particularly useful for automotive professionals involved in interior vehicle design.

We hope this publication will bring you more insight into the areas of seating comfort, rear vision, and human factors in driving. As both technology and customer expectations advance, a continuous refreshment of mind with creative problem-solving skills is essential. This publication will certainly help you.

Wenqi Shen
Delphi Interior and Lighting Systems

Roger Veldman
Donnelly Corp.

Alicia M. Vertiz
Delphi Interior and Lighting Systems

Gretchen Zobel
Ford Motor Co.

Session Organizers

TABLE OF CONTENTS

980011   **The Driver Model and Identification of Its Parameters** 1
    Andrzej Reński
        Warsaw University of Technology
        Institute of Vehicles

980012   **Development of an Improved Driver Eye Position Model** 9
    Miriam A. Manary, Carol A. C. Flannagan,
    Matthew P. Reed, and Lawrence W. Schneider
        University of Michigan Transportation Research Institute

980013   **Electronic Sound in Automobile and Sound Feeling** 17
    Hiroaki Kosaka, Takeshi Shigematsu and Kajiro Watanabe
        College of Engineering, Hosei University

980015   **Situation for Occurrence of Traffic Accidents at Dusk as Seen from the Standpoint of the Viewing Environment** 23
    Kazumoto Morita
        Traffic Safety and Nuisance Research Institute

980016   **Data Processing Method of Finger Blood Pulse for Estimating Human Internal States** 33
    Hiroshi Hashimoto and Tsuyoshi Katayama
        Japan Automobile Research Institute

980017   **Design of a Hybrid Driver Model** 37
    R. Majjad, U. Kiencke, and H. Körner
        Institute for Industrial Information Systems, University of Karlsruhe

980651   **An Improved Seating Accommodation Model with Application to Different User Populations** 43
    Carol A. C. Flannagan, Miriam A. Manary,
    Lawrence W. Schneider, and Matthew P. Reed
        Transportation Research Institute, University of Michigan

980653   **Seating Physical Characteristics and Subjective Comfort: Design Considerations** 51
    Se-Jin Park
        Ergonomics Research Group
        Korea Research Institute of Standards and Science
    Young-Shin Lee and Yoon-Eui Nahm
        Department of Mechanical Design Engineering
        Chung-Nam National University
    Jung-Woo Lee and Jin-Sun Kim
        Department of Industrial Engineering
        Jeon-Ju University

980654	**Evaluation of Comfort Properties with Covering Textiles of Car Seats**	63
	Yoon-Sook Hur and Se-Jin Park	
	Korean Research Institute of Standards and Science	
980655	**A Comparison Test of Transmissibility Response from Human Occupant and Anthropodynamic Dummy**	71
	Yi Gu	
	Lear Corporation	
980656	**Automotive Seating Comfort: Investigating the Polyurethane Foam Contribution-Phase 1**	75
	G. R. Blair, R. So, A. Milivojevich, and J. D. van Heumen	
	The Woodbridge Group	
980657	**The Influence of Polyurethane Foam Dynamics on the Vibration Isolation Character of Full Foam Seats**	97
	Mark R. Kinkelaar and Brian L. Neal	
	ARCO Chemical Company	
	Guy Crocco	
	ARCO Chemical Products Europe, Inc.	
980658	**The Effects of Regional Compliance and Instantaneous Stiffness on Seat Back Comfort**	105
	Eric C. Hughes, Wenqi Shen and Alicia Vértiz	
	Delphi Interior and Lighting Systems	
980659	**Noise Absorption of Automotive Seats**	117
	Pusheng Chen and Gordon Ebbitt	
	Lear Corporation	
980660	**Dynamic Ride Quality Investigation for Passenger Car**	123
	Se-Jin Park and Wan-Sup Cheung	
	Korea Research Institute of Standards and Science	
	Young-Gun Cho and Yong-San Yoon	
	Korea Advanced Institute of Science and Technology	
980661	**Stirling Air Conditioned Variable Temperature Seat (SVTS) & Comparison with Thermoelectric Air Conditioned Variable Temperature Seat (VTS)**	131
	Steve Feher	
	Feher Research Co.	
980916	**A Field Study of Distance Perception with Large-Radius Convex Rearview Mirrors**	139
	Michael J. Flannagan, Michael Sivak, Shinichi Kojima and Eric C. Traube	
	The University of Michigan	

980917	**Measuring Curvature of Mirrors using Image Analysis** 147	
	Dorothy J. Helder	
	Donnelly Corporation	

980918	**Binocular Disparity in Aspherical Mirrors** 153	
	Stephen M. O'Day	
	Ford Motor Company	

980919	**Acceptance of Nonplanar Rearview Mirrors by U.S. Drivers** 165	
	Carol A. C. Flannagan and Michael J. Flannagan	
	The University of Michigan	

980920	**The 'Double Objective' Milner Prismatic Exterior Rear View Mirror** 171	
	Peter J. Milner	
	de Montfort, UK	
	Richard E. Berg	
	de Montfort, US	

980921	**An Advanced Optic Rear Vision Device for Motor Vehicles** 181	
	S. Li	
	Su Li Patent	
	S. LukSang	
	Su's Autoptic Limited	

980922	**Added Feature Automotive Mirrors** 191	
	Niall R. Lynam	
	Donnelly Corporation	

980923	**School Bus Visibility: Driver's Field of View and Performance of Mirror Systems on a Conventional Long-nosed School Bus** 201	
	Paul Lemay and Alex Vincent	
	Transport Canada	

980011

The Driver Model and Identification of Its Parameters

Andrzej Reński
Warsaw University of Technology
Institute of Vehicles

Copyright © 1988 Society of Automotive Engineers, Inc.

ABSTRACT

The interaction between the driver and the vehicle plays a significant role in all problems of active safety. Various schemes for vehicle-driver-environment systems have been considered in many papers.

Elaborated driver model is characterised by three parameters: Aim Point Distance, Response Delay and Steering Angle Gain.

To reduce a danger and research cost for identification of the driver model parameters, instant of the road tests, a drum type test stand was used. The stand allows to simulate a curved ride of a car, for instance: the double lane change manoeuvre (ISO/TR 3888). The ride is controlled by a driver according to the changing road pictures presented simultaneously on the computer screen.

Several drivers took part in the tests. The identification of the driver parameters consisted in comparison of the test results (diagrams of lateral position, steering angle, yaw velocity and lateral acceleration versus time) with the results of computer calculations obtained for the vehicle-driver-environment model with different driver data sets.

INTRODUCTION

Active safety problem is a part of vehicle dynamic study. Computer simulation of the curved ride and directional control of the car plays a significant role in this area. Analysis based on the advanced dynamic vehicle model only is not satisfactory in many cases. A control response of the driver model, co-ordinated with visual stimuli and related do the dynamic vehicle control model can potentially give possibility the correct simulation of the car motion. Traffic safety problems involving interaction between the driver and the vehicle should also regard highway environment. Therefore the system of these three elements and the vehicle-driver-environment model should be taken into consideration.

CONCEPTS OF VEHICLE-DRIVER-ENVIRONMENT SYSTEM

Various schemes for vehicle-driver systems have been considered and the driver-vehicle model structure has been developed and improved over the last years in several papers (e.g. more elaborate list of publications in [1]). Many of these systems consist three components: the first represents the vehicle (road-vehicle kinematics and vehicle dynamics), the second - the driver (his perception and response), and the third - the environment and the impact of different disturbances.

Interdependence of these three components are shown in a form of a block-diagram in Fig. 1. In the "driver" block the following driver's functions are shown as separate elements: "perception and valuation" and "control". The "perception and valuation" element represents the driver's function which consists in receiving any information about the car's movement (e.g. lateral displacement, yaw angle, vibration) and comparing them with assumed ride parameters (e.g. desired path. desired speed). The driver takes also into consideration various environment influences such as weather, unexpected events, and last but not least, the traffic regulations. Taking all these into consideration the driver (in his "control" element) corrects the car motion elaborating signals for control system of the car (steering wheel angle, accelerator or brake pedal position). Thus the car control system influences the car motion changing its speed and/or yaw velocity. On the other hand the car as a dynamic system (the

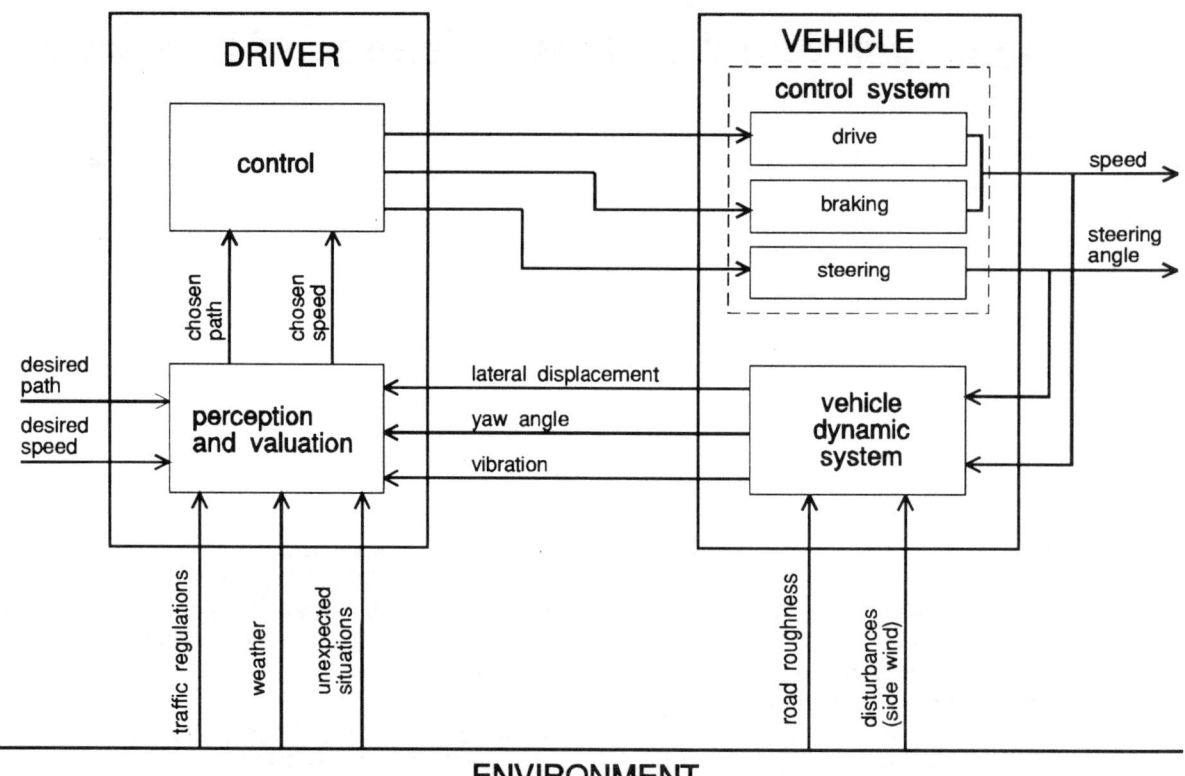

Fig. 1. Block diagram of the vehicle-driver-environment system

"vehicle dynamics" element in the diagram) is affected by such environment disturbances as road roughness, side wind and others. The result is a new vehicle path (described by the lateral displacement and the yaw angle), which usually is different from the desired path. In the same way the resulting speed differs from the desired speed.

From the block diagram in Fig. 1 a number of feed back loops can be isolated. One of the most important is a feed back loop for directional control, which is the subject of this paper. The driver-vehicle system applied here consists of the driver model and a vehicle model. The vehicle component generates dynamic output based on the drivers steering actions. Mathematical description of the vehicle is relatively easy. In this case a simple bicycle vehicle model can be here although there is no restriction for using more complicated non-linear models.

Much more complicated is elaboration of the driver model. In order to describe the function of the driver mathematically, the driver model should be reduced to several conceptually simple parameters for closed loop compensatory vehicle control.

For the study of steerability and directional stability of cars different driver's activity schemes are used:

- A driver holds a steering wheel in a constant position (fixed control). Applied in the study of a steady state turning of the car or an influence of disturbances (e.g. side wind) on the car's ride.
- A driver turns a steering wheel in a desired way (open loop control). Applied in the study of transient responses of the car on a step sinusoidal or random steering input.
- A driver controls a ride of the car using the steering wheel to keep the vehicle as precisely as possible on the desire path (closed loop control). Applied in the study of the whole vehicle-driver system. E.g. tests of the double lane change manoeuvre (ISO-TR 3888 [2]) and the wind gust manoeuvre.

In the two first cases (fixed control and open loop) a car model is adequate for computer simulation and only car data are to be delivered. However, for computer simulation of the close loop manoeuvres the car model should be extended by a driver model which samples the car position in relation to desired path and realises a control with a feedback of position error.

The method used for dynamic response analysis of the driver-vehicle model depends on a type of a model. For linear model, frequency domain

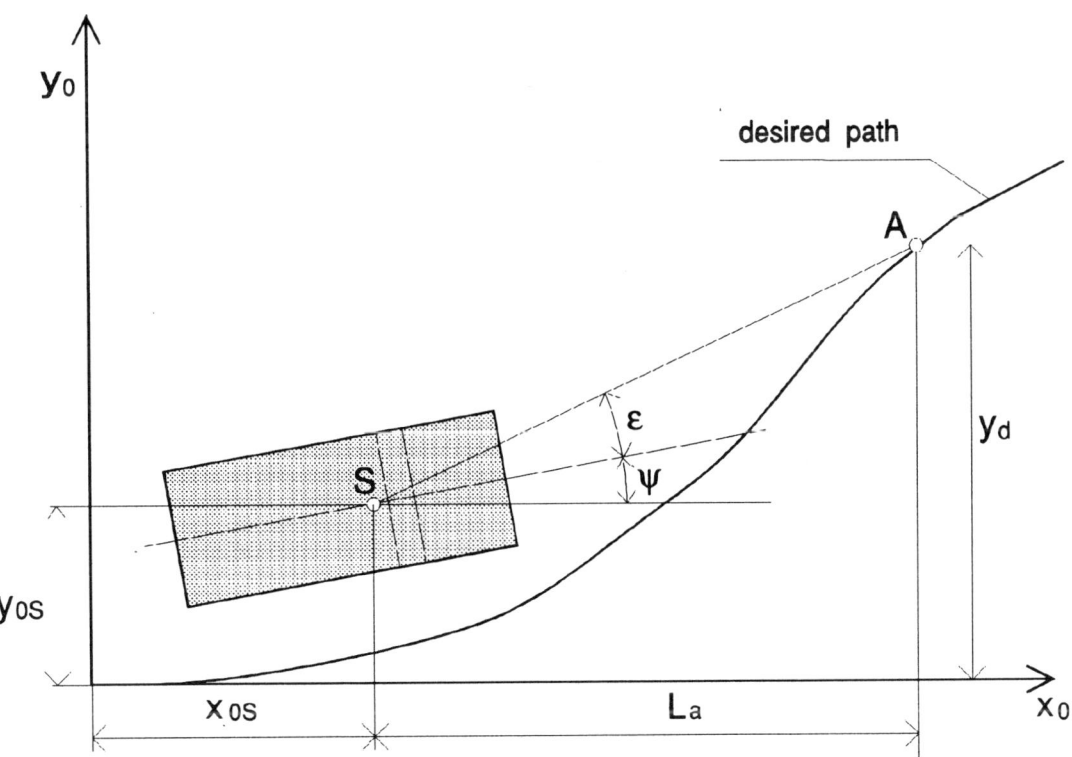

Fig. 2. Driver's steering control law used in the driver model

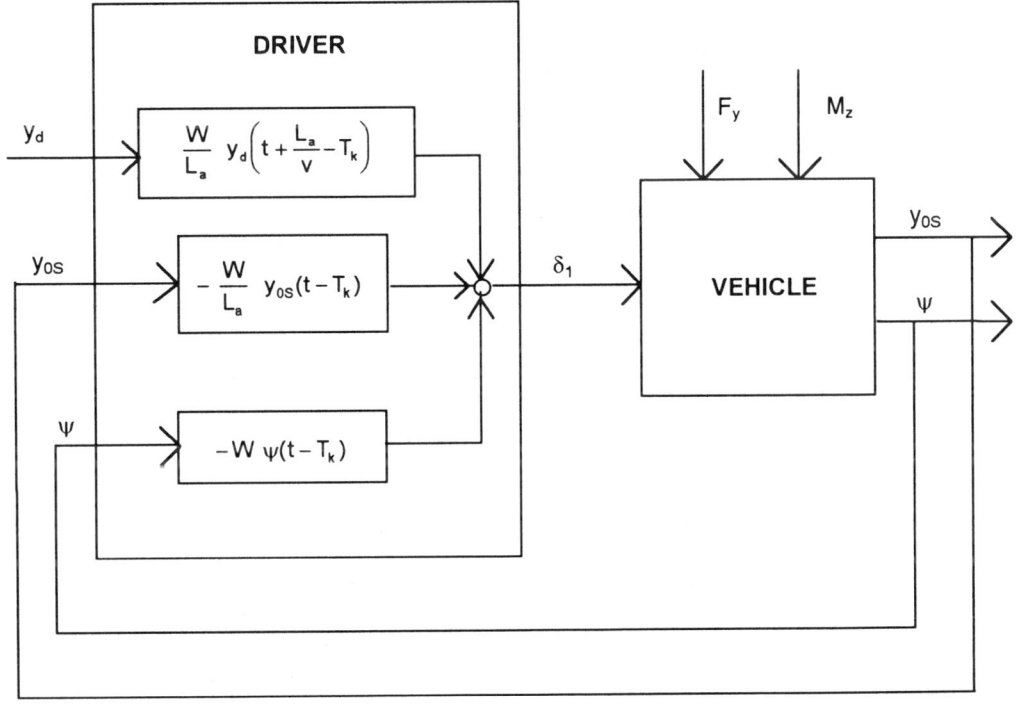

Fig. 3. Block diagram of the vehicle-driver model.

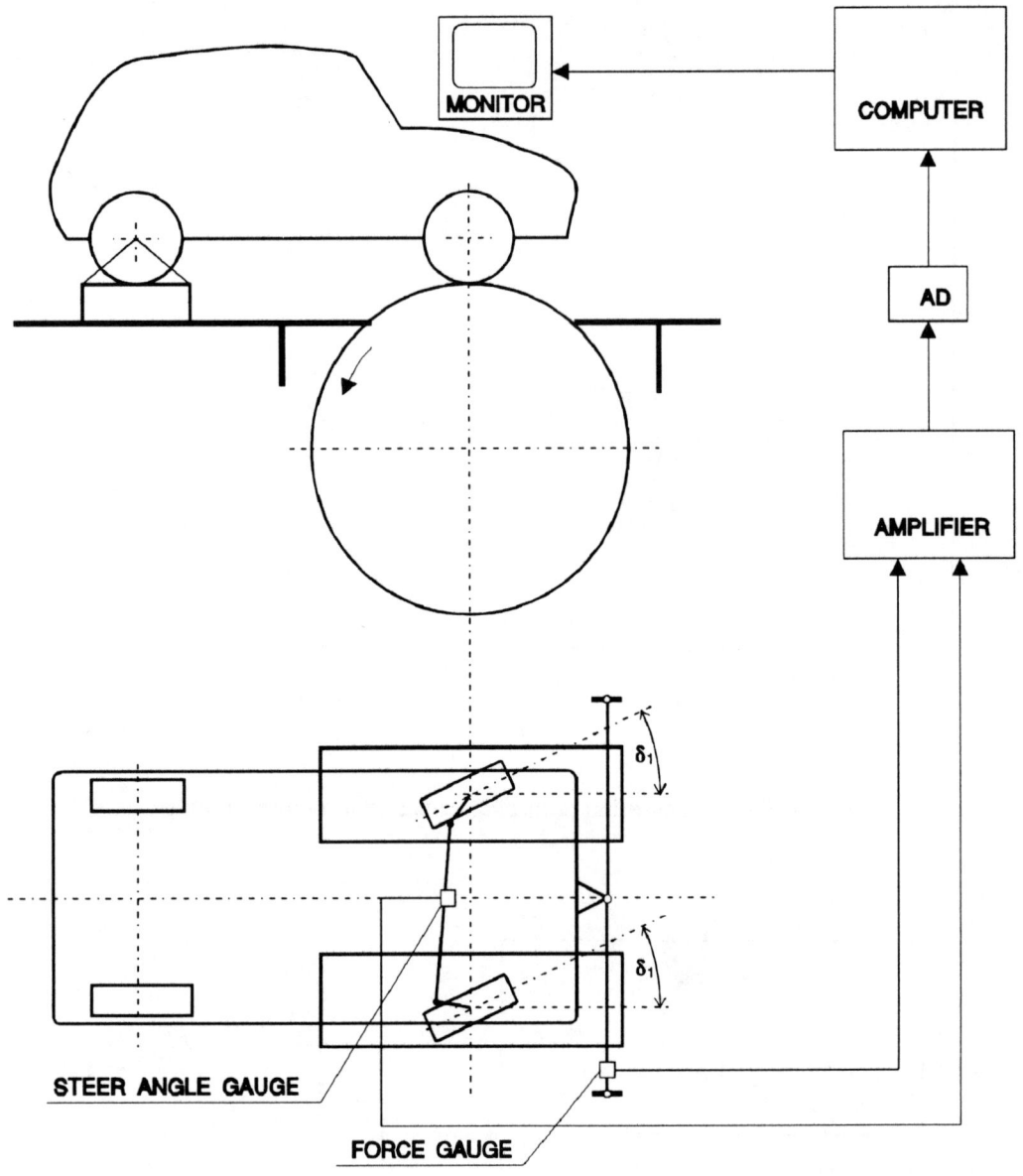

Fig. 4. Drum type test stand for simulation of the curved ride of the car

procedures can be used to study dynamic modes, stability conditions and transient response to various forcing functions like road disturbances, wind excitations, path profiles, obstacle avoidance. Time domain techniques are also useful and are necessary where non-linearities (such as e.g. tire characteristics under high lateral acceleration) are significant.

DRIVER MODEL

The aim of the work was to elaborate the driver model, which could be described by a possibly small number of parameters. It was assumed that for directional control the most important is visual signal. Thus the driver's steering control law, used in this model, is limited to visual feed back loop and is defined as follows (see Fig. 2): A driver observes an

Aim Point A which is situated on the desired path at a distance L_a down the road - Aim Point Distance. The driver sees this point at an angle ε to longitudinal axle of the car. This angle is equal to

$$\varepsilon(t) = \frac{y_d(x_{0S} + L_a) - y_{0S}(x_{0S})}{L_a} - \psi(x_{0S}) \quad (1)$$

where:
$x_{0S} = v\,t$ - longitudinal position, a way covered by the car down the road,
v - constant speed of the car,
y_{0S} - lateral position of the car,
y_d - desired path deviation from x_0-axis,
ψ - heading angle.
Steering angle δ_1 controlled by the driver is proportional to the ε-angle (Steering Angle Gain: W). The driver Response Delay between position perception and steering response is T_k. Then

$$\delta_1(t) = W\,\varepsilon(t - T_k) \quad (2)$$

Finally

$$\delta_1(t) = \frac{W}{L_a} y_d\!\left(t + \frac{L_a}{v} - T_k\right) - \frac{W}{L_a} y_{0S}(t - T_k) - W\,\psi(t - T_k) \quad (3)$$

Thus the driver model is characterised by three parameters: Aim Point Distance L_a, Response Delay T_k and Steering Angle Gain W. The block diagram of the model is shown in the Fig. 3.

TEST STAND

To reduce a danger and study costs for identification of the driver model parameters, instead of the road tests, a drum type test stand [3] was used (Fig. 4). The stand simulates a curved ride of a car taking into consideration the forces acting between tyres and road surface and in a realistic way the activity of the driver during the test manoeuvre, such as the double lane change manoeuvre.

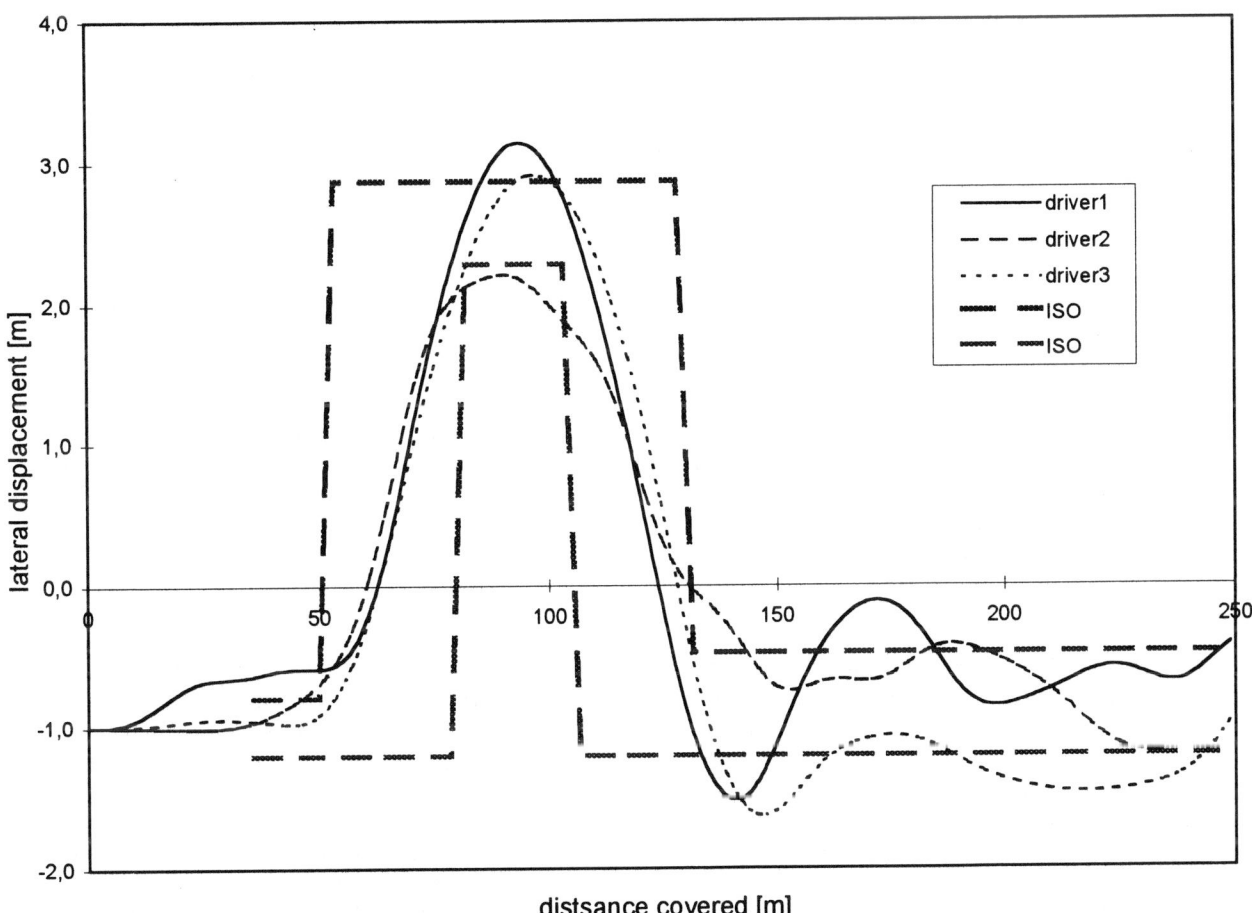

Fig. 5. Diagram of the lateral displacement obtained on the test stand by 3 drivers

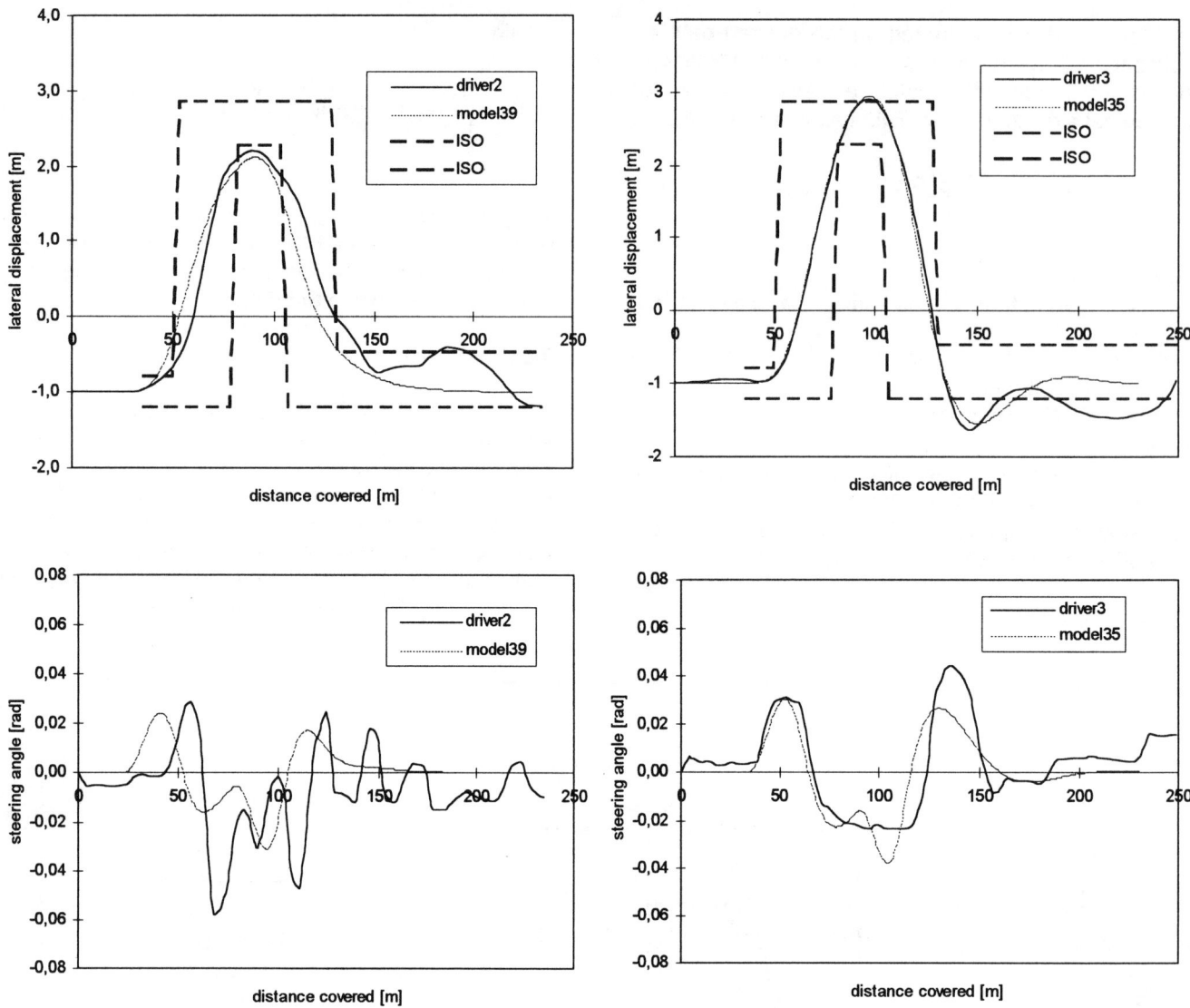

Fig. 6. Comparison of the diagrams of the lateral displacement and the steering angle obtained on the stand („driver2" and „driver3") and as a result of computer calculation for driver models with appropriately chosen parameters.
Model parameters: „model39" - L_a = 18m, T_k = 0,3s, W = 0,3;
„model35" - L_a = 30m, T_k = 0,4s, W = 0,3;

Front wheels of the medium size saloon car are positioned on the drums. The car is fastened to the stand by the ties. In one of the side tie a force gauge is mounted. The wheels rolling on the drums produce a side force, which is measured by this gauge. Since on the stand the slip angles of the front tyres are equal to steering angles, in order to obtain the same side forces as on the road the tyres cornering stiffness should be properly changed by increase or decrease of air pressure.

The car movement along the road is simulated in real time by a computer using the car model. The measured side force is an input signal for the calculation. As an output the car position in relation to the road (lateral displacement and speed, heading angle and yaw velocity) is obtained. Depending on this position the computer produces on the screen simultaneously changing pictures of the road in front of the car. Parameters describing the car's ride such as lateral displacement, steering angle, yaw velocity, lateral acceleration etc. are recorded during the test.

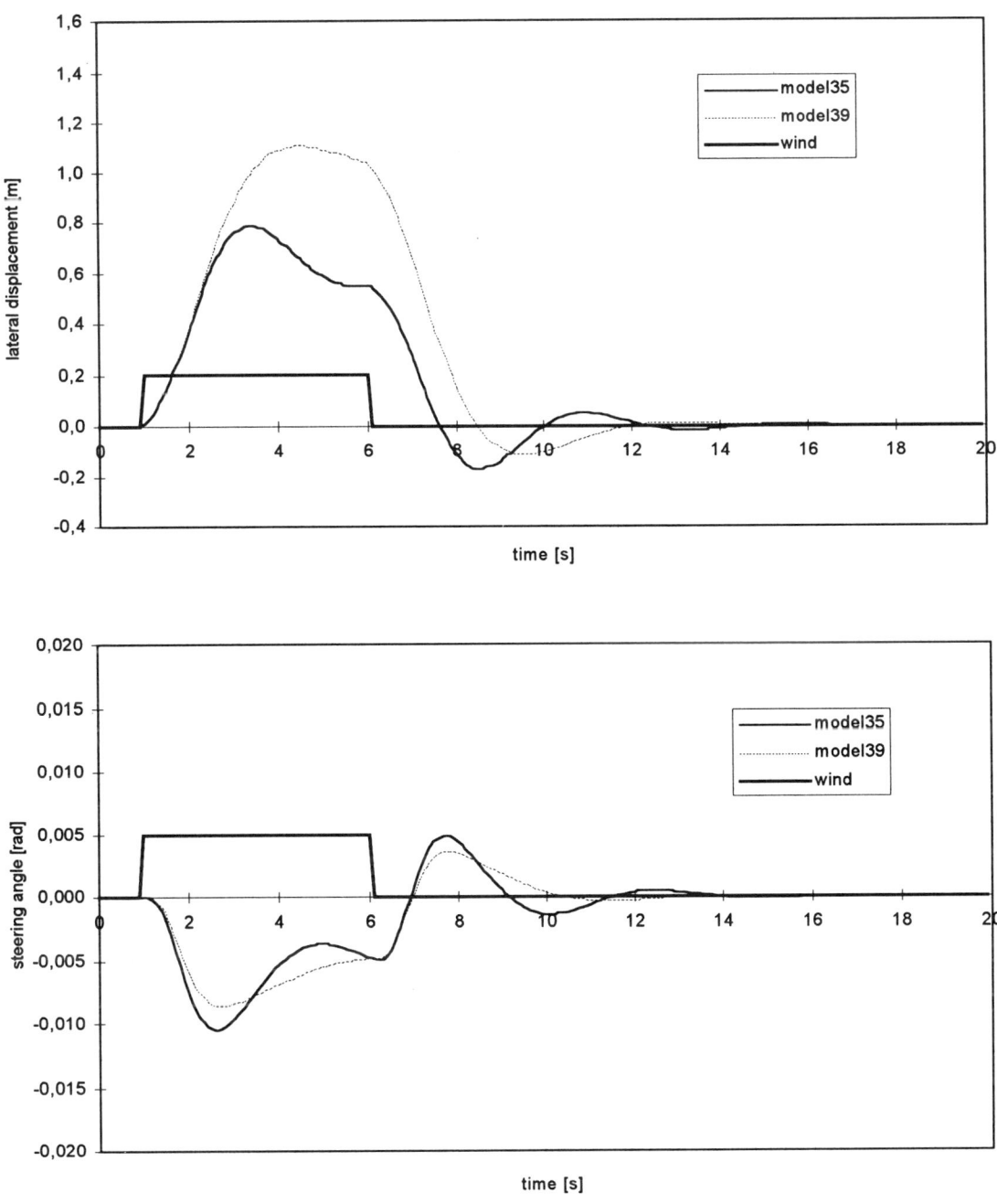

Fig. 7. Comparison of two driver models („model35" and „model39") reaction to side wind.

The task of the driver is to control by the steering wheel the ride of the car according to the changing road pictures presented on the monitor placed in front of the windscreen. Shown in Fig. 5 are the examples of diagrams of lateral displacement versus the way covered by the car, realised by 3 drivers in the dual line manoeuvre The curves are compared with the test tolerance borders (track width minus car width) according to ISO-TR 3888.

IDENTIFICATION OF THE DRIVER MODEL PARAMETERS

The test results were used for the identification of the driver parameters. The identification consisted in a comparison of the test results (diagrams of lateral position, steering angle, yaw velocity and lateral acceleration versus x-distance) with the results of computer calculations obtained for the vehicle-driver-environment model. The dual line manoeuvre was carried both on the stand and in the computer simulation. The same mathematical car model was used for the computer calculations and for the stand simulation. A number of drivers took part in the investigations.

CONCLUSION

A complete identification of the driver model requires a lot of road test carried with a considerable number of drivers of various age, sex and skills. The application of the drum type stand instead of road tests makes it possible to test a large number of drivers at relatively small costs, although all differences between both methods should be taken into consideration.

The shapes of curves from the computer simulation and from the stand test are similar, therefore presented driver model, though relatively simple, can be used in computer simulations.

On the basis of many calculations it can be concluded that for the simulation of different manoeuvres different driver parameters should be used.

The elaborated driver model can be used in the studies of vehicle handling and stability when closed loop procedures are needed. (As an example diagrams of reaction on side wind of 2 different drivers are shown in Fig. 7.) The model can be used also for studies of influence of different car parameters on stability of driver-vehicle system and in this way for better matching of car dynamic properties to driver characteristics.

REFERENCES

[1] Guo K., Guan H.: Modelling of Driver/Vehicle Directional Control System. *Vehicle System Dynamics* 22(1993)

[2] ISO-TR 3888. Road Vehicles - Test Procedure for a Severe Lane - Change Manoeuvre. 1975

[3] Reński A., Pokorski J., Lozia Z., Stegienka I.: Simulation of the curved ride of the car on the drum type stand. *Warsaw University of Technology. Institute of Vehicle Papers* 2(14)/1995 (in the Polish language)

980012

Development of an Improved Driver Eye Position Model

Miriam A. Manary, Carol A. C. Flannagan,
Matthew P. Reed, and Lawrence W. Schneider
University of Michigan Transportation Research Institute

Copyright © 1988 Society of Automotive Engineers, Inc.

ABSTRACT

SAE Recommended Practice J941 describes the eyellipse, a statistical representation of driver eye locations, that is used to facilitate design decisions regarding vehicle interiors, including the display locations, mirror placement, and headspace requirements. Eye-position data collected recently at University of Michigan Transportation Research Institute (UMTRI) suggest that the SAE J941 practice could be improved. SAE J941 currently uses the SgRP location, seat-track travel (L23), and design seatback angle (L40) as inputs to the eyellipse model. However, UMTRI data show that the characteristics of empirical eyellipses can be predicted more accurately using seat height, steering-wheel position, and seat-track rise. A series of UMTRI studies collected eye-location data from groups of 50 to 120 drivers with statures spanning over 97 percent of the U.S. population. Data were collected in thirty-three vehicles that represent a wide range of vehicle geometry. Significant and consistent differences were observed between eye-position data collected before and after driving, indicating that actual driving is important protocol feature for accurate measurement of driver eye position. In six vehicles, eyellipses obtained with two-way and six-way seat-track travel were only slightly different. Comparisons between mean preferred and design seatback angles show that design seatback angle does not accurately predict mean driver-selected seatback angle. On average, drivers select seatback angles that are about 1.6 degrees more upright than design. Stepwise regression techniques were used to identify the vehicle variables that have important effects on the distribution of driver eye locations.

INTRODUCTION

The expected range of locations of drivers' eyes is critical information used to design the interior of vehicles to accommodate vision requirements. Mean expected eye location is also used to anchor the headroom curves that define the headspace envelope required for typical driving conditions (SAE J1052). SAE J941 provides formulae to calculate the position and orientation of different percentile eyellipses given the vehicle seating reference point (SgRP), seat-track length (L23), and design seatback angle (L40). The resulting ellipsoids represent driver eye locations in the driver workspace(1).[*] Separate predictions are available for the right and left eyes and for the cyclopean eye, a reference point at the midpoint of the line segment connecting the right and left eyes.

SAE J941 is based on several studies of driver eye position and interpretation of the data for automotive design. The bulk of the data for the passenger-car condition were collected in a study described by Meldrum (2), in which 2300 drivers were tested in three convertibles (about 775 subjects per vehicle) with fixed seatback angles under static conditions to determine driver eye position and construct percentile eyellipses. Subsequent work (3, 4) expanded on the available data to include vehicles with seatback angle adjustment, driver head turn, definitions of the visual field, and methods to implement the eyellipse in automotive design practices.

Recent studies of automotive driver posture and driver-preferred seat position and seatback angle have raised several potential concerns about the use of design seatback angle (L40) as an input to the SAE J941 eyellipse. Typical automotive design practice leaves the selection of design seatback angle to the discretion of the designer. Most design seatback angles are between 20 to 27 degrees, as measured with the J826 manikin, and are typically selected to have an inverse relationship with seat height. Previous studies of driver-preferred seat position and seatback angle have shown that mean preferred seatback angles are generally two to four degrees more upright than design, and little relationship between preferred seatback angle and seat height has been found (5). Previous work at UMTRI has also shown only a weak relationship between manikin-measured seatback angle and driver torso orientation (6), and only small effects of vehicle geometry, including seat height, on driver torso posture (7). These findings suggest that design seatback angle is not a good predictor of driver eye position in vehicles with adjustable seatback recline.

In previous UMTRI studies of driver position and posture, statistically significant differences between the parameters of the observed eyellipses and those predicted by SAE J941 were observed (5). A typical result is shown in Figure 1, which compares the SAE-predicted eyellipse and the observed eyellipse for a 1987 Chevrolet Camaro. The empirical

---

[*] Numbers in parentheses designate references at the end of the paper.

eyellipse has a centroid that is positioned more rearward and higher than the SAE-predicted centroid, a longer X-axis, and shorter Y and Z axes. These differences could be attributed to changes in vehicle features since the early 1970s, including increased range and type of seat adjustments, firmer and more contoured seats, and increased seat-track travel. The differences between J941-predicted and observed eyellipses could also originate in the differences in test conditions between the two investigations. Meldrum tested parked convertibles while the UMTRI studies collected data after on-road driving in vehicles with roofs.

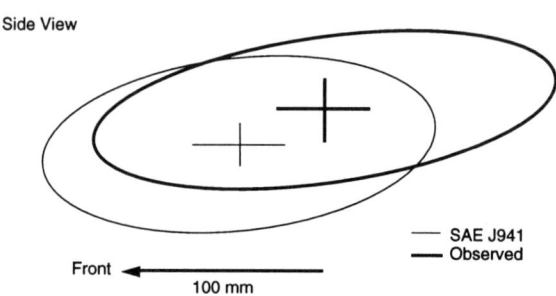

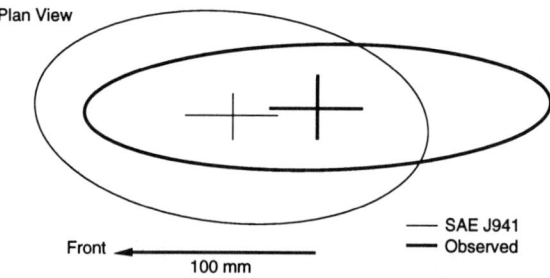

Figure 1. Side- and plan-view comparison of observed and J941 eyellipses for 1987 Chevrolet Camaro.

The consistent differences observed between J941 and experimental eyellipses generated interest in further study of driver eye position in late-model vehicles. The current investigation was initiated to:

1. compile previous data and collect additional data to create a comprehensive eye-position database that includes vehicles equipped with six-way seat-track travel (adjustable height, angle, and fore-aft position),
2. compare design and driver-selected seatback angles,
3. quantify the differences between eye-position data collected under static and dynamic driving conditions,
4. determine the effects of increased seat-track adjustability (six-way vs. two-way) on driver eye positions,
5. determine if the new eye-position database supports the need for changes to SAE J941,
6. identify the vehicle variables that significantly affect driver eye positions, and
7. suggest the form of a new eye position model.

To address these goals, driver eye-position data were collected for 60 to 120 subjects in sixteen vehicles, to increase the UMTRI database to include eye-position data collected in thirty-three vehicles. Experimental eyellipses were calculated for all vehicles and compared to those predicted by J941. Analysis of the empirical eyellipses provides an improved eye-position model that predicts the three-dimensional eyellipse centroid location, three eyellipse axis lengths, and two eyellipse offset angles, given the vehicle seat height, seat-track rise and steering wheel location relative to Ball of Foot (BOF), a pedal reference point defined in SAE J1516 (8).

METHODS

Test Conditions

Driver eye positions were measured in the thirty-three vehicles listed in Table 1, along with vehicle package dimensions of interest. In thirty of the vehicles, eye-position data were collected immediately after subjects had driven. For three vehicles, including the Econoline, Firebird and Taurus L, subjects had previously driven the vehicles but the eye-position data were collected in a laboratory buck study in which the subject-selected seat position and seatback recline angles from dynamic testing were established in a seating buck.

Subject Populations**

Due to the duration of the eye-position test sessions and the importance of the data from the anthropometric extremes of the population, representative sampling was rejected in favor of a stratified sampling strategy was used whereby short and tall drivers are oversampled relative to their representation in the driving population. The resulting eye-position data can be weighted to represent many different population stature distributions (e.g., U.S., Asian, European) and/or different gender mixes of a defined target population, with assurance that an adequate number of subjects are available to define the extreme percentiles.

Subjects were selected to fill twelve gender/stature groups described in Table 2. These groups include subjects who are shorter than the 5th-percentile U.S. female and taller than the 95th-percentile U.S. male based on the 1974 HANES survey data (9). An effort was made to select subjects so that weight, body proportion, and age spanned a wide range. Previous analyses have shown that this sampling strategy is effective in obtaining adequate ranges of these secondary anthropometric variables.

** The rights, welfare, and informed consent of the volunteer subjects who participated in this study were observed under guidelines established by the U.S. Department of Health, Education, and Welfare (now Health and Human Services) on Protection of Human Subjects and accomplished under medical research design protocol standards approved by the Committee to Review Grants for Clinical Research and Investigation Involving Human Beings, Medical School, The University of Michigan.

Table 1
Test Vehicle Summary

Vehicle	Seat Height: H30 (mm)	Seat Cushion Angle: L23 (degrees)	Horizontal Wheel-to-BOF (mm)	Design Seatback Angle: L40 (degrees)	Seat-track Adjuster Type	Seat-track Rise (degrees)	Seat-track Length (mm)
Firebird†	154	15.5	650	27	2-way	6.5	180
TransAm†*	165	14.0	623	24	2-way	8.7	256
Camaro*	177	13.0	616	26	2-way	11.5	293
Avenger 2	189	16.6	577	24	2-way	5.5	230
Avenger 6	189	adjustable	577	24	6-way	5.5	230
Laser 2*	197	11.3	550	25	2-way	5.0	180
Laser 6*	197	adjustable	550	25	6-way	5.0	180
Neon†	212	22.5	565	24	2-way	8.6	210
Probe	216	adjustable	576	24	6-way	4.7	232
Acclaim 2 *	220	17.7	559	24	2-way	9.0	188
Acclaim 6*	220	adjustable	559	24	6-way	9.0	188
Monte Carlo*	231	11.0	597	27	2-way	8.0	262
Mazda 626†	234	18.0	561	25	2-way	5.1	232
Cadillac DeVille*	240	8.0	590	26	2-way	9.0	262
Grand Prix	250	12.0	623	24	2-way	8.0	282
LHS 2	250	17.7	597	24	2-way	8.2	221
LHS 6	250	adjustable	597	24	6-way	8.2	221
Grand Prix*	254	adjustable	610	24	6-way	7.9	279
Taurus SHO†	257	12.0	557	24	2-way	7.0	184
Pontiac 6000*	266	16.0	583	26	2-way	7.5	196
Acclaim 0*	278	9.0	555	24	2-way	0.0	188
Blazer*	288	7.5	591	23	2-way	11.1	231
Grand Cherokee 2	298	11.3	607	24	2-way	9.4	192
Dakota Pickup†	298	12.0	600	22	2-way	2.9	135
Grand Cherokee 6	298	adjustable	607	24	6-way	9.4	192
CK Pickup Truck*	303	12.5	570	21	2-way	5.7	310
Voyager 2	326	14.0	504	22	2-way	7.6	190
Voyager 6	326	adjustable	504	22	6-way	7.6	190
Cherokee Sport†	333	adjustable	589	24	6-way	9.9	189
RAM†	346	13.0	512	20	2-way	5.8	189
Windstar	349	adjustable	504	26	6-way	3.3	181
APV*	381	12.0	518	24	2-way	0.0	300
Econoline†	420	9.5	447	21	2-way	4.5	130

* indicates vehicles that were modified from design to meet specific criteria of the studies.
  All package dimensions were measured directly from the vehicle to assure accuracy.
† denotes manual-transmission vehicles

Table 2
Subject Groups

Group	Gender	Percentile Stature Range by Gender (9)	Stature Range (mm)
0	Female	< 5th	under 1511
1	Female	5-15	1511 - 1549
2	Female	15-40	1549 - 1595
3	Female	40-60	1595 - 1638
4	Female	60-85	1638 - 1681
5	Female	85-95	1681 - 1722
6	Male	5-15	1636 - 1679
7	Male	15-40	1679 - 1727
8	Male	40-60	1727 - 1775
9	Male	60-85	1775 - 1826
10	Male	85-95	1826 - 1869
11	Male	> 95th	over 1869

Test Protocol

The same general procedure was used for all testing, although the eye-position data were collected in six separate studies. Each subject completed a consent form, health questionnaire, and survey asking about their current vehicle and driving habits and a set of twenty standard anthropometric measures were taken. In each test session subjects were tested in 2 to 6 vehicles and the vehicles were tested in random sequence. The initial positions of the seat, seatback, and steering wheel prior to testing in each vehicle were the same for every subject. The subject was instructed on the operation of the seat, seatback, and steering wheel adjustments and was asked to experiment extensively with the adjustments while driving over a 10- to 20-minute road route. During the drive the subject was asked to find the most comfortable driving position and to note the posture of his/her head in straight-ahead driving. Immediately after returning from the drive, driver eye position was measured while the subject maintained a relaxed, normal driving position. While remaining in the vehicle, s/he rated the position of the primary controls and selected seat parameters using a standardized questionnaire.

Methods for Measuring Eye Position

Stereophotogrammetry- Three-dimensional eye locations in fourteen of the vehicles were collected immediately after the subject's return from the drive using two-camera stereophotogrammetry. Direct Linear Transformation (DLT) techniques (10) were used to calibrate eye space inside the vehicle using a set of high-contrast targets, whose locations were precisely known. The targets were attached to the exterior of each vehicle surrounding the driver seating space. A pseudo-eye target of known location was established inside the vehicle and used to confirm the accuracy of the stereophotogrammetric measurements.

To collect eye position data, two cameras were positioned about 70 degrees apart in the parking lot so that all the calibration targets on each vehicle and the left eye of each test subject were visible from both cameras. The cameras were triggered simultaneously to collect a front and side-view image of the driver. The resulting slide film was processed, cut, and mounted between two glass plates. Each pair of images was projected on a tablet digitizer and the position of the calibration targets and the left eye of each subject were manually recorded using a PC equipped with software for processing the 2-D image coordinates into vehicle X, Y, Z coordinates. Location data from the pseudo-eye targets were processed for each subject in each vehicle as a check on the accuracy and consistency of the data-acquisition and processing protocol. This method locates the driver eye within 4 mm of its actual position.

Sonic Digitizer- A Science Accessories Corporation, Inc. sonic digitizer was used to measure eye position in three vehicles whose package geometry was simulated in a laboratory seating buck. The sonic digitizing system uses a fixed array of four microphones to detect the three-dimensional locations of sonic emitters within 2 mm. A sonic probe, with two emitters is placed on a palpated body landmark and the emitters are fired in rapid sequence. The three-dimensional coordinates of the probe tip are calculated from the measured locations of the probe emitters. For each trial, the subject's corner of eye, infraorbitale landmark and glabella were digitized and the location of the pupil was calculated using measured inter-eye anthropometric data.

FARO Arm - A portable, articulated arm for coordinate measurement, manufactured by FARO Technologies, Inc., was used to measure eye position in sixteen vehicles. The FARO arm is a three-link mechanical pointing device instrumented at each of six joints with rotary transducers. The joint angles and the lengths of the three articulated links are used to calculate the position of the probe tip.

To measure subjects' eye positions, the FARO arm, attached to a rolling platform, was positioned next to the vehicle immediately after the subjects returned from the drive and braced rigidly against the vehicle body. Subjects were asked to maintain their driving posture while the FARO arm apparatus was aligned to the data-collection coordinate system by digitizing three reference points on the vehicle body. The FARO arm was then used to record the driver's corner of eye, infraorbitale and glabella locations, followed by seventeen other subject and vehicle landmarks. The digitization was completed in approximately thirty seconds. The FARO arm accuracy under the data collection conditions was determined to be ± 2 mm.

RESULTS

Pre-drive versus Post-drive Eye Position

In six of the vehicles, eye position was measured both before and after the subjects drove to determine the effects of driving on eye position. Table 3 lists the *p*-values resulting from paired *t*-tests of pre- and post-drive eye position in six test conditions. In every vehicle there is a significant difference between pre- and post-drive eye positions, particularly for eye height. Subjects' eyes are an average of 9 mm lower after the drive. The average differences in the X and Y directions are smaller and not always significant, but the eyes tend to be further rearward and further inboard after the drive.

Table 3
Comparison of Pre- and Post-Drive Eye Positions (Post- minus Pre-Drive)
(mm)

Vehicle	Seat-track Condition	X		Y		Z	
		p-value	mean diff.	*p*-value	mean diff.	*p*-value	mean diff.
Acclaim	2-way	0.0589	3.6	0.0003	4.2	0.0000	-8.8
Acclaim	6-way	0.0402	4.4	0.0000	6.0	0.0000	-8.4
LHS	2-way	0.0006	6.9	0.0000	5.3	0.0000	-9.7
LHS	6-way	0.0060	5.2	0.0000	5.6	0.0000	-8.4
Voyager	2-way	0.0426	4.2	0.0430	2.2	0.0000	-9.2
Voyager	6-way	0.0927	3.4	0.5819	0.7	0.0000	-10.2

Design Seatback Angle versus Mean Preferred Seatback Angle

Subject-selected seatback angle was recorded after the drive and compared to the design seatback angle for each vehicle. In Figure 2, symbols below the diagonal line indicate vehicles in which the mean selected seatback angles are more upright than design. In twenty-eight of thirty-three vehicles, the mean preferred seatback angle is more upright than the design seatback angle. The average observed difference is 1.6 degrees and the average absolute error is 2.0 degrees.

Table 4
Average Parameter Values for Empirical and
SAE J941 95th %ile Eyellipses

Parameter	Coordinate or view	Eyellipse		
		SAE	Empirical	Difference
Centroid	X	914	934	20
(mm from	Y	0	7	7
BOF/SCL/AHP)	Z	892	902	10
Axis length	X	199	212	13
(mm)	Y	105	59	-46
	Z	86	100	14
Angle of offset	Side	6.4	9.3	2.9
(deg)	Plan	-5.4	0.3	5.7

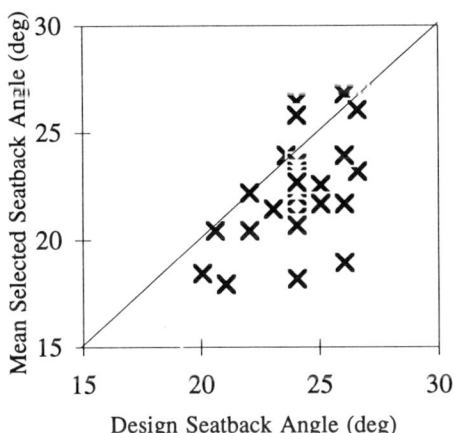

Figure 2. Design seatback angle versus mean selected seatback angle.

SAE J941 versus Experimental Eyellipses

The eye-position data collected for one eye were converted to cyclopean eye using anthropometric data from each subject. The data were weighted according to the fraction of the U.S. population represented by each subject group. The weighted data were used to compute cyclopean side- and plan-view eyellipses following the procedures given by Hammond and Roe (3). The X-, Y-, and Z-coordinates of the centroid of the eyellipses are expressed relative to BOF, seat centerline, and Accelerator Heel Point (AHP).

Table 4 summarizes the observed differences. Every parameter of the empirical eyellipse is significantly different from the corresponding SAE J941 parameter (paired *t*-tests, $p<0.01$). In general, observed eyellipse centroids are rearward and above J941 predictions. Also, the observed X and Z axes are longer and the Y-axis is shorter than J941. In addition, the side-view angle is steeper and the plan-view angle is shallower. The mean differences are illustrated in Figure 3.

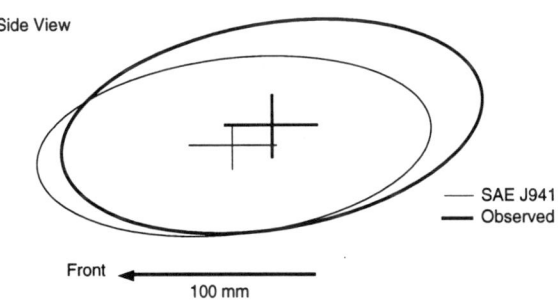

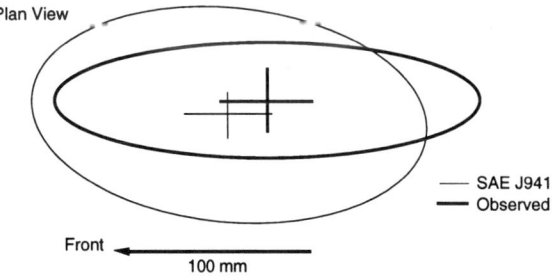

Figure 3. Side and plan-view comparison of mean observed and J941 eyellipses.

Effect of Six-way vs. Two-way Seat Adjustment on Driver Eye Position

Six of the vehicles were tested both as two-way and six-way seat-track vehicles and the resulting data afford a direct comparison while minimizing the possible effects of other variables. Table 5 shows the differences in the eyellipse parameters between the six-way and two-way seat-track conditions. In general, the differences are quite small and only one difference, less vertical variability of eye position with six-way seats, is common to all vehicles. Paired t-test analysis shows that both the Z-axis difference and the decrease in side-view angle inclination are significant with $p<0.01$.

Table 5
Differences in Eyellipse Parameters
due to Seat-track Adjustability
(two-way seat-track minus six-way seat-track)

Vehicle	Eyellipse Centroid (mm)			Eyellipse Axes (mm)			Offset Angle (degrees)	
	X	Y	Z	X	Y	Z	Side	Plan
Avenger	-4	0	-8	9	6	4	3	0
Laser	-4	3	-11	-10	3	9	6	-2
Acclaim	-8	0	-6	-15	-5	3	5	0
LHS	1	2	-2	-1	-5	2	3	-2
Gd. Cherokee	-3	-1	6	-5	3	5	3	1
Voyager	0	-1	2	0	-4	6	0	-2
Mean	-3	0	-3	-4	0	5	3	-1

Predicting Eyellipse Parameters

The consistent differences between the SAE J941 and empirical eyellipses indicate that further analysis with regard to the relationship between vehicle dimensions and eyellipse parameters is warranted. Eight stepwise regressions were performed, one for each of the eyellipse parameters (three centroid coordinates, three axis lengths, and two angles of inclination). The centroid coordinates were expressed relative to package reference points (BOF in the X direction, AHP in the Z direction, and seat centerline in the Y direction). Each eyellipse parameter was regressed on seat height (H30), steering-wheel-to-BOF distance, seat-cushion angle, seat-track rise, seat-track length, SgRP-to-BOF distance, and design seatback angle. Two different seat position predictions were included to determine if eye location could be best predicted with respect to seat position. The 50th-percentile seat position predicted by the new Seating Accommodation Model (11) and J1517 50th-percentile predicted seat position were included for each vehicle. Since J941 includes design seatback angle as an input variable, the mean subject-selected seatback angle was also used to determine if an accurate predictor of seatback angle would be useful in predicting eye location. Finally, two dummy variables were included, one for seat-track type (two- vs. six-way) and one for transmission type (auto vs. manual). Due to the preliminary nature of this model, only its general form is provided.

For each regression, the results of each step were inspected for improvement in prediction and mechanistic plausibility. For present purposes, there was a bias toward simpler equations and predictors with strong effects. As the model is honed in the future, it is possible that new components will be added.

The results of this analysis indicated that steering wheel position is the best predictor of X centroid location, accounting for over 90% of the variance. The Y-coordinate of the eyellipse centroid is best described by a constant, as in SAE J941. Also like J941, the Z-coordinate is a function of seat height (H30). The form of the three prediction equations are given below.

$$c_x = C_1 + C_2 w$$
$$c_y = C_3$$
$$c_z = C_4 + h$$

where, h = seat height (H30) in mm, and
w = fore/aft distance from steering wheel center to BOF in mm.

The regression analysis indicated that the X- and Y-axis lengths are not substantially related to any of the potential predictors and so are best described by constants. The Z-axis length, however, was related to seat-track rise, with steeper track rises associated with decreased vertical variability. The prediction equations for the axis lengths take the form:

$$l_x = C_5$$
$$l_y = C_6$$
$$l_z = C_7 - C_8 r$$

where, r = seat-track rise in degrees.

The ranges of angles of inclination in side and plan views are small and the angles are not strongly related to any vehicle variables. Thus side- and plan-view angles of inclination are best predicted by constants, based on these data.

$$a_{xz} = C_9$$
$$a_{xy} = C_{10}$$

DISCUSSION

For six vehicles, driver eye height was measured before and after the drive and a significant difference in eye height, ranging from 8-10 mm, was observed in all vehicles. These consistent differences between pre- and post-drive eye height show that the driving experience is important to the study of eye position. At the same time, very few differences were observed between pre- and post-drive seat fore/aft position and seatback angle. This suggests that the difference in eye height originates in a change in driver posture during the drive or a change in the amount of seat deflection due to increased load time for the seat cushion.

Data from six vehicles tested with both two-way and six-way seat-track travel provide for direct comparison of the effects of the additional seat adjustability on eye position. The differences in the eyellipse centroid parameters are small and suggest that the same eyellipse model is appropriate for vehicles equipped with two-way or six-way seat-track travel. However, six-way adjustability significantly decreased the vertical range of eye position and decreased side-view offset angle. These differences suggest that the added vertical adjustability allows drivers to deviate from the path of the nominal seat-track travel so that eye height is more consistent across statures.

Comparisons between the empirical and J941 eyellipses show large and consistent differences and suggest that J941 can be improved. Stepwise regressions demonstrate that the distribution of driver eye positions is influenced by seat height, steering-wheel-to-BOF distance and seat-track rise. Although it makes intuitive sense that the X-coordinate of the eyellipse centroid might be most strongly related to measures of driver-selected seat position, steering wheel position emerged as the best predictor. However, research at UMTRI has shown that wheel position is influential to both selected seat position (11) and driver-preferred torso posture (7) which are in turn hypothesized to be closely related to driver eye location.

As expected, the Z component of the centroid is related to seat height, with the centroid a fixed distance above SgRP. The Z-axis length, a measure of the vertical variability in eye locations, is lower in vehicles with higher seat track angles. One potential interpretation of this finding is that the steeper rise raises the eye position of shorter drivers who sit toward the front of the seat track, reducing the eye height difference between short and tall drivers. However, seat-track rise is not well-defined for six-way travel seats and may be undesirable as an eye position predictor.

Design seatback angle and seat-track length are inputs to the J941 eyellipse location procedure, but did not significantly influence eye position in the current study. Mean subject-selected seatback angles were, on average, 1.6 degrees more upright than design seatback angles. Subject-selected seatback angle was explored as a potential input to the eye location model, under the assumption that a mechanistic relationship might exist between seatback angle and eye location. However, no significant relationship was found, suggesting that seatback angle should not be used as a predictor of driver eye location when the driver is provided with an adjustable seatback angle.

SAE J941 provides for two ranges of seat-track length, 100 to 133 mm and over 133 mm. The range of seat-track lengths in the sample was 130 to 411 mm, with all but two vehicles over 180 mm. A seat track length of 170-230 mm is typical in most late-model passenger cars. If more vehicles with restricted track length were included in the study, censoring of seat and eye position would likely result in decreased X-axis length and increased Z-axis length, but testing for these conditions would only be warranted if reduced track lengths were likely to become more prevalent.

The work described in this paper will be expanded and used to suggest improvements to SAE J941. In the future, topics such as Class B vehicles and reference points for the J1052 headroom curves will be addressed. Methods for generating eyellipses for populations with different gender mixes and different anthropometry will be considered.

SUMMARY AND CONCLUSIONS

Previous studies of driver position conducted at UMTRI have found substantial differences between SAE J941-predicted and empirical eyellipses in late-model vehicles. The current investigation combined existing data for seventeen vehicles with newly-collected data for sixteen vehicles to provide an extensive database to study driver eye position. Analysis of these data support the following conclusions:

1. Manufacturer-designated design seatback angles are, on average, 1.6 degrees more reclined than mean subject-selected seatback angles.

2. Eye position collected after subjects drove over a road route was consistently lower than that measured before the drive.

3. Increased seat-track adjustability (i.e., six-way power seats) have little effect on the eyellipse location, size or orientation. A small, though significant, trend for decreased vertical variability and side-view offset angle was associated with the presence of six-way power seats.

4. Large and consistent differences were found between SAE J941-predicted and empirical eyellipses.

5. Stepwise regression techniques show that the distribution of driver eye positions is significantly influenced by seat height, wheel position, and seat-track rise.

ACKNOWLEDGMENTS

This work was sponsored by the American Automobile Manufacturers Association and greatly benefited from the contributions of the AAMA Human Factors Task Force, which includes Howard Estes of Chrysler Corporation, Gary Rupp of Ford Motor Company, and Debra Synetka of General Motors Corporation. Many UMTRI Biosciences Division staff contributed to the success of this project. Ron Roe provided valuable insight into industry automotive practices and helped focus the study objectives. Cathy Harden, Bethany Eby, Stacy Harden and Lynn Langenderfer tested hundreds of subjects to collect the data reported in this paper. Brian Eby and James Whitley fabricated and repaired the test fixtures needed for the study and Stewart Simonette provided technical electronics support for the equipment and vehicles. Tracey Melville was responsible for the stereophotogrammetry equipment and provided photo documentation of all phases of the project.

REFERENCES

1. Society of Automotive Engineers (1997). Recommended Practice J941, *Automotive Engineering Handbook.* Society of Automotive Engineers, Warrendale PA..

2. Meldrum, J. (1965). *Driver eye position.* Technical report no. S-65-3. Ford Motor Company, Automotive Safety Office, Dearborn, MI.

3. Hammond, D.C., and Roe, R.W. (1972). *Driver head and eye positions.* SAE paper no. 720200. Society of Automotive Engineers, Warrendale PA.

4. Devlin, W.A., and Roe, R.W. (1968). *The Eyellipse and Considerations in the Driver's Forward Field of View.* SAE Technical Paper 600105. Society of Automotive Engineers, Warrendale PA.

5. Schneider, L.W., and Manary, M.A. (1991). *An Investigation of Preferred Steering Wheel Location and Driver Positioning in Late-Model Vehicles*, UMTRI Technical Report UMTRI-91-29, Ann Arbor, MI.

6. Manary, M.A., Schneider, L.W., Flannagan, C.A.C., and Eby, B.A.H. (1994). *Evaluation of the SAE J826 3-D Manikin Measures of Driver Positioning and Posture.* SAE Technical Paper 941048. Society of Automotive Engineers, Warrendale PA.

7. Reed, M.P. (1998). *Statistical and Biomechanical Prediction of Automobile Driving Posture.* Unpublished doctoral dissertation, University of Michigan, Ann Arbor, MI.

8. Society of Automotive Engineers (1997). Recommended Practice J1516, *Automotive Engineering Handbook.* Society of Automotive Engineers, Warrendale PA..

9. Abraham, S., Johnson, C.L., and Najjar, F. (1979). Weight and height of adults 18-74 years of age. *Vital and Health Statistics*, Series 11, Number 211.

10. Abdel-Aziz, Y.I., and Karara, H.M. (1971). Direct linear transformation from comparator coordinates into object-space coordinates in close-range photogrammetry. In *Proc. of ASP Symposium on Close Range Photogrammetry.* Urbana, IL.

11. Flannagan, C.A.C., Schneider, L.W., and Manary, M.A. (1996). *Development of a New Seating Accommodation Model.* SAE Technical Paper 960479. Society of Automotive Engineers, Warrendale PA.

980013

Electronic Sound in Automobile and Sound Feeling

Hiroaki Kosaka, Takeshi Shigematsu and Kajiro Watanabe
College of Engineering, Hosei University

Copyright © 1988 Society of Automotive Engineers, Inc.

ABSTRACT

The electronic sound is used as one of the very common means of human-machine interface. Few investigations of how human feels to such the simple sounds noisy or comfortable, exist. The electronic sound must be designed by considering how or in what situation it is used. This research is aimed at investigating how human feels to a variety of electronic sound and describes the relation between human feeling and physical characteristics of sound. We investigated how physical parameters of sound effect to the human feeling via evaluation experiments. We found relations between the human auditory feeling and the physical characteristics of the sound. We showed a guideline of how to design a variety of electronic sound that satisfy the given human feeling specifications.

1 INTRODUCTION

For advanced vehicle systems, one should consider human sense and feeling in the human-vehicle interface. Electronic sounds in automobile, i.e., car reverse alarm and the alarm in missing to remove the ignition key for example, are one of human interfaces where feeling must be considered. A research that investigated the quality of sound in a cellular [1] and an investigation about the relation between electronic sound and it's signal function [2] exist. But the few investigation from the point of view of sound feeling exist. Design problems of electronic sounds were experimentally solved in each case. All electronic sound must take human feeling into account in designing. This study aims at obtaining quantitative relation between physical parameters like basic frequency and human sound feeling.

2 PROBLEM DESCRIPTIONS

2.1 ELECTRONIC SOUND

Figure 1 shows 1 cycle wave form of typical electronic sound. This sounds "Pi Pi Pi Pi...". Figure 2 shows spectrum of the sound shown in Fig. 1. The peak frequency of this sound is 2kHz. In most cases, electronic sounds in automobile can be designed using

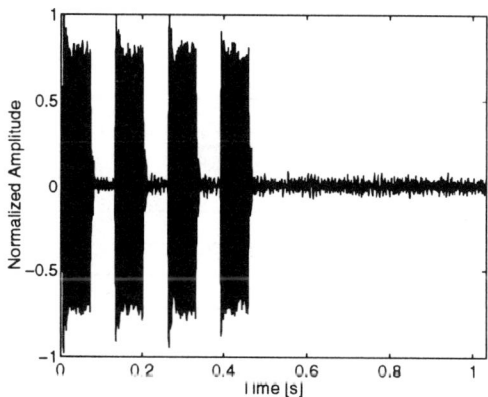

Figure 1: Typical vehicle electronic sound

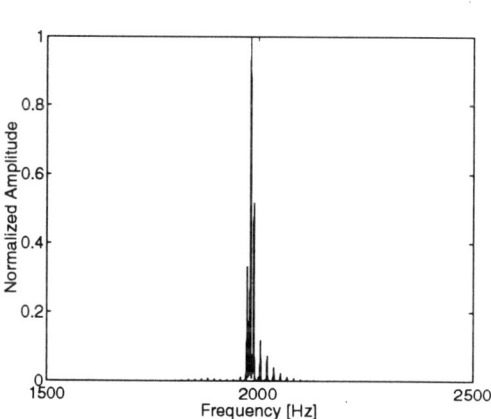

Figure 2: Spectrum of typical vehicle electronic sound

physical parameters in **Figure 3**. In **Table 1**, we defined parameters necessary to form the electronic sound in Figure 3.

We set 4 combinations of 8 physical parameters. **Table 2** shows the values of the combinations. All existing electronic sound can be given by the 4 combinations. **Figure 4** shows the envelope of square, sin, triangle, sawtooth in **Table 2**.

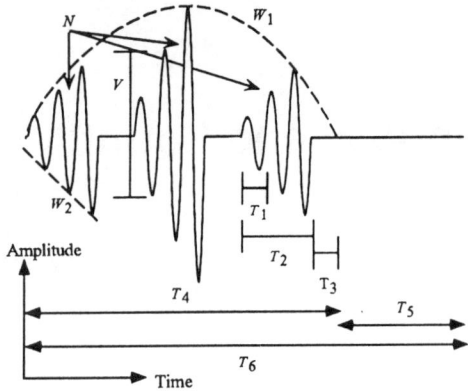

Figure 3: Generalized form of electronic sound wave

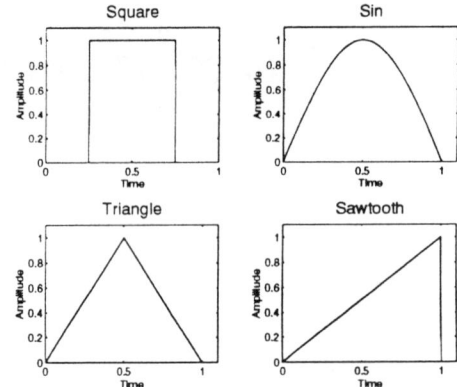

Figure 4: Envelope forms(square, sin, triangle and sawtooth)

Table 1: Definition of physical sound parameters

Basic Frequency	$F = T_1^{-1}$ [kHz]
Outside Envelope Form	W_1 [type]
Inside Envelope Form	W_2 [type]
Outside Cycle	T_6 [s]
Duty Ratio(Outside)	$R_1 = T_2 / T_3$
Duty Ratio(Inside)	$R_2 = T_4 / T_5$
Number of Inside Wave	N [wave]
Volume	V [dB]

Table 2: Physical values of electronic sound

Case	1	2	3	4
F	0.5	1.0	1.5	2.0
W_1	square	sin	triangle	sawtooth
W_2	square	sin	triangle	sawtooth
T_6	0.5	0.8	1.1	1.4
R_1	0.286	0.444	0.556	0.714
R_2	0.286	0.444	0.556	0.714
N	1	2	3	4
V	60	64	68	72

2.2 PROBLEM DESCRIPTION

We consider the following problems.

(P1) Select the words to express the human feeling.
(P2) Investigate the way to decrease the load of a subject.
(P3) Obtain the relation between the physical sound parameters and human feeling.

3 SELECTION TO EXPRESS HUMAN FEELING

Here we consider about problem (P1).

Few studies about the human feeling expressing words to evaluate electronic sound feeling exist. It is desirable that the electronic sound feeling is expressed or evaluated by few words as possible. The purpose of this section is to find the few words appropriate to evaluate and/or express the feeling of electronic sounds. First, we collected 83 words related with sound from dictionaries and related publication [3]. Among those, we selected 14 from the 83 words removing the words which were not appropriate to evaluate or to express the human feeling of electronic sounds clearly. The 14 words are listed in **Table 3**.

Table 3: Words to evaluate the electronic sound human feeling

Bright	Clean	Emergency
Pleasant	Dim	Loud
High-pitched	Artificial	Harsh
Smooth	Clear	Conspicuous
Soft	Comfortable	

Questionnaire tests by the rating method [4] about 14 words above were carried out by the subjects of 15 engineering students. For example, the electronic sound feeling for the word "Bright" was evaluated by the rating scale in which the degree of, "Bright" to "Not Bright" is plotted. The rating scale was 5 grades. The subjects evaluate electronic sounds by selecting one of the 5 grades. We let subjects understand the meanings of the words in Table 3 and let them heard the 8 sounds. The factor analysis to the averages of questionnaire grades was applied. The 14 points in **Figure 5** show the results of the factor analysis [5]. In order to cluster to several groups, cluster analysis [5] was applied. The 6 groups in Figure 5 were clustered by the cluster analysis. We named 6 group from G_1 to G_6.

The 6 words on behalf of groups are shown in **Table 4**. After this, we investigated about the 6 words(groups) as evaluation words for electronic sound.

4 DECREASE LOAD OF SUBJECT IN EVALUATION TESTS

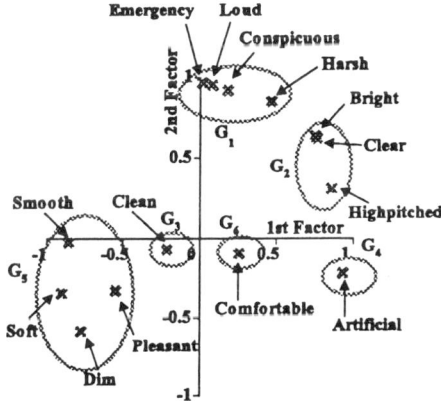

Figure 5: Results of factor analysis and cluster analysis

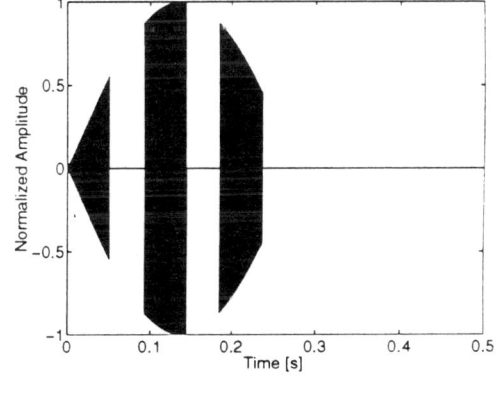

Figure 6: Electronic sound No.6 in Table 6

Table 4: Selected words from experimental results

G_1	Emergency
G_2	Clear
G_3	Clean
G_4	Artificial
G_5	Soft
G_6	Comfortable

Here we consider about problem (P2).

We set 4 combinations of 8 physical parameters of electronic sound. We evaluated electric sounds by the 6words. If we were to evaluate electronic sound for all the combinations, we would have to evaluate $4^8 \times 6 = 393{,}216$ times, very many cases. Therefore, we needed a way to decrease the number of evaluation tests.

4.1 REDUCTION OF THE NUMBER OF TESTS

We applied the design of experiments (Taguchi method) [6] to problem (P2). The design of experiments assesses that varying factors have the most effect to a desired outcome. And we can decrease the number of the evaluation tests if we determine the combination of physical parameters of electronic sound that we must evaluate by the table of orthogonal arrays.

Thus, it seemed appropriate to use the Taguchi method.

4.2 ASSIGNMENT

We designed the combination of physical parameters using the table of orthogonal arrays $L_{32}(2^{31})$. The individual factors in $L_{32}(2^{31})$ are varied between two values each. Since we designed 4 cases of 8 factors(physical parameters), we used two column per factor in designing. We can design 4 cases using two column per factor. We assigned F to the column no.8 and 16, W_1 to 9 and 19, W_2 to 10 and 20, T_6 to 11 and 23, R_1 to 12 and 17, R_2 to 13 and 18, N to 14 and 21 and V to 15 and 22 in $L_{32}(2^{31})$. The results of the assignment is shown in **Table 5**. The number of row of Table 5 is the number of sound we must evaluate. Therefore we only have to evaluate 32 electronic sound.

Here we focus on the first row in Table 5. The cases of sound no.1 are all 1. Then, values of physical parameters of sound no.1 are the values of case 1. Physical parameters of 32 sound are shown in **Table 6**. The graph of wave of Sound No.6 is shown in **Figure 6** as a example of designed sound.

Table 5: A part of orthogonal array

No.	F	W_1	W_2	T_6	R_1	R_2	N	V
1	1	1	1	1	1	1	1	1
2	2	2	2	2	2	2	2	2
3	3	3	3	3	3	3	3	3
4	4	4	4	4	4	4	4	4
5	1	1	2	2	3	3	4	4
6	2	2	1	1	4	4	3	3
7	3	3	4	4	1	1	2	2
8	4	4	3	3	2	2	1	1
9	1	2	3	4	1	2	3	4
10	2	1	4	3	2	1	4	3
11	3	4	1	2	3	4	1	2
12	4	3	2	1	4	3	2	1
13	1	2	4	3	3	4	2	1
14	2	1	3	4	4	3	1	2
15	3	4	2	1	1	2	4	3
16	4	3	1	2	2	1	3	4
17	1	4	1	4	2	3	2	3
18	2	3	2	3	1	4	1	4
19	3	2	3	2	4	1	4	1
20	4	1	4	1	3	2	3	2
21	1	4	2	3	4	1	3	2
22	2	3	1	4	3	2	4	1
23	3	2	4	1	2	3	1	4
24	4	1	3	2	1	4	2	3
25	1	3	3	1	2	4	4	2
26	2	4	4	2	1	3	3	1
27	3	1	1	3	4	2	2	4
28	4	2	2	4	3	1	1	3
29	1	3	4	2	4	2	1	3
30	2	4	3	1	3	1	2	4
31	3	1	2	4	2	4	3	1
32	4	2	1	3	1	3	4	2

Table 6: Physical values

No.	F	W_1	W_2	T_6	R_1	R_2	N	V
1	0.5	square	square	0.5	0.286	0.286	1	60
2	1.0	sin	sin	0.8	0.714	0.714	2	64
3	1.5	sawtooth	sawtooth	1.1	0.444	0.444	3	68
4	2.0	triangle	triangle	1.4	0.556	0.556	4	72
5	0.5	square	square	0.8	0.444	0.444	4	72
6	1.0	sin	sin	0.5	0.556	0.556	3	68
7	1.5	sawtooth	sawtooth	1.4	0.286	0.286	2	64
8	2.0	triangle	triangle	1.1	0.714	0.714	1	60
9	0.5	square	sin	1.4	0.286	0.714	3	72
10	1.0	sin	square	1.1	0.714	0.286	4	68
11	1.5	sawtooth	triangle	0.8	0.444	0.556	1	64
12	2.0	triangle	sawtooth	0.5	0.556	0.444	2	60
13	0.5	square	sin	1.1	0.444	0.556	2	60
14	1.0	sin	square	1.4	0.556	0.444	1	64
15	1.5	sawtooth	triangle	0.5	0.286	0.714	4	68
16	2.0	triangle	sawtooth	0.8	0.714	0.286	3	72
17	0.5	square	triangle	1.4	0.714	0.444	2	68
18	1.0	sin	sawtooth	1.1	0.286	0.556	1	72
19	1.5	sawtooth	sin	0.8	0.556	0.286	4	60
20	2.0	triangle	square	0.5	0.444	0.714	3	64
21	0.5	square	triangle	1.1	0.556	0.286	3	64
22	1.0	sin	sawtooth	1.4	0.444	0.714	4	60
23	1.5	sawtooth	sin	0.5	0.714	0.444	1	72
24	2.0	triangle	square	0.8	0.286	0.556	2	68
25	0.5	square	sawtooth	0.5	0.714	0.556	4	64
26	1.0	sin	triangle	0.8	0.286	0.444	3	60
27	1.5	sawtooth	square	1.1	0.556	0.714	2	72
28	2.0	triangle	sin	1.4	0.444	0.286	1	68
29	0.5	square	sawtooth	0.8	0.556	0.714	1	68
30	1.0	sin	triangle	0.5	0.444	0.286	2	72
31	1.5	sawtooth	square	1.4	0.714	0.556	3	60
32	2.0	triangle	sin	1.1	0.286	0.444	4	64

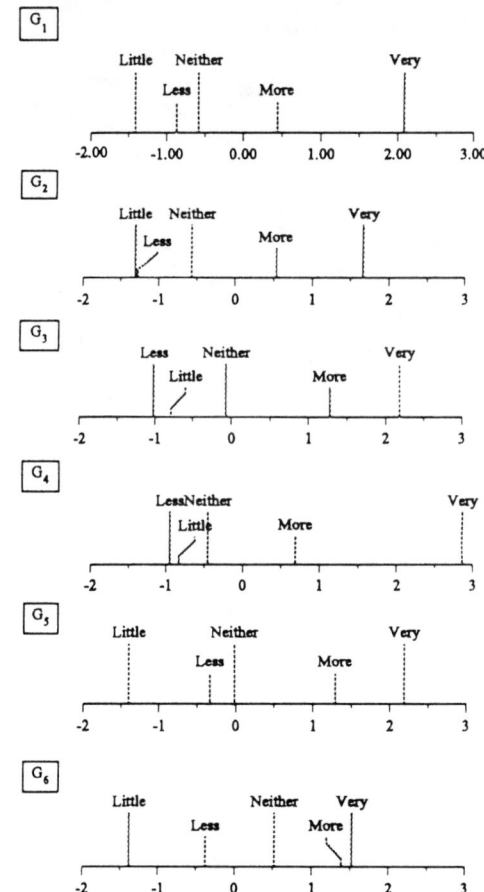

Figure 7: Dual scales for 6 evaluation words

5 PHYSICS OF ELECTRONIC SOUND AND HUMAN FEELING

Here we consider about problem (P3).

5.1 QUESTIONNAIRE

Questionnaire tests by the rating method of 6 feeling evaluation words were carried out by the subjects of 34 engineering students. They feeling-evaluated electronic sound by selecting one of the 5 grades.

5.2 RESULTS

ANALYSIS OF VARIANCE

In order to investigate how sound physical parameters affect to human feeling, questionnaire data were analyzed by their variances [6]. Independent physical variables effective to each human feeling were obtained from the analysis of variance. Independent physical variables are shown in **Table 7**. Values in parenthesis are the percentage of the effect to sound feeling.

DUAL SCALING

To apply the multiple regression analysis, the categorical data are quantified by the dual scaling [7]. The dual scales obtained are shown in **Figure 7**. Application of χ^2 test to the dual scales yielded that all scales were significant within 1% level.

MULTIPLE REGRESSION ANALYSIS

From the analysis of variance, the relation between the human feelings and the individual physical parameters can be estimated linear. Thus, linear multiple regression analysis can be applied to the relations. The multiple regression models are in eq. (1) to eq.(6). The minimum value of multiple regression coefficient was 0.814. Application of F test to the estimated multiple regression model yielded that all regression coefficients were significant within 1% level. From these equations, we can estimate the degree of human feeling.

SOUND DESIGN BASED ON HUMAN FEELING

For automobile alarm sound, the degree of emergency must be considered by the distance between the automobile and obstruction. The electronic sounds are required as one example that we feel "very emergent" and "a little emergent". Such the electronic sound can be

Table 7: Significant physical parameters and ratios of contribution

Parameter	1	2	3	4	5	6	7
G_1(Emergency)	V(100)	T_6(51.1)	W_2(19.2)	F(15.3)	W_1(11.6)	R_2(6.8)	R_1(6.6)
G_2(Clear)	V(100)	W_1(42.3)	W_2(36.7)	F(6.6)	T_6(4.6)	R_1(3.8)	
G_3(Clean)	F(100)	W_1(29.5)	V(8.4)	T_6(5.7)			
G_4(Artificial)	F(100)	V(31.8)	N(20.6)				
G_5(Soft)	V(100)	W_2(76.2)	T_6(40.5)	W_1(13.9)			
G_6(Soft)	T_6(100)	N(44.5)	V(26.1)	W_2(12.4)	W_1(8.7)	F(5.4)	R_1(2.7)

$$G_1 = -0.1733F - 0.1157W_1 - 0.1568W_2 - 0.6081C + 0.4924R_1 + 0.5200R_2 + 0.7741 \times 10^{-1}V - 4.8210 \quad (1)$$

$$G_2 = 0.3799F - 0.5199 \times 10^{-1}N + 0.2234 \times 10^{-1}V - 1.8069 \quad (2)$$

$$G_3 = -0.4040 \times 10^{-2}F - 0.2662W_1 - 0.2553W_2 - 0.2483C + 0.1780R_1 - 0.1044N + 0.8951 \times 10^{-1}V$$
$$-5.49478 \quad (3)$$

$$G_4 = -0.9103 \times 10^{-1}W_1 - 0.1964W_2 - 0.3830C - 0.4370 \times 10^{-1}V + 2.5203 \quad (4)$$

$$G_5 = 0.7269F - 0.2311W_2 - 0.1220C - 0.1844 \times 10^{-1}V + 0.1925 \quad (5)$$

$$G_6 = -0.9052 \times 10^{-1}F - 0.1109W_1 - 0.1369W_2 + 1.066C - 0.2644R_1 - 0.1961N - 0.3733 \times 10^{-1}V + 2.186 \quad (6)$$

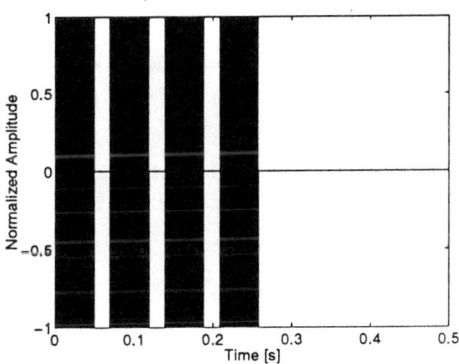

Figure 8: Electronic sound No.1 in Table 8

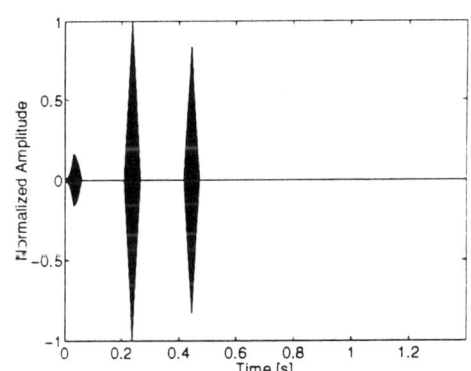

Figure 9: Electronic sound No.3 in Table 8

designed by the design of experiments.

Here we designed 18 electronic sounds that we feel {very, neither, little} about the given human feeling. The feelings and the physical parameters of 18 sounds are shown in **Table 8**. The parameters not influence to the feeling in Table 8 were selected randomly. A electronic sound wave form no.1("Very Emergent" feeling) in Table 8 is shown in **Figure 8**, electronic sound wave form no.3("Little Emergent" feeling) in Table 8 is shown in **Figure 9**.

6 RECONFIRM BY QUESTIONNAIRE TESTS

Questionnaire tests by the rating method of 18 sounds with the 6 feeling words were carried out by subjects of 15 engineering students. The subjects evaluated electronic sound by selecting one of the 5 grades. The 15 subjects did not join to the questionnaire tests described in **5.1**.

6.1 RESULTS OF EXPERIMENTS

The degree of human feeling were obtained by substituting values of independent variables in Table 8 into eq. (1) to eq. (6). These degree are shown on the upper side of scales in **Figure 10**. The averages of the scaled questionnaire grades are shown on the under side of the scales.

From Figure 10, the scale on the upper side and the scale on the under side in G_1 coincided. G_2, G_5 and G_6 coincided on order between the upper side and the under side. G_3 and G_4 did not coincide. The discords are probably from the following reasons:

- Variance in evaluation in questionnaire test was big.
- The tendencies of the evaluation between the questionnaire in **5.1** and **6** did not coincide.

Table 8: Feeling aim and physical values about the designed 18 sounds

Group	No.	Aim	F	W_1	W_2	T_6	R_1	R_2	N	V
G_1	1	Very Emergency	1.0	square	square	0.5	0.556	0.714	4	72
	2	Neither	1.5	triangle	sin	1.1	0.286	0.714	2	64
	3	Little Emergency	0.5	sin	triangle	1.4	0.444	0.286	3	60
G_2	4	Very Clear	1.5	square	square	0.5	0.556	0.444	2	72
	5	Neither	0.5	triangle	triangle	1.1	0.444	0.714	1	64
	6	Little Clear	1.5	sin	triangle	1.4	0.556	0.444	2	60
G_3	7	Very Clean	2.0	square	triangle	0.5	0.714	0.714	1	72
	8	Neither	1.0	sin	square	1.1	0.556	0.714	1	64
	9	Little Clean	0.5	sawtooth	sawtooth	0.8	0.714	0.444	3	68
G_4	10	Very Artificial	2.0	sin	square	0.5	0.556	0.444	2	72
	11	Neither	0.5	square	sin	1.4	0.714	0.444	2	68
	12	Little Artificial	0.5	square	square	1.1	0.556	0.556	2	60
G_5	13	Very Soft	1.5	triangle	sin	0.8	0.444	0.714	1	64
	14	Neither	1.0	sin	sawtooth	1.1	0.714	0.714	2	60
	15	Little Soft	1.0	square	square	0.5	0.444	0.714	1	72
G_6	16	Very Comfortable	2.0	square	sawtooth	1.4	0.714	0.444	1	64
	17	Neither	0.5	triangle	sin	1.4	0.286	0.714	1	60
	18	Little Comfortable	1.0	square	square	0.5	0.556	0.714	4	72

7 CONCLUSIONS

This study investigated the relation between the degree of human feeling expressed by the 6 words and the 32 electronic sounds via questionnaire. The physical parameters to design electronic sound and the human feeling evaluation words were obtained.

We applied the design of experiments to reduce the number of evaluation tests. The quantitative relation between the physical sound parameters in Table 1 and the human feeling evaluation words in Table 4 were obtained by the questionnaire tests. The results of the questionnaire are as follows:

- The analysis of variance to obtain the independent parameters and the percentage of contribution was applied for each feeling evaluation word.
- The 18 electronic sounds were designed. Each sound was designed so as to have the human feeling "very emergency", for example. We carried out questionnaire test to reconfirm the 18 sounds. From Figure 10, the scale on the upper side and that on the under side in G_1 coincided. G_2, G_5 and G_6 coincided on order.

REFERENCES

1. Arai Y. and K.Miura, ``Sound Quality Evaluation of Cordless telephones by the three-way scaling method, and the correlation analysis between psychological measure and physical measure,'' 43.88.Si, THE JOURNAL OF THE ACOUSTICAL SOCIETY OF JAPAN, Shibuyaku, Tokyo, 1994.
2. Arai Y. and K.Miura, ``Evaluation on the Signal Function of Frequency Modulation Sounds,'' 321/322, 1990 Acoustic Society of Japan Lecture Meeting, Shibuyaku, Tokyo, 1990.
3. Sakai H. and T.Nakayama, Auditory Sense and Sound Psychology, Corona Publishing Co., Ltd., 1978.
4. JUSE, Sensory Evaluation Handbook, JUSE Publishing, 1973.
5. Okuno T., et al., Multivariate Analysis, JUSE Publishing, 1971.
6. Taguchi G., et al., Design of Experiments, Maruzen Publishing, 1976.
7. Nishizato S., Scaling of Categorical Data, Asakura Press, 1982.

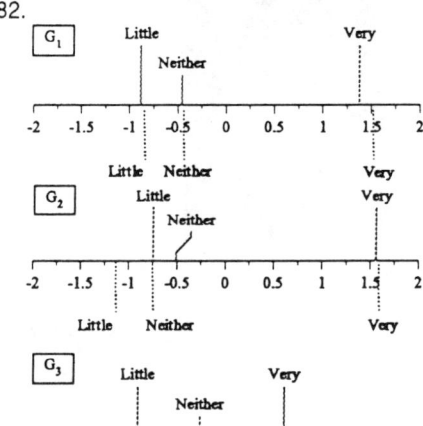
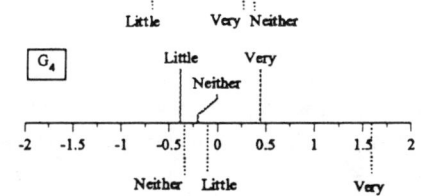
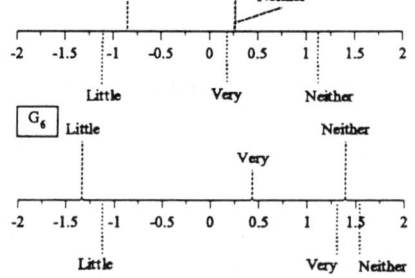

Figure 10: Dual scales about 6 evaluation words
(upper: predicted, lower: experimented on each scale)

980015

Situation for Occurrence of Traffic Accidents at Dusk as Seen from the Standpoint of the Viewing Environment

Kazumoto Morita
Traffic Safety and Nuisance Research Institute

Copyright © 1988 Society of Automotive Engineers, Inc.

ABSTRACT

This paper deals with an investigation on whether the change in the viewing environment at dusk has a connection with the occurrence of traffic accidents by statistical analysis of data on traffic accidents in Japan. In this paper, dusk is considered to be a time period of civil twilight which is about 30 minutes in Tokyo. The target area is limited so that the time period for dusk would be basically the same time. As a result, the author could not recognize a tendency for traffic accidents to occur with especial frequency in the period of dusk. In addition, there was analysis in terms of the ages of the drivers (drivers in their 20s, 40s and 65 years or over). Judging from data on traffic accidents, the author could not confirm a tendency toward more frequent traffic accidents at dusk in any of the age groups, including drivers of advanced age.

INTRODUCTION

When driving in the evening we often experience how the surrounding brightness becomes darker and how objects in front of us are encountered as difficult to see. In this dusk period, the change in the brightness of our periphery is striking. This sudden change in the viewing environment could have an effect on the vision of the driver, possibly leading to traffic accidents. However, there have been insufficient analyses up to now regarding the occurrence of traffic accidents at dusk. Thus, the purpose of this study is to determine on hand of statistical data on traffic accidents whether changes in the viewing environment at dusk have a major influence on the occurrence of traffic accidents or not.

We can easily obtain statistical data on traffic accidents in Japan which clarifies the situation for occurrence of traffic accidents in time periods of two hours each (See Figure 1. Based on data in Reference [1].) This figure shows that, in the course of a day in the period from 4 PM to 6 PM, there is a frequent occurrence of traffic accidents. (In the case of Japan, one type of Japan Standard Time is used which is based on an eastern longitude of 135 degrees).

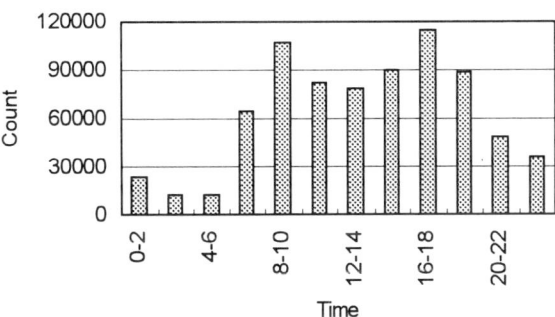

Fig.1 Number of traffic accidents with respect to time of day

As possible reasons for this we can cite the sudden change in the viewing environment at dusk as well as an increase in traffic volume accompanying the end of work. However, in order to verify whether this change in viewing environment is connected as a factor or not, the statistics shown in Figure 1 are insufficient. As reasons for this we can cite how there are no considerations of the region of occurrence, season or other factors. In other words, in order to accurately determine the situation for occurrence of traffic accidents at dusk we need to consider the difference in the time when sunset occurs. For example, if we compare Sapporo (43° 04' north, 141° 21' east) located in the northern part of Japan, and Naha (26° 13' north, 127° 41' east) in southern island in Japan, we notice that sunset occurs 98 minutes earlier in Sapporo (comparison of sunset times on January 1, 1996 [2]). Thus, if we simply make a totalization for all of Japan according to time period, the differences in the time for dusk according to region will not appear. Moreover, because the time for sunset is different in the summer and winter, we also have to consider that point. In addition, when considering time periods at dusk when the change in the viewing environment is particularly pronounced, we must consider a shorter time period than two hours. Considering these various points, the author attempted in this study to collate statistical data on traffic accidents from the standpoint of the viewing environment.

CONSIDERATION ABOUT DUSK

In examining the situation for occurrence of traffic accidents at dusk, we need to determine which time period forms the target of study. For this purpose, we first make considerations about twilight.

DURATION OF TWILIGHT - Generally speaking, concerning the duration of twilight, this is considered as the period between the moment when the central height of the sun is 18 degrees below the horizon and sunrise or sunset. This is defined as the duration of astronomical twilight [2,3]. In other words, regarding evening, the time period from sunset until when the center of the sun is 18 degrees below the horizon line is known as duration of twilight. However, the time when the sun's center is 18 degrees below the horizon line at evening is the period when a star of the sixth magnitude can be observed. Although this has significance in terms of astronomy, when considering normal road traffic, it is actually closer to the night period than the twilight period. On the other hand, the period from sunset until the sun's center is 6 degrees below the horizon is defined as the duration of civil twilight [3]. The time when the sun's center becomes 6 degrees below the horizon is said to be the time when a star of the first magnitude can be observed.

MEASUREMENT OF BRIGHTNESS AT DUSK - Regarding the time period in the evening, the author carried out an experiment to actually measure changes in the sky illuminance (horizontal illuminance) and the road surface luminance on 11 January 1996. The test site was a test course (36° 09' north, 139° 18' east) in Kumagaya City, Saitama Prefecture, Japan. Regarding sky illuminance, the receptor of an illuminance meter was pointed toward the zenith for measurement. Regarding road surface luminance, there was measurement, with a luminance meter, of the luminance of an asphalt surface which was about 50 meters away from the measurer. As for measurement results (Figure 2), the horizontal axis is considered to be time and the vertical axis is considered to be sky illuminance and the road surface luminance. Although the vertical axis shown in Figure 2 shows the sky illuminance and the road surface luminance with the same calibration, regarding the individual units, the sky illuminance is expressed as lx (lux) and the road surface luminance is expressed as cd/m^2. The weather was clear. The age of the moon was 20.0 and the time of moonrise was 21:51, so there was no influence from the brightness of the moon. When expressing the vertical axis as a logarithm, we notice that the level of change in the sky illuminance and road surface luminance is large starting at sunset. The reason why the vertical axis is considered a logarithm is because human perception is generally said to be related to the logarithm of the stimulus value.

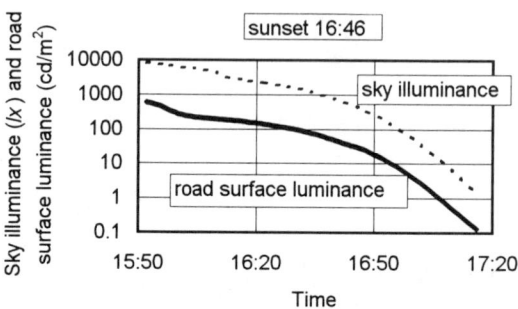

Fig.2 Measurement results of brightness at dusk

On the day of the experiment, the time of sunset was 16:46 and the end of civil twilight was 17:16. An examination of brightness at the time of sunset shows that the sky illuminance was about 440 lx and the road surface luminance was about 30 cd/m^2. This is considered sufficiently bright. An examination of the brightness at the time when civil twilight finished shows that the sky illuminance was about 1.6 lx and the road surface luminance was about 0.14 cd/m^2.

Noguchi et al. investigated the lighting conditions of automobiles at evening on normal roads where there were no street lamps [4]. Results show that, on normal roads, the road surface luminance where the compositional ratio of the parking lamps being lit was maximum was 3.0 to 4.8 cd/m^2 (average of 4.1 cd/m^2). In addition, the road surface luminance on normal roads where the compositional ratio of the headlamps being lit was 50% was a figure of 1.4 to 2.2 cd/m^2 (average of 1.8 cd/m^2).

Regarding the day of the experiment in the present study, an investigation of the time when the road surface luminance was 4.1 cd/m^2 and 1.8 cd/m^2 produced the figures such as 16:59 and 17:04 respectively. We see that they are within the time period for civil twilight.

Regarding a time period when the road surface luminance becomes about 2 cd/m^2, because the headlamps of the vehicle are turned on or the street lamps are starting to go on, a state of natural light disappears. According to "Street Lighting Facilities Installation Standards of the Ministry of Construction [5]", although there are differences according to conditions, it is determined, for example, in such a way that the minimum value of road surface luminance on a principal roadway is 1.0 or 0.7 cd/m^2.

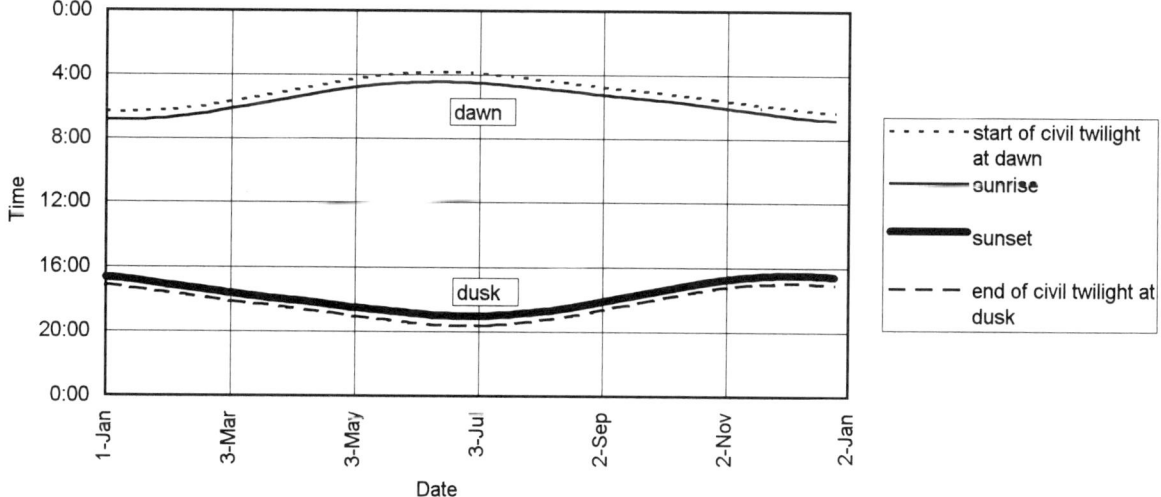

Fig.3 Cange in time of civil twilight in Tokyo

DEFINITION OF TWILIGHT IN THIS STUDY - Taking these various facts into consideration, in this study, dusk is considered to be a time period of civil twilight in which there are complex changes in the viewing environment due to a rapid change of sky illuminance, lighting of auto lights and lighting of street lights. Moreover, the word evening is used here in a slightly broader sense that includes dusk as it is commonly understood.

Figure 3 shows the change over a year period in the time of civil twilight in Tokyo (35° 39' north, 139° 45' east) (compiled according to data in Reference [2]). Although this time period fluctuates according to the season, it is generally considered as the period of 30 minutes starting with sunset. Because sunset is later in summer and earlier in winter, we can confirm a tendency for the starting time of dusk to fluctuate along with this change. The difference in the time for sunset in summer and winter is about 2 hours and 30 minutes.

TARGET TRAFFIC ACCIDENTS

In the present study there is analysis by using the number of traffic accidents involving death or injury as the target data. This traffic accident data makes use of data which is recorded and stored by the Institute for Traffic Accident Research and Data Analysis. In the present study, traffic accidents under the following conditions are used as the subject of analysis.

FIRST PARTY CONCERNED - The purpose of the present analysis is to clarify the relationship between the vision of the driver and the occurrence of the traffic accident. For this reason the author selected traffic accidents with the condition that the first party concerned was the driver of the vehicle. In this case, "vehicle" refers to a personal vehicle, a commercial vehicle and a special vehicle. Also, the first party concerned refers to the party on whose side lies the serious negligence in the accident regarding the drivers of the vehicles or pedestrians initially involved in the traffic accident. When the level of negligence is about the same, it refers to the party whose injuries are less serious.

The number of traffic accidents in which the first party concerned was the driver of the vehicle was 665,172 cases (based on data for 1995 in Reference [1]). This is about 87% of a total traffic accident amount of 761,789 cases [1].

TARGET AREA - As for the target area, it was limited to regions where the time of dusk does not change to a large degree. Moreover, it was necessary to select regions with a high rate of traffic accident occurrence. Based on this approach, Tokyo and three adjacent prefectures (Kanagawa Prefecture, Chiba Prefecture and Saitama Prefecture, see Figure 4) were chosen in which there is a concentration of population in Japan. The total area of this region is about 13,554 square kilometers. This corresponds to about 3.6% of total land area in Japan (377,829 square kilometers). The total population of the region is about 32,575 thousand (as of 1995). This is about 26% of the total population of Japan (125,569 thousand). The total number of traffic accidents in the target area was 187,961 cases (based on data for 1995 [1]). This amounts to about 25% of all traffic accidents in Japan (761,789 cases [1]). As this shows, this satisfies the conditions of being a limited area in which there is a high rate of traffic accident occurrence. Moreover, this region includes small islands far removed from the main archipelago. Thus, in order to achieve strictness in analyzing subsequent data on traffic accidents, these islands are eliminated from the above-mentioned target area. If we express in terms of latitude and longitude the general range of this region after eliminating these islands, we obtain a range of about 1.4 degrees for latitude (34° 54' to 36° 15' north) and a range of about 2.2 degrees for

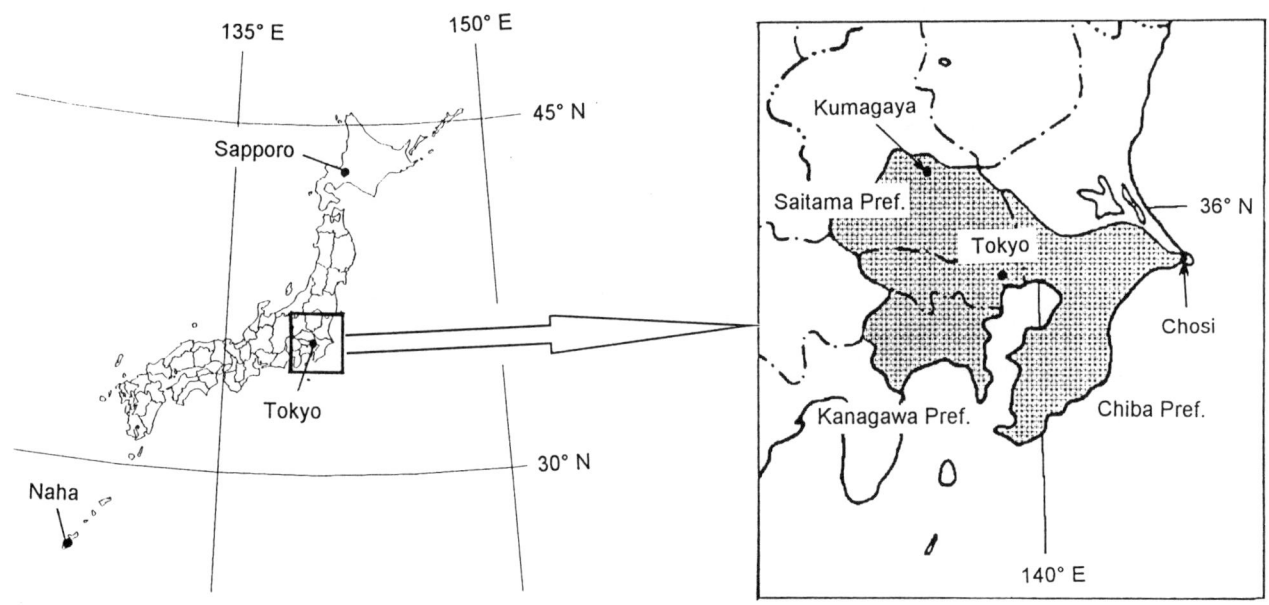

Fig.4 Target area

longitude (138° 43' to 140° 53' east).

As for the range of this region, if we consider this from the standpoint of the time of sunset, we obtain the following. For example, if we take Chosi in Chiba Prefecture (35° 44' north, 140° 52' east) which is the furthest point east in relation to Tokyo (35° 39' north, 139° 45' east), we notice that there is a difference of only about 4 minutes in comparison with Tokyo regarding the time of sunset (based on data for January 1, 1996, although there are seasonal changes in the difference). If we take Kumagaya (36° 09' north, 139° 23' east) in Saitama Prefecture which is the furthest point north in relation to Tokyo, we obtain the same sunset time as in Tokyo (January 1, 1996).

Thus, considering the time for sunset, for the purpose of the present analysis, this is considered to be a sufficiently limited region. The following involves analysis based on the time of sunset in Tokyo which is located roughly in the center of the target region.

WEATHER CONDITION - In addition, it is necessary to make the weather conditions the same. In the present study, the author adds the condition that the weather is clear in totaling the data on traffic accidents. The total number of traffic accidents in Japan under the condition of clear weather was 483,461 cases (based on data for 1995 [1]). This corresponds to about 63% of all traffic accidents (761,789 cases [1]).

TARGET YEAR - In this study, the target years are the four years from 1992 to 1995. The total for all of Japan was 2,911,266 cases (total for four year period 1992-1995).

RESULTS OF ANALYSIS

SITUATION FOR OCCURRENCE OF TRAFFIC ACCIDENTS IN THE TARGET TIME PERIODS FOR ONE DAY - In making the present analysis, it is necessary to consider seasonal fluctuations in the starting time of dusk. Thus, for this reason there is totalization of the number of traffic accidents according to individual months. Because there are differences in the time of sunset even in the same month depending on whether it is the beginning or end of the month, in order to minimize that difference, there is totalization in terms of the beginning of the month (1st to 10th), middle of the month (11th to 20th) and end of the month (21st to last day). (For example, although the time of sunset on January 1, 1996 was 16:38, the time for sunset on January 31 was 17:06, a difference that cannot be ignored).

Regarding the number of target accidents under the conditions listed in the above Chapter, there is totalization of the number of traffic accidents according to month (beginning, middle and end) and time period for one day.

The results are shown in three dimensions in Figure 5. A plane figure is shown in Figure 6. As was mentioned before, considering that the time period for dusk is about 30 minutes, regarding the time in one day there is totalization for each 30 minutes. The total number of target traffic accidents for the four year period in this study was 375,002 cases. As was mentioned before, the total for all of Japan was 2,911,266 cases (total for four year period 1992-1995). That means there is analysis of data for about 12.9% of that total.

As figures 5 and 6 show, generally speaking, there are more traffic accidents occurring in the daytime than at night. Seen in terms of month, the number of cases around June is fewer and the number of cases around December is higher. However, we must pay attention here,

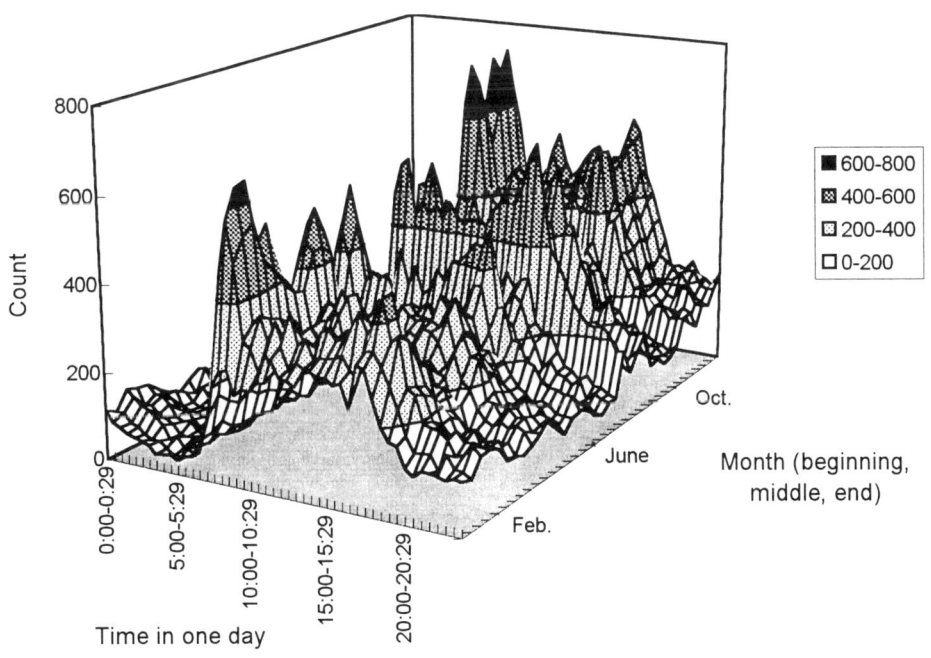

Fig.5 Situation for occurrence of traffic accidents (total for 1992-1995)

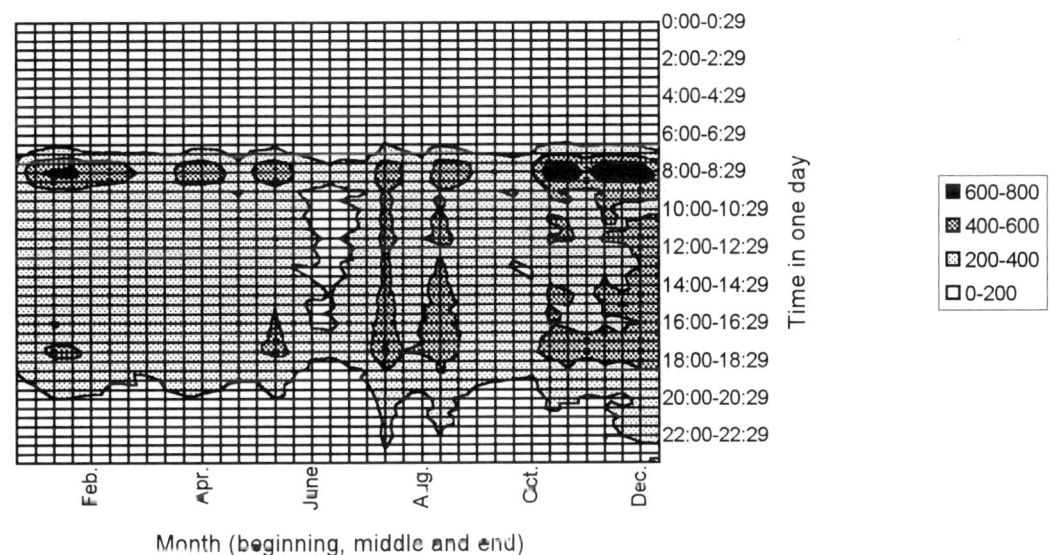

Fig.6 Plan view of the situation for occurrence of traffic accidents

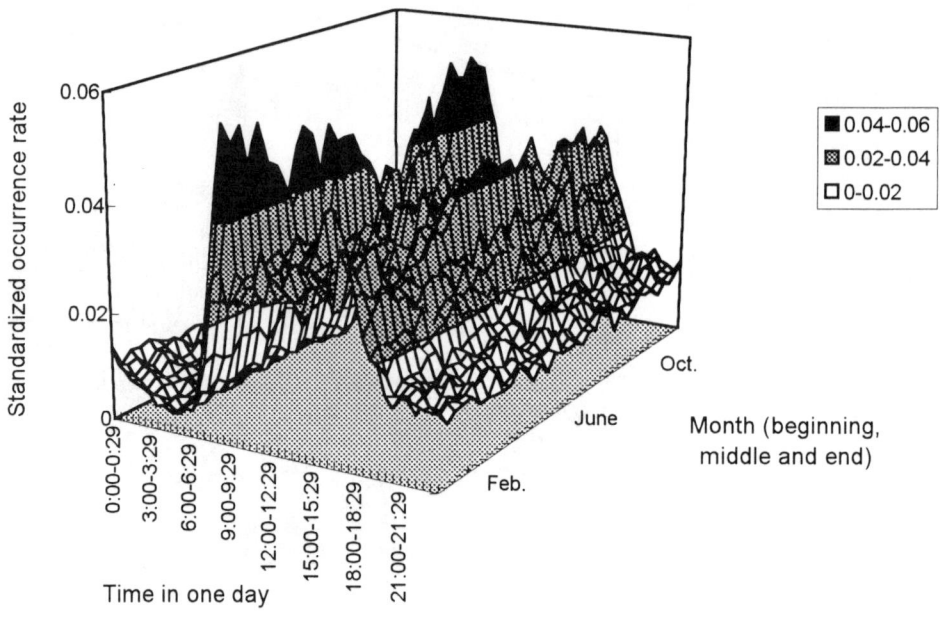

Fig.7 Standardized occurrence rate of traffic accidents

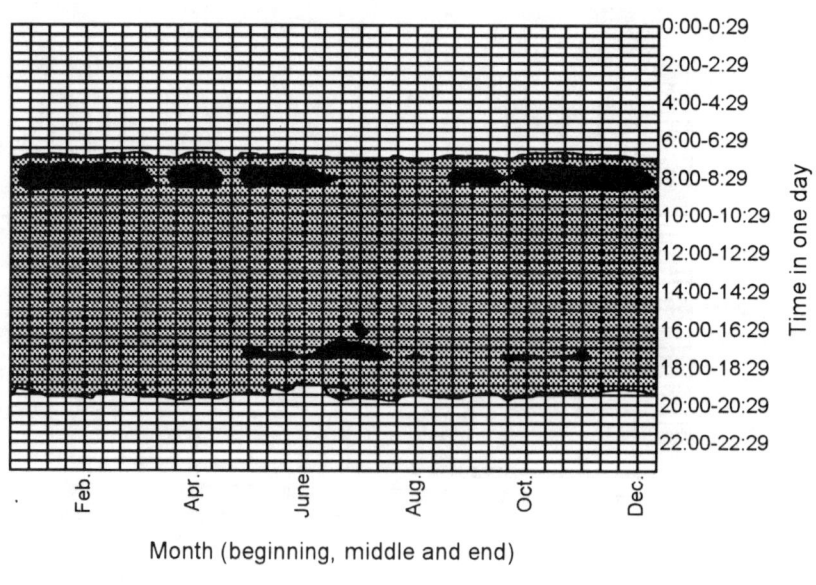

Fig.8 Plan view of the standardized occurrence rate of traffic accidents

as was mentioned above, to the fact that totalization is only made for days of clear weather. This is one reason why the number of cases is lower in June, a period of rainy weather. Regarding figures 5 and 6, we should not make a comparison for each month but rather in terms of time periods in one day.

As a result, in order to make the comparison of time periods in one day easier to understand, the following processes are carried out. There is a totalization of traffic accidents for a 24-hour period for each month (that is, for each beginning, middle and end). The number of cases for each time period in one day are divided in that totalization to achieve the standardized occurrence rate for traffic accidents. By carrying out such a process, it is possible to investigate the situation for occurrence of traffic accidents in each time period in one day for each month (beginning, middle and end). The results are shown in Figures 7 and 8 (plane figure). Both figures clearly show that the standardized occurrence rate of traffic accidents is higher in the period from 7:30 to 9 AM and the period from 5 PM to 6 PM. If there is a high rate of occurrence of traffic accidents in the period of dusk, there should be seasonal fluctuation by which the rate becomes higher at a later period in summer and at an earlier period in winter. However, we cannot find any such tendencies in the charts.

Also, regarding the time periods of 30 minutes each in the evening, if we totalize the number of traffic accidents throughout the year we obtain results as in Figure 9. At the same time, we take the time period of 30 minutes from sunset to totalize the number of traffic accidents in this time period for the entire year. This is shown in the same figure. As was mentioned above, this 30-minute period corresponds more or less to the time period for dusk. As the results in Figure 9 also show, we cannot say that there is any particularly high occurrence of traffic accidents during the period of dusk.

Judging from the above facts, we can say that the reason why the rate of occurrence of traffic accidents is high is because it is the time period for start and finish of work. Thus, the factor of an increase in traffic volume during those time periods plays a more important role.

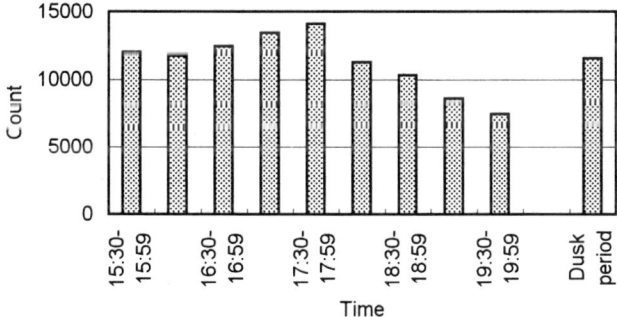

Fig.9 Totalization of the number of traffic accidents in 30-minute period

INFLUENCE OF THE AGE OF THE DRIVER - Regarding the age of the driver, it is expected that there will be a change in the driver's response to changes in the viewing environment. Particularly in the case of drivers of advanced age, because there is a decline in the visual functions, there is a possibility that such drivers are more susceptible to the influence of changes in the viewing environment. As a result, we can expect that the situation for occurrence of traffic accidents will also change in comparison with the other age groups.

For this reason, regarding the target traffic accidents, there is totalization for drivers in different age groups (20s, 40s and 65 or over). If we obtain the standardized occurrence rate mentioned before, we obtain Figure 10 (20s), Figure 11 (40s) and Figure 12 (65 or over). Each case shows a plane figure. The number of target accidents for the group in their 20s is 137,318 cases. This is 36.6% of the figure of 375,002 cases of target accidents as shown in Figure 5. In the same way, regarding the number of target accidents for the group in their 40s, the figure is 70,966 cases (18.9% of total). The figure for the group aged 65 years or over is 12,982 cases (3.5% of total).

Regarding drivers in their 20s and 40s, as was true for the results in the former section, it is recognized that the occurrence rate for traffic accidents is high in the time periods from 7:30 AM to 9 AM and 5 PM to 6 PM. There is not tendency recognized by which traffic accidents occur more frequently in the period of dusk.

Regarding older drivers of 65 years or over, there is a tendency spread over daytime period, but no tendency is recognized for traffic accidents to occur with particular frequency during the dusk period. As reasons for the distribution over the daytime period we can say regarding persons of 65 years or older that, because they have generally retired from work, the attendant tendency for accidents to occur in the periods for starting and finishing work no longer exists in their case.

As this shows, even seen in terms of age group, there is no tendency recognized for traffic accidents to occur with special frequency during the dusk.

TRAFFIC ACCIDENTS INVOLVING VEHICLES AND PEDESTRIANS - During the period of dusk, there is lighting of the lights in the vehicles with the result that the visibility of the vehicle is increased. The author and his colleagues have already carried out a quantitative analysis regarding how visibility increases in a vehicle with its lights on compared to a vehicle with no lights on [6]. Those results made clear experimentally that a vehicle with its lights on has a major increase in visibility compared to a vehicle with no lights. Conversely, regarding a vehicle with no lights on, the visibility decreases correspondingly so that there is a greater possibility of the vehicle being hit. Nevertheless, in the traffic accident data used in the present study, there is no record on what kind of lighting existed for the vehicles. Thus, it is not possible to carry out totalization regarding the lighting conditions of

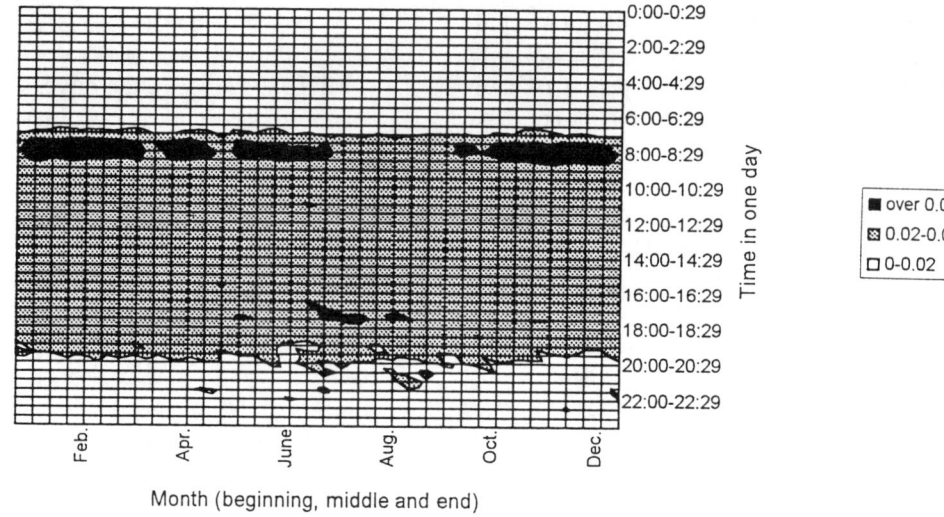

Fig.10 Influence of the age of the driver (drivers in their 20s)

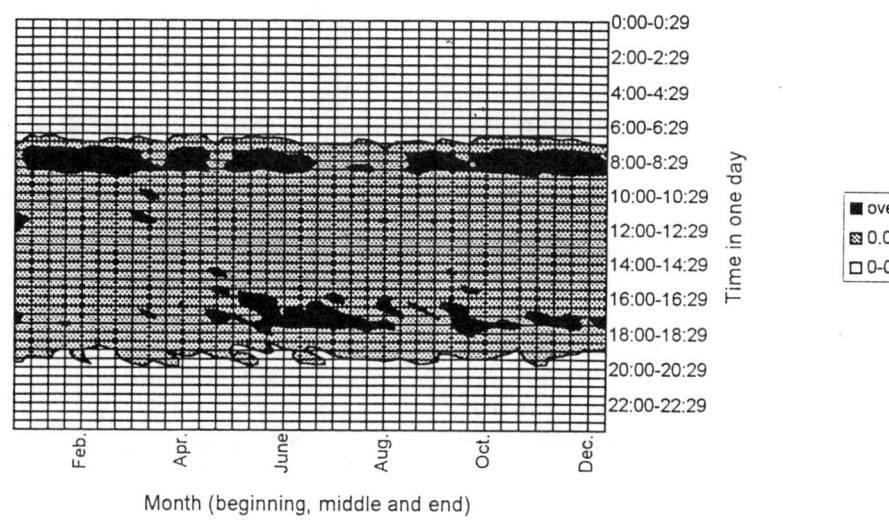

Fig.11 Influence of the age of the driver (drivers in their 40s)

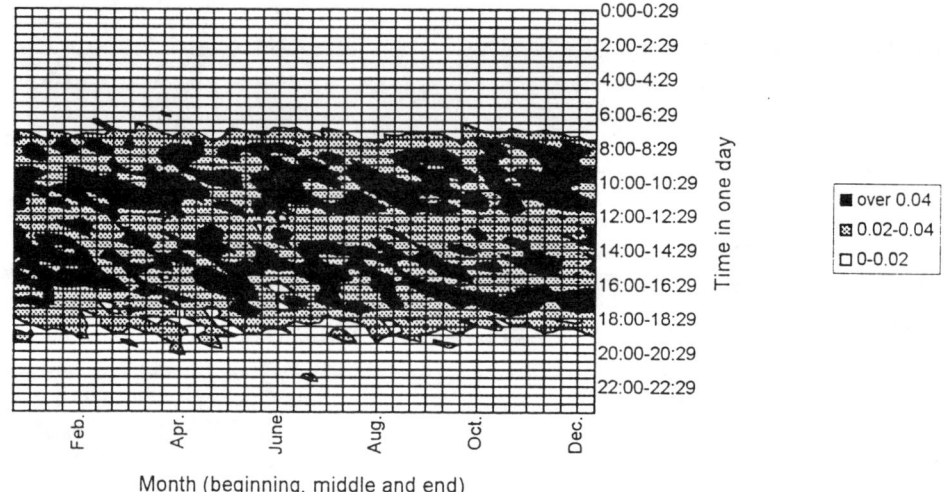

Fig.12 Influence of the age of the driver (drivers of 65 or over)

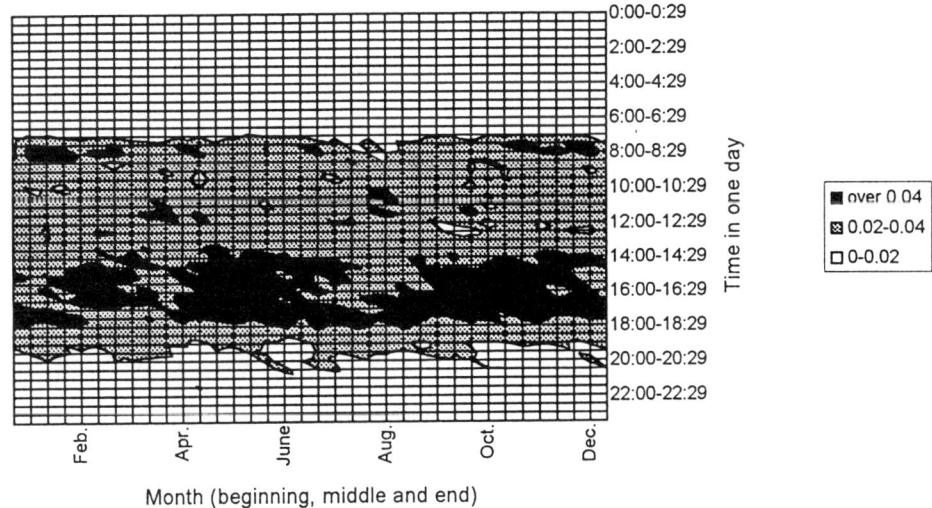

Fig.13 Results in case that the type of accident is vehicle vs. pedestrian

lighting on the vehicles which suffered collision.

Nevertheless, if the object of the collision is a pedestrian, the lights are not lit by the subject himself. If the surroundings have become dark in dusk, the visibility will be correspondingly inferior to that of a vehicle with lights.

Thus, among the totalization conditions in Chapter "Target Traffic Accidents" there is inclusion in totalization of the condition that the object of the collision is a pedestrian (i.e., the condition that the type of accident is vehicle vs. pedestrian). The total number of target traffic accidents in such a case is 36,503 cases. As was the case up to now, if we make a chart of the standardized occurrence rate we obtain a chart such as Figure 13. The authors can not recognize any tendency for accidents to occur with especial frequency in the period of dusk. The accidents occur with particular frequency in the time period from 2 PM to 6 PM. As reasons why a large number of accidents occur in that period, there is the fact that there are many accidents which involve children playing outdoors after school [7].

As this shows, even in the case of a target which has no lights lit, we cannot conclude that traffic accidents occur with especial frequency in the period of dusk.

CONCLUSION

The author carried out an investigation on whether the change in the viewing environment at dusk has a connection with the occurrence of traffic accidents by statistical analysis of data on traffic accidents. In this paper, dusk was considered to be a time period of civil twilight. The target area was limited so that the time period for dusk would be basically the same time. In the present study, there was statistical investigation of the number of traffic accidents in the region in Japan consisting of the following areas: Tokyo, Kanagawa Prefecture, Chiba Prefecture and Saitama Prefecture.

As a result, the author could not recognize a tendency for traffic accidents to occur with especial frequency in the period of dusk. Seen in terms of time period for one day, it is confirmed that the occurrence rate for traffic accidents is high in the time periods from 7:30 to 9 AM and from 5 to 6 PM throughout the year. In other words, the reason why the occurrence rate for traffic accidents is high is because these are considered to be time periods related to the start and finish of work, thus leading to an increase in the traffic volume during those times. In addition, there was analysis in terms of the ages of the drivers (drivers in their 20s, 40s and 65 years or over). This was used to investigate differences in the situation for occurrence of traffic accidents in terms of age differences. Particularly in the case of older drivers, there is said to be a decline in the visual function. However, judging from data on traffic accidents, the author could not confirm a tendency toward more frequent traffic accidents at dusk in any of the age groups, including drivers of advanced age. It is also believed that the visibility of pedestrians declines at dusk compared to vehicles with their lights on. The author also investigated data on traffic accidents concerning pedestrians. However, in this case as well, it could not be said that the number of traffic accidents occurring at dusk was particularly high.

Judging from the above, we cannot conclude, at least in terms of statistics, that the sudden change in the viewing environment at dusk has a major relationship to occurrence of traffic accidents. However, these statistical results express a tendency concerning traffic accidents as a whole. In considering the individual examples we cannot deny the possibility that there was a problem concerning the visibility of the objects.

REFERENCES

[1] Traffic Bureau, National Police Agency, "Statistics '95 Road Accidents Japan", International Association of Traffic and Safety Sciences, (1996)
[2] "Astronomical Almanac 1996", Seibundo Shinkosha Publishing Co., Ltd., (1995) (in Japanese)
[3] Maritime Safety Agency, "1996 Nautical Almanac", Japan Hydrographic Association, p.458 (1995) (in Japanese)
[4] T. Noguchi et al., "Use of Car Lights at Dusk", J. Illum. Engng. Inst. Jpn. Vol.75 No.2, pp.16-20 (1991) (in Japanese)
[5] Japan Road Association, "Street Lighting Facilities Installation Standards of the Ministry of Construction", Maruzen, pp.25-27 (1981) (in Japanese)
[6] K. Morita et al., "Change in Automobile Visibility at Dusk", Journal of Light & Visual Environment, Vol.19,No.2, pp.20-26, Illum. Engng. Inst. Jpn. (1995)
[7] Traffic Bureau, National Police Agency, "Statistics of Road Accidents '95", p.176-179 (1996) (in Japanese)

About the Author

Kazumoto Morita was born in November,1951 in Japan. He received a M.Eng. from Kyoto University (1976). His research interests are concerned with the safety of automobiles, especially with the automobile lighting, visibility, navigation systems and elderly drivers performance.
Fax No. +81 422 41 3233,   E-mail i90393@simail.ne.jp

980016

Data Processing Method of Finger Blood Pulse for Estimating Human Internal States

Hiroshi HASHIMOTO, Tsuyoshi KATAYAMA
Japan Automobile Research Institute

Copyright © 1998 Society of Automotive Engineers, Inc.

ABSTRACT

It was found that the finger blood pulse shows various fluctuations in different driving conditions. The nature of the finger blood pulse fluctuations was used for estimating a driver's internal state. Indexes suitable for expressing the fluctuations were moment and density; these indexes were calculated by using a return-map. However these results were measured by an off-line system and were calculated after the experiment. So, an on-line (real-time) system was needed in order to construct a driver's internal state monitoring system. As a first step, an on-line system for estimating the human internal state was developed. This system is available for estimating the human internal state every 30 seconds.

INTRODUCTION

The importance of accident prevention is now recognized. Active safety of the vehicle is as important as passive safety. As one candidate for active safety, a monitoring system for a driver's internal state, such as arousal levels and emotional states, will be a powerful tool to decrease traffic accidents.

It is well known that the heart rate and the capillary pulse rate show various aspects of fluctuations under different mind and body conditions. But the detailed relationship between these fluctuations and internal states is not well known and is now under investigation [1][2].

The ultimate goal of this study was to construct a driver's internal state monitoring system. This paper describes the data processing method and the real-time (on-line) system. Also the relationship between the finger blood pulse and internal states is discussed. The finger blood pulse was measured during driving. Then the data was analyzed in terms of a return-map; which is familiar in the analysis of chaotic phenomenon.

This study shows that fluctuation of the finger blood pulse is available for estimating a driver's internal state. Appropriate indexes of the fluctuation are moment and density. A real-time system was developed based on these results.

DATA PROCESSING METHOD

Before developing a real-time system, the experiment was done during driving. The data was analyzed in terms of a return-map. This chapter describes the off-line measurement method and the way data was processed.

OFF-LINE MEASUREMENT METHOD

The data was analyzed by an off-line system. The data was recorded in a data recorder and the recorded data was processed after the experiment. This system consisted of the finger blood pulse sensor, the amplifier, the data recorder and the workstation. (Fg.1)

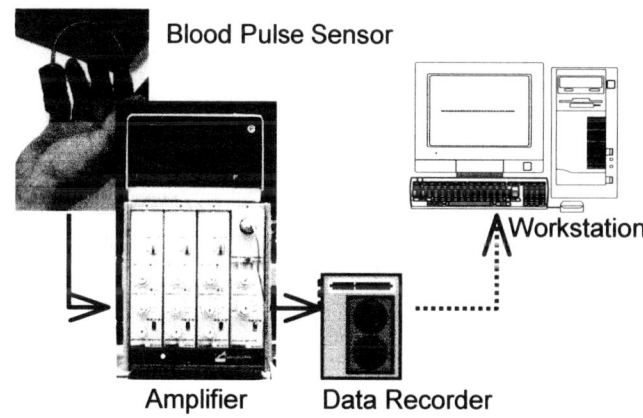

Figure 1: Off-line Measurement System

RETURN-MAP

The data was analyzed in terms of a return-map.

It was processed to convert the time series of finger blood pulses into a return-map. The dates were sampled at regular intervals. These data were plotted to the x-axis and the y-axis by turns. (Fg.2)

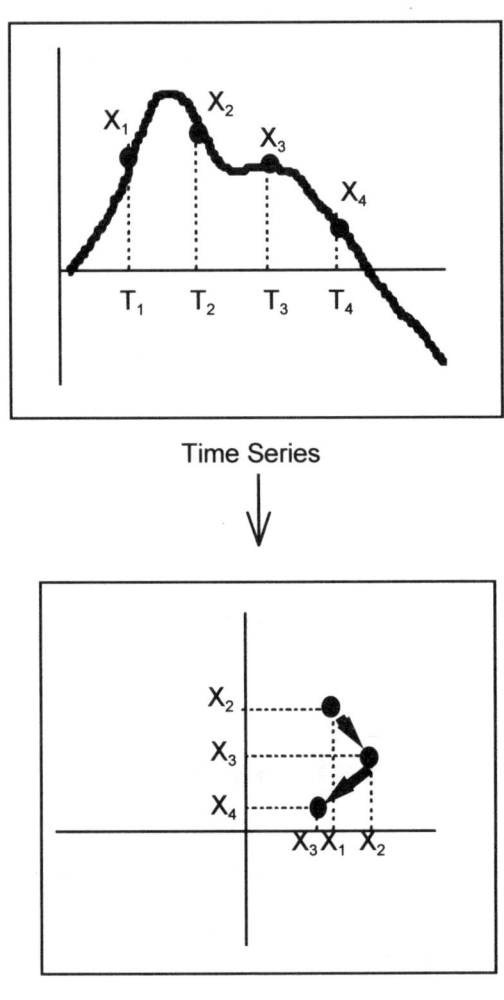

Figure 2: Conversion of the Time Series of the Finger Blood Pulse into the Return-Map

EXAMPLE OF THE RETURN-MAP

Figure 3 shows four examples of the return-map during driving. Subjects drove a car from Tsukuba to Tokyo. Public roads and highways were used as a test course.

"Eyes-Closed" is a return-map before driving. "Ordinary driving" is driving on a public road while relaxed. "Traffic jam" is a level of low arousal while driving. "Disorderly Highway Junction" is a high-tense situation when subjects pass through a disorderly highway junction. (A disorderly highway junction is a place where two highways merge and diverge and there is a traffic jam. So, some drivers go straight and some drivers cross lanes across the highway in order to merge with the other highway. In our study, the subjects had to cross the highway.)

These return-maps have two features. One feature is the size of the return-map. For example, an "Eyes-closed" map is large; on the other hand, a "Disorderly Traffic" map is small. The other feature is density. For example, on a "Ordinary Driving" map, the rings are bunched together and concentrated to the outside; whereas, on a "Traffic Jam" map, the rings are random and not concentrated. (Fg.3)

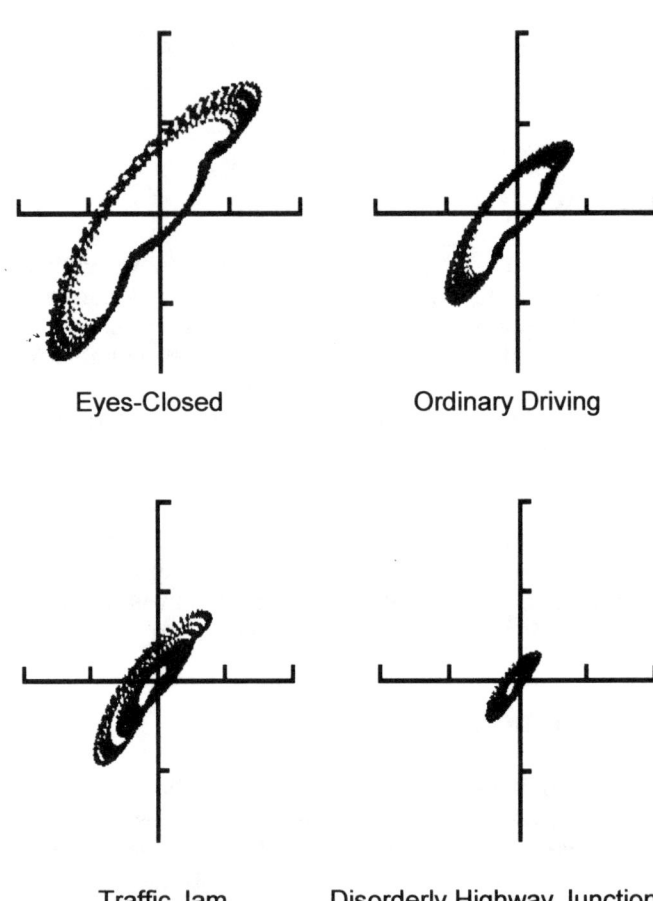

Figure 3: Examples of the Return-Map

INDEXES OF THE RETURN-MAP

Moment and density were utilized as an index. Moment reflects the size of the return-map. And density reflects the order of the return-map.

MOMENT

Moment is a polar moment around its center of gravity. This is an index of the size of the return-map and its numerical value is proportional to its size.

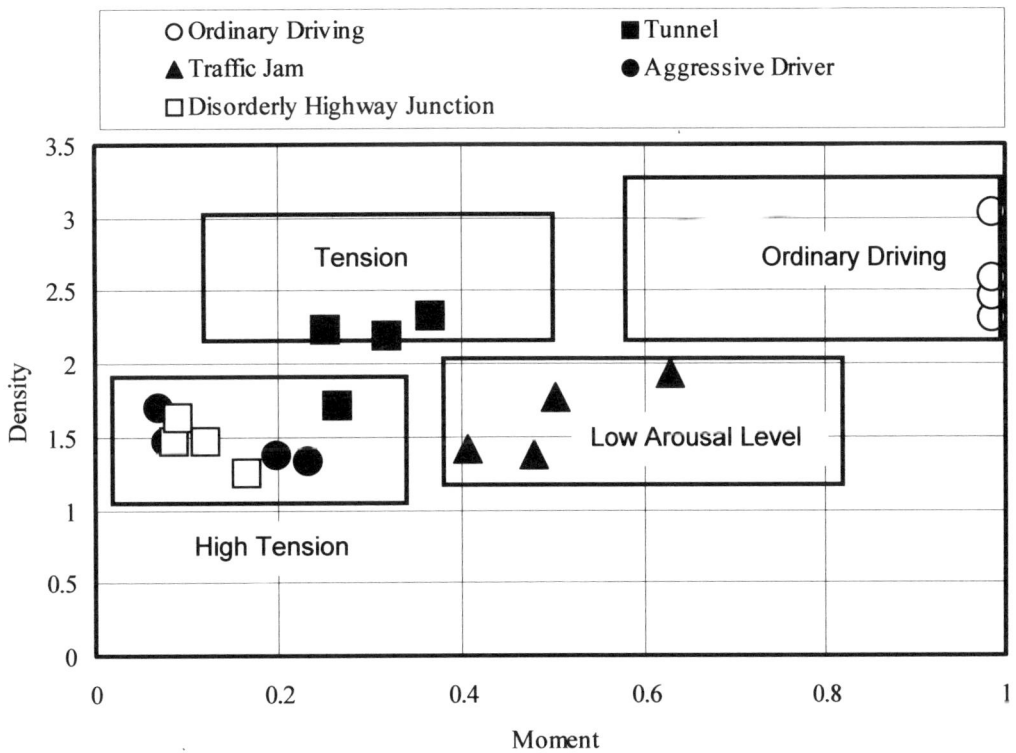

Figure 4: Moment-Density Map

$$M = \sum_{i=1}^{n-1}\left\{(x_i - x_c)^2 + (x_{i+1} - y_c)^2\right\}$$

M: Moment

x: Data (sampling at regular intervals)

x_c: Center of Gravity (x-axis)

y_c: Center of Gravity (y-axis)

DENSITY

Density is when moments normalize with a maximum value to the return-map. This is an index of the return-map, and a bigger density corresponds to a smaller fluctuation.

$$D = M / (A L_{max}^2)$$

D: Density

M: Moment

L_{max}: Maximum Length from Center of Gravity

A: Proportional Fixed Number

MOMENT-DENSITY MAP

The moment-density map was made based on moment and density related to each internal state. (Fg.4)

Only four of nine subjects data was plotted to the Moment-Density Map because they encountered an aggressive driver during the study.

Table 1 shows a feature of each internal state.

Subjects felt tension when they were in a "Tunnel" on the Highway. When subjects felt tension, the moment was smaller than the one in "Ordinary Driving" and the density was almost the same.

When subjects got through the long traffic jam, they felt tired, so they became sleepy (A level of low arousal). In this case, the moment was almost the same, and the density was smaller than the density in "Ordinary Driving".

When subjects encountered an aggressive driver and when they came to the disorderly highway junction, they became very tense. In these cases, both moment and density were smaller than the ones in "Ordinary Driving".

Table 1: Feature of each internal state

		Moment	
		Small	Big
Density	Big	Tension	Relaxed
	Small	High Tension	Low Arousal

REAL-TIME SYSTEM

A real-time system was developed based on the "Moment-Density Map". (Fg.5) This system consisted of the finger blood pulse sensor, the amplifier, A/D card (PC Card type) and the personal computer.

Today, the personal computer has been miniaturized and has higher performance with a lower price. As a result, it is possible to install a personal computer into a car.

Table 2 shows the system specifications. In this system, the data-sampling frequency was 1000Hz and the data-sampling time was 30-seconds per one process.

The time series data, a return-map and a Moment-Density Map were indicated in the monitor of the computer. (Fg.6)

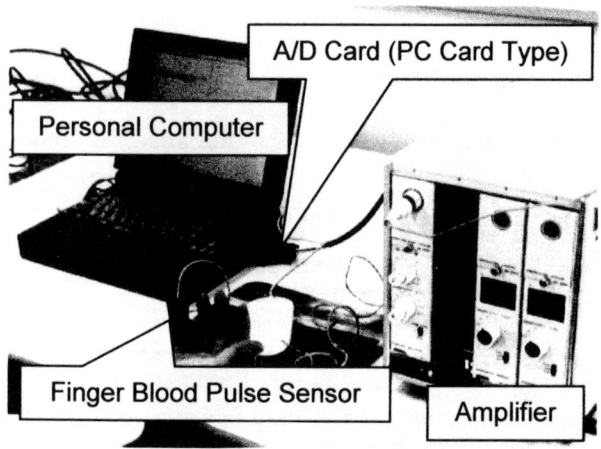

Figure 5: Real-time System

Table 2: System Specification

Device	Finger Blood Pulse Sensor
	Amplifier
	A/D Card (PC Card Type)
	Personal Computer (IBM Compatible)
Operating System	Windows95
Sampling Frequency	1000Hz
Sampling Time	30 seconds / process
Monitor	Finger Blood Pulse (Time Series)
	Return-Map (Every 30 second)
	Moment-Density map (Every 30 seconds)

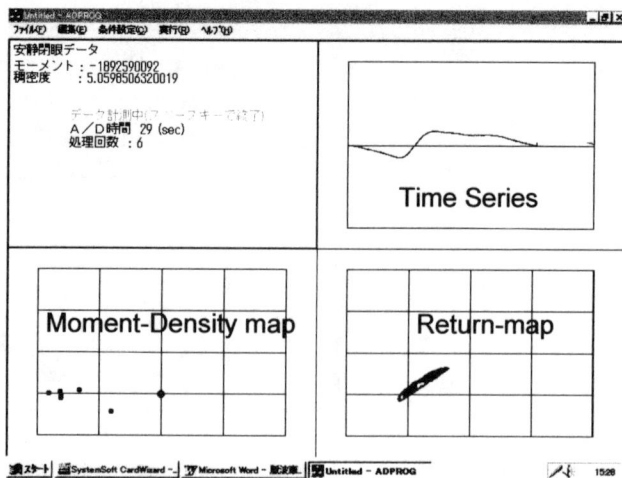

Figure 6: Monitor

CONCLUSION

In this study, finger blood pulses were measured in various internal states. The data was analyzed by means of a return-map. As a result, it was possible to estimate the human internal state by its location on the Moment-Density Map. Also a real-time system was developed based on this result. This system is available for estimating the human internal state every 30 seconds.

The future direction of this study will be to construct a driver's internal state monitoring system. As a next step, more data will be needed for more accurate estimating. Also a study of appropriate preventative measures will be needed; for example, a warning system when a driver is sleepy, etc.

Therefore, more progress into investigations concerning the finger blood pulse will make a greater contribution to the active safety of vehicles.

CONTACT

Hiroshi HASHIMOTO

E-mail: hhiroshi@jari.or.jp

http://www.jari.or.jp

REFERENCES

1. Keiichi Yamamoto and Shinichi Higuchi, A Study on Truck Driver's Arousal Level in Long-Distance Transportation, Journal of the Society of Automotive Engineering of Japan (JPN), Vol.46 (1992), p23-p28.
2. Yoshinaka Kawakami, Koichi Akamiya and Katsunori Hamatani, Lowering in Arousal Level of a Driver, Journal of the Society of Automotive Engineers of Japan (JPN), Vol.46 (1992), p408-p410.
3. Tsuyoshi Katayama and Kazuo Sakai, Fluctuation of Capillary Pulse as an Index for Driver's Internal States, Vehicle navigation & information Systems Conference Proceedings, (1994), p11-p14.

980017

Design of a Hybrid Driver Model

R. Majjad, U. Kiencke, H. Körner
Institute for Industrial Information Systems, University of Karlsruhe, Germany

Copyright © 1998 Society of Automotive Engineers, Inc.

ABSTRACT

In this article a hybrid driver model is described which has been developped at the Institute for Industrial Information Systems (IIIT) of the University of Karlsruhe. After a short introduction into the subject an overview over the human information perception and processing is given. A concept is presented which adapts the human behaviour to a realistic driver model. The developped driver model, which is composed of a queueing network and two GPC-controllers, is described. For the purpose of simulation three different driver types are defined and simulation results are discussed.

INTRODUCTION

In recent years the automobile industry has developed and improved many technical innovations in the scope of driver support systems, like ABS, electronic traction control (ETC), or Bosch FDR. These systems help the drivers to avoid critical driving situations which may lead to accidents.

For a fast and cost saving development of these systems suitable vehicle models are needed which reproduce the mechanics, the electric system and the hydraulics of a car sufficiently. These models supply realistic information about the vehicle dynamics with the help of simulations. On the other hand a realistic driver model is required that is able to manipulate the vehicle model and to imitate the human driving behaviour in order to achieve a closed loop simulation.

In order to validate the performance of driver support systems a test driver was needed so far who handled a vehicle simulator. This method is combined with a high computational effort. Another disadvantage is that a test driver is not able to simulate different types of drivers.

In earlier studies the human driving behaviour was modelled with the help of continuous controllers like the PID controller [Mit90]. These models are not sufficient in order to get enough information about a real driver because they carry out driving manoeuvres with more or less optimal performance.

Another driver model that was implemented as Fuzzy-controller [Wol97] performes nearly the human steering behaviour but is difficult to analyze and to adapt to different driver types.

It is common to all of these controllers that they only simulate the steering behaviour but not the control of the longitudinal motion. Further these models are not able to reproduce the information reception and processing of a human driver.

For this reason a realistic hybrid driver model was developped at the IIIT of the University of Karlsruhe. It is able to simulate both the human cognitive process during driving and the human controlling behaviour. Therefore the queueing theory is used to model the information reception and processing. The control of the lateral and the longitudinal dynamics is carried out by GPC-controllers (**G**eneralized **P**redictive **C**ontrol).

INFORMATION PERCEPTION AND PROCESSING

The process of getting information is composed of different stages [Wic84] as it can be seen in figure 1. In every stage of this process some kind of data transformation takes place which needs a certain amount of time. Sensorial defects of sense organs may affect the quality and the quantity of the percepted signals and, as a consequence, the following processes, too.

At first signals of the environment are percepted with the different sense organs. These stimuli are stored in the short-term sensory store for a short period of time. This time constant is less then a second for visual signals and up to a few seconds for signals of the other senses.

In the next stage the physical stimuli are assigned to certain perceptual categories. There exist different levels of complexity in the categorization task dependent on the task confronting the operator. The more physical dimensions are involved the more complex is the task.

In the following function block 'decision and response selection' a decision is made based on the percepted information and on the situation. This decision may be thoughtful as if to choose the traffic lane on a motorway or it can be reflex-like and almost automatically. Alternatively the perceived information may be stored in the working memory to be processed at a later moment. Informations that are presented again and again can enter the long-term memory in this way.

After the decision a suitable series of muscular movements is made to generate the desired actions. The most important tasks of the so called response execution while driving are steering, accelerating and braking. These actions are mostly trained and they proceed almost automatically. The results of the actions are finally fed back to the different senses.

Every stage of information processing needs a certain part of attention which is a limited ressource. If one special task requires more attention than there is left altogether, there is less available for the other processes. As a consequence their performance will descend.

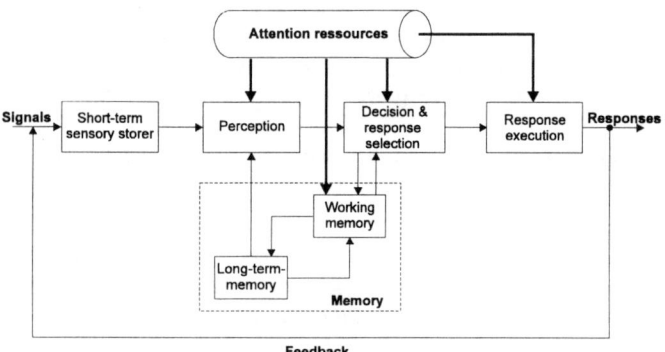

Figure 1: Human information processing

THE CONCEPT

PERCEPTION, MEMORY

The entire model including the driver and the vehicle is shown in figure 2. The block 'Model of the human sensory perception' represents the information perception and a part of the information processing which are modelled as a queueing network. Here the different stimuli, which are available as sensor data, are received. The percepted signals are then related to the informations which correspond to them.

In the function block 'Actualisation, memory' the suitable sensor data of the vehicle are actualised and stored until they are overwritten by new data.

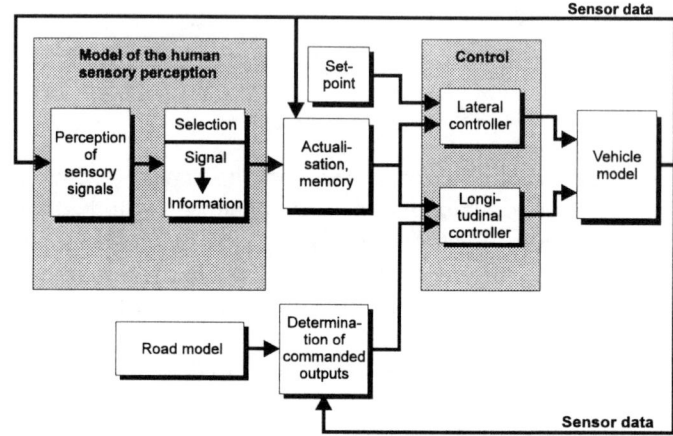

Figure 2: The entire concept

DECISION AND RESPONSE SELECTION

The set-point of the lateral controller is always chosen to zero because the task of this controller is to minimize the lateral displacement. The desired output of the acceleration controller is calculated in the function block 'Determination of commanded output'. Here the desired acceleration is determined depending on the road surface, the curvature and on the driver.

RESPONSE EXECUTION

The block 'Control' contains the lateral and the longitudinal controller which are implemented as GPC-controllers. The lateral controller uses the steering angle as output to minimize the lateral displacement. The output of the longitudinal controller, which is implemented as acceleration controller, are the angle of throttle valve and the force on the brake pedal. The task of this ontroller is to reach the desired acceleration. Therefore the output data manipulate the vehicle model and as a consequence new sensor data is produced.

THE HYBRID DRIVER MODEL

THE QUEUEING NETWORK

The human information processing is modeled with the help of queueing systems (figure 3): one for the visual and one for the vestibular perception. These senses are able to work in parallel as long as there are enough attention ressources for both types of perception left. The other forms of perception can be neglected because they are by far less important.

The **visual** perception is divided into the road fixation points, which supply the information that is relevant for the driving task, and the disturbances. The last ones represent the fixation points inside the car, like the radio, and do not yield any information about the traffic so that they disturb the relevant information perception.

Every time the 'driver' looks at one of these fixation points a transaction is generated in the corresponding source node. The interarrival time of a transaction of the kind 'road fixation point' can be calculated as follows:

$$t_h = \frac{30\frac{m}{s}b_1 + 0\frac{m}{s^2}b_2 + 1.7\frac{\circ}{s}b_3 + 4.5\frac{\circ}{s^2}b_4 + 90^\circ b_5}{\left(\frac{1}{3}\frac{s}{m}|a_x| - 0.001401\frac{s}{m^2}v^2 - 0.09384\frac{1}{m}v + 7\right)*\left(b_1 v + b_2|a_y| + b_3|\dot{\psi}| + b_4|\ddot{\psi}| + b_5|\Delta\psi|\right)*p_h}$$

The parameters $b_1 \ldots b_5$ and p_h are characteristic for each road fixation point and have to be chosen appropriately. The interarrival times of the fixation points inside the car can be determined with the help of [Wie97].

Each generated transaction is corresponded to a characteristic priority and a service time. The priority signifies the importance of the fixation point and the service time reflects the amount of time the driver glances at a fixation point.

Once arrived at the queue, customers with higher priorities are served before the ones with lower priorities. Even if a customer is in service it can be interrupted by another one with higher priority. When this second customer has finished its service the server resumes the interrupted service with the service time decreased by the amount of time already spent on the customer before. Between transactions of the same priority the firs-come-first-served principle is valid.

This queueing system includes the short-term sensory store, that is all customers which wait in the queue for more than one second will be removed.

In the **vestibular** perception transactions are generated when the input signals exceed a specific perception level. The vestibular queueing system includes three parallel servers because it is possible to perceive several stimuli at the same time [Liu94]. The service times of the different vestibular transactions are shorter than those of the visual transactions since the information processing is less complex.

If a customer receives service in one of the queueing systems, the information which is suitable to the percepted signals is selected and then actualized during the whole time of service. When the service of this customer has ended the information will be stored until a transaction that carries the same information receives service. The stored values in the memory which may be not actual serve as input data for the lateral and the longitudinal controller. If a disturbance is in service, the memory is not actualized so there is no actual data available.

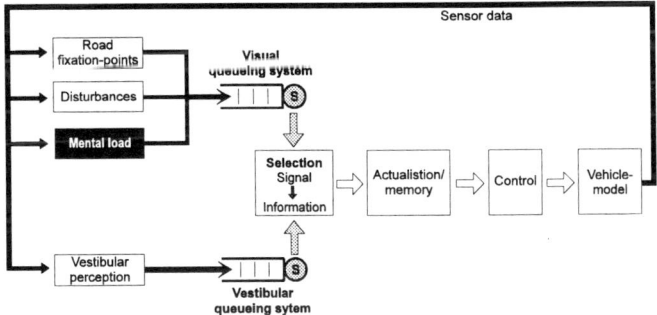

Figure 3: Queueing network of the information processing

The model is also able to reproduce the mental load of a driver which depends on the traffic situation, the velocity and on the experience of a driver. The higher the velocity, the shorter the interarrival time of a transaction of the categorie 'Mental load'. It is possible to vary the experience of a driver with the help of the service time.

If a transaction of the categorie 'Mental load' is in service, no driving relevant information will be actualized. As the mental load increases less actual data is available for the controllers so the control performance gets worse. As a cosequence the event rate of the information reception increases.

DETERMINATION OF THE DESIRED OUTPUT OF THE LONGITUDINAL CONTROLLER

The desired vehicle acceleration is determined with the help of an automata graph [Kra97]. Dependent on the current driving state the suitable acceleration is calculated subject to the road surface, the curvature and to the driver type.

THE CONTROLLERS

The response execution is carried out by two GPC-controllers, one for the lateral and one for the longitudinal control. The characteristic of the GPC-algorithm is that at any time the plant's output (the lateral displacement respectively the deviation of the desired acceleration) is predicted over several steps based on assumptions about future control actions [Cla87]. This corresponds to the behaviour of a human driver who estimates the deviation of the desired output in advance.

Another particularity of the implemented GPC-controllers is the fact that they only need one input parameter in contrast to many other lateral and longitudinal controllers used in earlier studies. Nevertheless the control performance is robust in regard of variations of the velocity and of road surfaces.

The **longitudinal** controller consists of two separate parts: one for the engine and one for the brake system. Dependent on the desired acceleration the suitable controller is switched on and the other one is switched off. If a negative acceleration is demanded, a comparison is made between the desired acceleration and the one that can be produced with the help of the braking effect of the engine. If this effect is big enough the braking controller remains inactive because the braking system is not needed.

DEFINITION OF DIFFERENT DRIVER TYPES

In order to examine the performance of the hybrid driver model three different driver types have been defined: a good driver, a fast one and a novice driver. Therefore three velocity shapes have been developed in order to imitate the speed behaviour of the drivers.

The lateral controller has been modified, too (figure 4). A switch with hysteresis and memory has to be added so that different types of steering behaviour could be simulated. This modification is due to the fact that a human driver doesn't regulate the lateral displacement continuously but event driven.

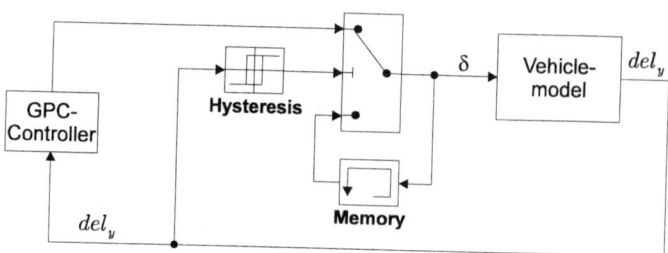

Figure 4: Modified lateral control

If the current lateral displacement exceeds a certain level, the controller output is connected with the vehicle model and the control is active. As soon as the lateral displacement reaches a second lower level the values that are stored in the memory are taken as controller output.

SIMULATION RESULTS

For the simulations a road model is used with different curvature and road conditions (wet, dry). This test path is shown in the first diagram of figure 5. The second diagram shows the different velocity shapes of the three drivers. It can be seen that the driver types differ very much as far as the speed level is concerned.

In figure 6 the lateral displacement and the steering angle during a simulation **without mental load** are presented. The good driver shows the best behaviour of all. The greatest lateral displacement can be observed for the novice driver who also has the biggest steering angles during the simulation.

Another simulation shows the behaviour of the driver types **with mental load** (figure7). It can be noticed that the lateral displacement increases for all drivers because the controller performance decreases since the controllers get less current traffic information. The driver types 'good' and 'fast' are still stable but the novice driver is not able to get enough information and so he leaves the road.

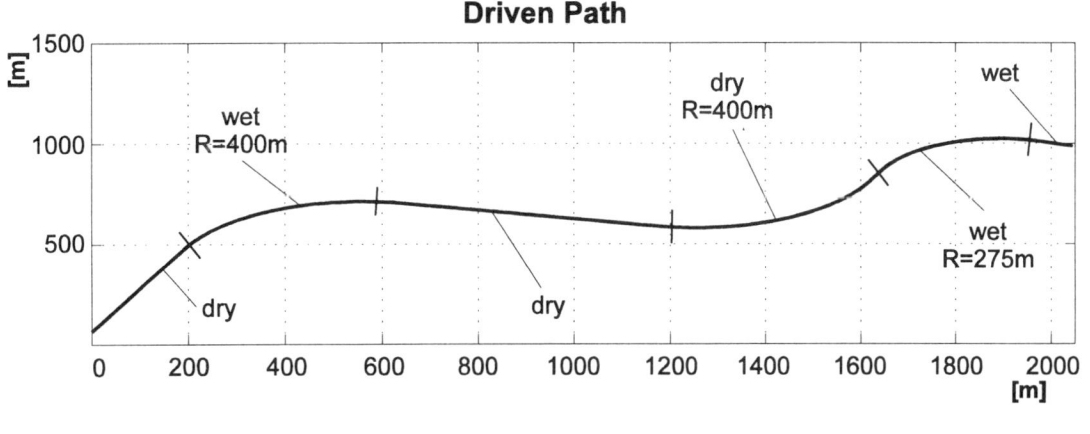

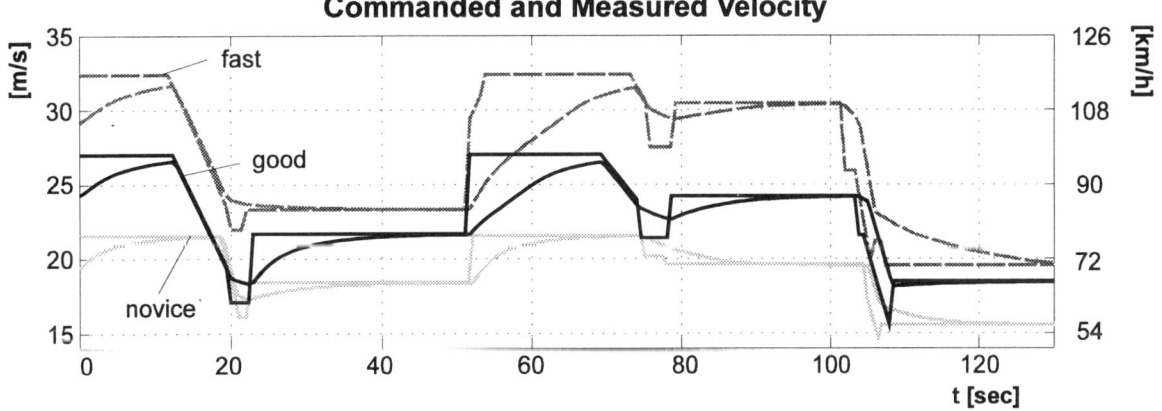

Figure 5: Test path and velocity shapes of three different driver types

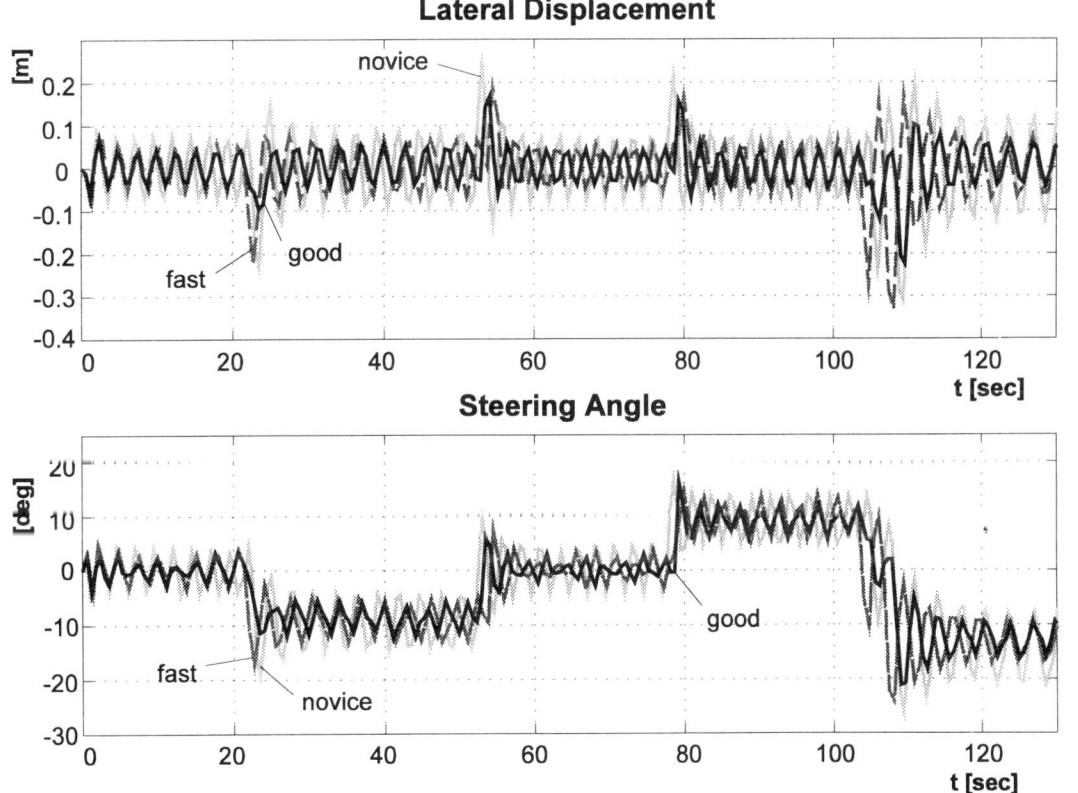

Figure 6: Lateral displacement and steering angle during simulations **without** mental load

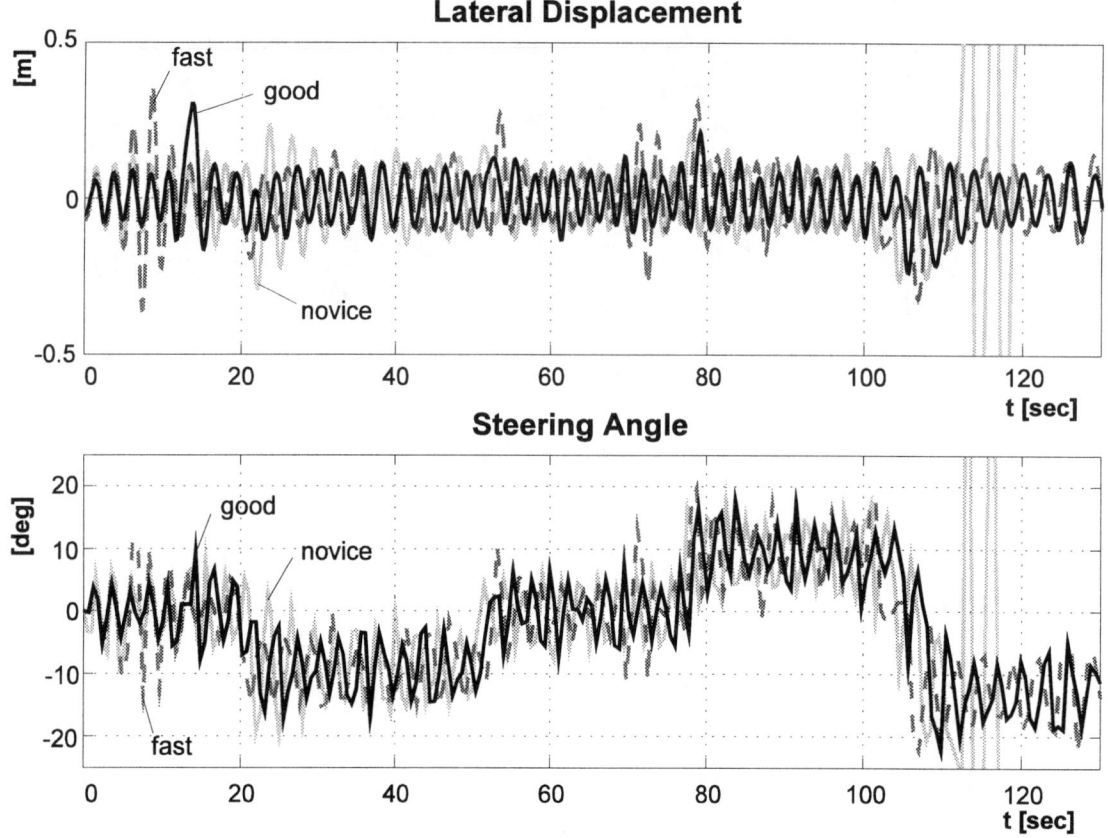

Figure 7: Lateral displacement and steering angle during simulations **with** mental load

REFERENCES

[Cla87] Clarke, D.W., Mohtadi, C., Tuffs, P.S., *Generalized Predictive Control*, Part I and II, Automatica, Vol. 23, No. 2, pp137-160, 1987

[Liu94] Liu, Y., *A Queueing Network Model of Human Performance of Concurrent Spatial and Verbal Tasks*, IEEE Transactions on Systems, Man and Cybernetics, S.2761-2766, 1994

[Mit90] Mitschke, M., *Dynamik der Kraftfahrzeuge*, Band C, Fahrverhalten, Springer Verlag, 1990

[Wic84] Wickens, C.D., *Engineering Psychology and Human Performance*, Charles E. Merill Publishing Company, S. 11-16, 1984

[Wie97] Wierwille, W.W., Tijerina, L., *Darstellung des Zusammenhangs zwischen der visuellen Beanspruchung des Fahrers im Fahrzeug und dem Eintreten eines Unfalls*, Zeitschrift für Verkehrssicherheit, Nr. 43, S. 67-74, 1997

[Wol97] Wolter, T.-M., Jürgensohn, T., Willumeit, H.-P., *Ein auf Fuzzy-Methoden basierendes Situations-Handlungsmodell des Fahrerverhaltens*, ATZ Automobiltechnische Zeitschrift 99, Nr. 3, S.142-147, 1997

980651

An Improved Seating Accommodation Model with Application to Different User Populations

Carol A. C. Flannagan, Miriam A. Manary, Lawrence W. Schneider, and Matthew P. Reed
Transportation Research Institute, University of Michigan

Copyright © 1988 Society of Automotive Engineers, Inc.

ABSTRACT

A new approach to driver seat-position modeling is presented. The equations of the Seating Accommodation Model (SAM) separately predict parameters of the distributions of male and female fore/aft seat position in a given vehicle. These distributions are used together to predict specific percentiles of the combined male-and-female seat-position distribution. The effects of vehicle parameters—seat height, steering-wheel-to-accelerator pedal distance, seat-cushion angle, and transmission type—are reflected in the prediction of mean seat position. The mean and standard deviation of driver population stature are included in the prediction for the mean and standard deviation of the seat-position distribution, respectively. SAM represents a new, more flexible approach to predicting fore/aft seat-position distributions for any driver population in passenger vehicles. Model performance is good, even at percentiles in the tails of the distribution.

INTRODUCTION

Prediction of drivers' seat-positions is important for many aspects of vehicle design, including comfort and safety. The current recommended practice for predicting population percentiles of driver-selected seat position is given in SAE J1517 (1).

Seat-position models have evolved over a period of more than ten years. Philippart et al. (2) used regression equations to predict each of seven percentiles of the seat-position distribution for a 1:1 male-female U.S. driver population. Each percentile was predicted using a second-order function of seat height (H30), obtained from empirical percentile values calculated for each of the vehicles in their database.

This approach directly fits the data at each percentile, thereby avoiding any assumptions about the form of the distribution. However, its use is restricted to the seven percentiles for which equations are available (though these are a good set), and a 50%-male, U.S.-driver population.

By adding the assumption that seat-position is normally distributed, a more flexible seat-position model was created by Flannagan et al. (3). Flannagan et al. also conducted studies to determine how certain vehicle factors influence seat position. They found that across many vehicles, the relationship between H30 and seat position is adequately represented by a linear, rather than a quadratic function. In addition, they found that horizontal steering-wheel-to-pedal distance (measured to the ball-of-foot reference point), seat-cushion angle, and transmission type all influence seat position independently.

Flannagan et al. (3) started with the assumption that the seat-position distribution for a typical driver population can be reasonably described as a normal distribution, and generated equations to predict the two parameters of the normal distribution, the mean and standard deviation. Means and standard deviations of seat position were calculated for each of a number of vehicles. The means were regressed on driver population stature, H30, wheel-to-BOF distance, seat-cushion angle, and transmission type. The standard deviations were regressed on the percentage of males in the driver population (fit with a quadratic function). To calculate predicted percentiles of the seat-position distribution, the user entered the vehicle factors and population factors, and generated predicted mean and standard deviation of the seat-position distribution. Either a normal distribution table or a computer could be used to predict any percentile of the seat-position distribution.

This approach represented an important advancement in modeling seat position because it included population mean stature and gender mix, so that seat position could be predicted for any target driver population.

Since the 1996 paper, an extensive modeling effort has shown that the assumption that the seat-position distribution can be described as a single normal distribution with two parameters leads to systematic errors in predicting tail percentiles. The effort to improve prediction accuracy in the tails of the distribution led to a new, fundamentally different approach to prediction of seat position. This new approach is the topic of the present paper.

MODEL DEVELOPMENT

General Approach

Modeling seat position presents interesting problems whose solutions could be applied in a wide variety of contexts. Two features of the modeling context are particularly challenging. First, because each within-gender stature distribution is approximately normal and because stature and seat position are strongly related, the seat-position distribution is best described as a mixture of two normal distributions. Second, the tail percentiles of the distribution are of greatest interest because these determine the level of accommodation provided by different seat track locations and lengths. However, the tails of any distribution are difficult to estimate because, by definition, there are fewer observations to be made there.

As described above, Flannagan et al. (3) proposed a model of seat position in which a single normal distribution was used to approximate the normal mixture. Although such an approach works well for estimating the mean of the distribution, the errors are larger and biased in the tails. Figure 1 illustrates a single-normal approximation to a normal mixture typical of a seat-position or stature distribution. The approximating distribution is close to the normal mixture, but the greatest deviation is in the tails.

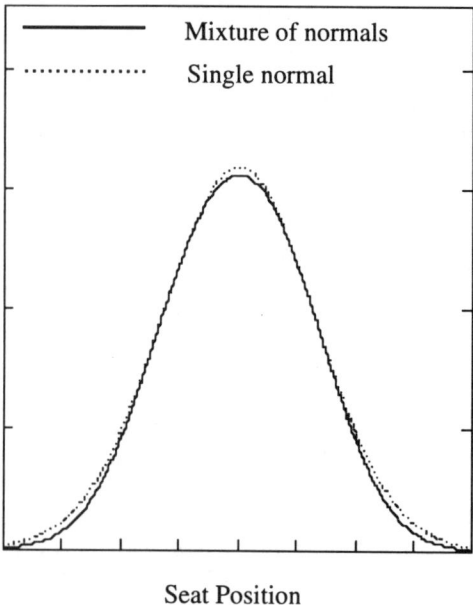

Figure 1. Hypothetical normal-mixture distribution (solid line) and single-normal approximation (dotted line).

Although it might be possible to choose a normal approximation that fits better at the tails at the expense of the center, a model that maintains separate predicted seat-position distributions for males and females was chosen instead. The two predicted distributions are then combined at the end of the prediction process. There are several advantages to this approach. First, the model form matches the form of the presumed true distribution, making artifactual errors of the sort shown in Figure 1 less likely. Second, population means and standard deviations of stature from anthropometric surveys are generally reported separately for males and females. Thus, the seat-position model can make use of these numbers directly. Third, asymmetrical seat-position distributions for populations with male-female ratios other than 1:1 can be better predicted than with a symmetrical single normal distribution. Finally, using separate normal distributions for males and females allows for a new approach to modeling the effect of population stature on seat position.

The new approach is based on certain basic properties of the normal distribution. First, the convolution of a normal distribution with any linear function results in another normal distribution with new parameters. In other words, a set of normally distributed inputs to a linear function will produce a normally distributed output. Second, the sum of two normal distributions is also normal, with mean and variance equal to the sums of the two original means and variances (4). These properties are described mathematically in Equations 1-6.

Suppose that X in Equation 1 represents the distribution of male stature in the driver population. Male stature is distributed normally with mean, μ_x, and variance, σ_x^2. The variable, y', in Equation 2, represents idealized seat position, perfectly linearly related to stature. Equation 3 gives the distribution of y', which is the convolution of the normal distribution (stature) and the linear relationship between stature and seat position. Note that the mean of the new distribution is simply equal to the linear function of the input mean, while the variance has been multiplied by the square of the slope. If the relationship were perfect, this would be sufficient for modeling. However, there is variance in seat position that is not explained by stature, and that variance is represented by ϵ, or error. Under the standard regression model, error is assumed to be normally distributed with mean zero and some variance, σ_ϵ^2 (Equation 4). Equation 5 shows the relationship between (non-idealized) seat position, y, and stature and error. Finally, Equation 6 gives the distribution of seat position as the sum of the two normal distributions from Equations 3 and 4, obtained by adding the error distribution to the idealized seat position distribution.

$$X \sim N(\mu_x, \sigma_x^2) \tag{1}$$

$$y' = mx + b \tag{2}$$

$$Y' \sim N(m\mu_x + b, m^2\sigma_x^2) \tag{3}$$

$$\mathcal{E} \sim N(0, \sigma_\epsilon^2) \tag{4}$$

$$y = mx + b + \epsilon \tag{5}$$

$$Y \sim N(m\mu_x + b, m^2\sigma_x^2 + \sigma_\epsilon^2) \tag{6}$$

Figures 2 and 3 illustrate this transformation with hypothetical relationships between stature and seat-position. In both figures, the normal distribution shown along the horizontal axis represents the stature distribution for one gender, e.g., males. The normal distribution shown along the vertical axis represents the predicted seat-position distribution based on convolution of the stature distribution with the linear relationship. Note that in Figure 2, the slope is relatively shallow, and in Figure 3, it is relatively steep. Ignoring the unexplained error in either relationship, the steeper slope leads to a seat-position distribution with greater variance than does the shallow slope.

The two elements described above, 1) separating prediction for males and females, and 2) using the seat-position/stature relationship to incorporate the effect of population stature in the model, form the basic approach embodied in the new Seating Accommodation Model (SAM). This modeling approach was used with an expanded seat-position database.

Seat-Position Database

Flannagan et al. (3) described the original UMTRI database of seat positions in 11 vehicles, a laboratory buck study involving 18 conditions, and a study of four pairs of vehicles matched on key interior dimensions but differing in transmission type. Since then, subject-selected seat positions have been measured in seventeen additional vehicles using the same basic methodology described in Flannagan et al. (3).*

Each subject sample was stratified by stature as indicated in Table 1 which shows the composition of subject groups. In nine of the vehicles, groups 0 and 11 were not used. For each vehicle, the same number of subjects was selected from each stature group, though that number was different for different vehicles. For example, 60 subjects were tested in a Neon, 5 from each group, while 120 subjects were tested in an Acclaim, 10 from each group. Subjects at the tails of the stature distribution were oversampled in each vehicle to ensure that the extremes were represented by more than one or two subjects. Each subject's stature was measured before driving the vehicle. Subjects drove each vehicle over a specified route for at least 15 minutes and seat position was recorded upon the subject's return. A number of other measures of driver position and posture were taken, but they are not relevant to the modeling of seat position.

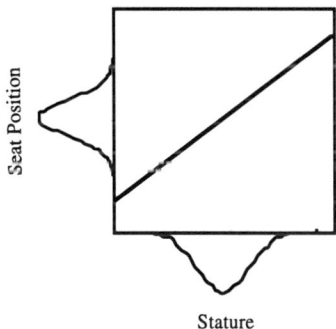

Figure 2. Hypothetical distributions of stature and seat position resulting from a linear relationship with shallow slope.

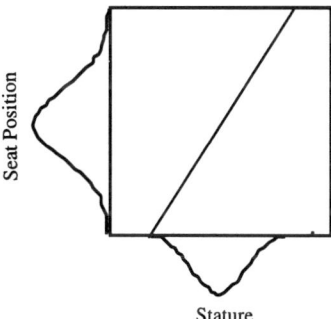

Figure 3. Hypothetical distributions of stature and seat position resulting from a linear relationship with a steep slope.

---

* The rights, welfare, and informed consent of the volunteer subjects who participated in this study were observed under guidelines established by the U.S. Department of Health, Education, and Welfare (now Health and Human Services) on Protection of Human Subjects and accomplished under medical research design protocol standards approved by the Committee to Review Grants for Clinical Research and Investigation Involving Human Beings, Medical School, The University of Michigan.

Table 1
Composition of Subject Groups

Subject Group	Gender	Stature Range	
		mm	Percentile*
Group 0	Female	< 1511	< 5
Group 1	Female	1511-1549	5-15
Group 2	Female	1549-1595	15-40
Group 3	Female	1595-1638	40-60
Group 4	Female	1638-1681	60-85
Group 5	Female	1681-1722	85-95
Group 6	Male	1636-1679	5-15
Group 7	Male	1679-1727	15-40
Group 8	Male	1727-1775	40-60
Group 9	Male	1775-1826	60-85
Group 10	Male	1826-1880	85-95
Group 11	Male	> 1880	> 95

*by gender from Abraham et al. (5)

The UMTRI database now contains seat positions for 36 to 120 subjects of varying statures in each of 36 vehicles and 18 buck conditions. The vehicles, along with key dimensions and number of subjects tested, are listed in Table 2.

Checking Basic Assumptions

The first step was to check the basic assumptions in Equations 1 through 6. Population stature is assumed to be normally distributed within gender, based on data from Abraham et al. (5). Stature and seat position are assumed to be linearly related as reported previously by Flannagan et al. (3) for a laboratory buck study, and there is no evidence of a nonlinear relationship between stature and seat position in any of the vehicles.

The third assumption is that the distribution of unexplained error is the same across all values of the independent variable, stature. This assumption was checked by regressing seat position on stature separately for each vehicle, and inspecting the residuals for signs of heteroscedasticity. In particular, it was hypothesized that people at the extremes of the stature distribution might be less variable in their selection of seat position than people in the middle of the distribution. However, only six vehicles showed any evidence of decreased variance at either tail. Furthermore, data from the laboratory buck study, in which subjects had ample seat track travel, showed no signs of heteroscedasticity. It was concluded that equal variance is a reasonable assumption to make for modeling purposes.

In addition to these basic assumptions, the data support additional simplifying assumptions that make both the modeling process and the end result more straightforward. First, stature does not interact with vehicle variables, such as seat height or steering-wheel position, in its effects on seat position. Second, vehicle variables do not affect the variability of seat position. Third, the effect of stature on seat position is the same for males and females.

Table 2
Vehicles, Dimensions, Seat Conditions and Sample Sizes in UMTRI Database

Vehicle	Number of Subjects	Seat Height (mm)	Wheel-BOF (mm)	Seat-Cushion Angle (degrees)	Transmission Type
Camaro	55	177	616	13	Automatic
Blazer	55	288	591	7.5	Automatic
Cadillac	55	240	590	8	Automatic
Oldsmobile	55	250	564	18	Automatic
Monte Carlo	55	231	597	11	Automatic
Pontiac 6000	55	266	583	16	Automatic
APV	55	381	518	12	Automatic
CK Truck	55	303	570	12.5	Automatic
Econoline	120	420	447	9.5	Manual
Firebird	120	154	650	15.5	Manual
Camry	50	240	567	13	Automatic
Civic	50	232	547	18	Automatic
Grand Am	50	230	588	15	Automatic
Grand Am	50	234	587	17	Manual
Celica auto	36	215	587	13	Automatic
Celica man	36	215	587	13	Manual
Taurus L	120	257	557	12	Automatic
Taurus SHO	120	257	557	12	Manual
C/K Pickup	50	327	551	10.5	Automatic
Yukon	50	327	551	10.5	Manual
Trans Am	100	165	623	14	Manual
Mazda 626	100	234	561	18	Manual
Grand Prix	100	250	623	12	Automatic
Transport	100	250	530	16	Manual
Tercel	100	324	478	13	Automatic
Previa	100	381	504	10	Automatic
Laser	100	188	555	19	Automatic
Voyager	120	326	504	14	Automatic
Acclaim	120	220	559	18	Automatic
LHS	120	250	597	18	Automatic
Avenger	120	189	577	17	Automatic
Cherokee	120	298	607	12	Automatic
Laser B	120	197	550	11	Automatic
Dakota	60	298	600	12	Manual
Ram Pickup	60	346	512	13	Manual
Neon	60	212	565	23	Manual
Lab. buck 1	100	180	650	11	Automatic
Lab. buck 2	100	180	600	11	Automatic
Lab. buck 3	100	180	550	11	Automatic
Lab. buck 4	100	180	650	18	Automatic
Lab. buck 5	100	180	600	18	Automatic
Lab. buck 6	100	180	550	18	Automatic
Lab. buck 7	100	270	600	11	Automatic
Lab. buck 8	100	270	550	11	Automatic
Lab. buck 9	100	270	500	11	Automatic
Lab. buck 10	100	270	600	18	Automatic
Lab. buck 11	100	270	550	18	Automatic
Lab. buck 12	100	270	500	18	Automatic
Lab. buck 13	100	360	550	11	Automatic
Lab. buck 14	100	360	500	11	Automatic
Lab. buck 15	100	360	450	11	Automatic
Lab. buck 16	100	360	550	18	Automatic
Lab. buck 17	100	360	500	18	Automatic
Lab. buck 18	100	360	450	18	Automatic

These three important results make it possible to separate the modeling of the effects of vehicle variables from the effects of stature. Specifically, vehicle variables need only be considered in predicting mean seat position. In addition, the same pair of equations can be used to predict parameters of the seat position distribution for both males and females.

Modeling the Effects of Vehicle Variables

Because the sample of drivers in each vehicle was stratified by stature, it is necessary to weight each observation according to its likelihood of occurrence in the population. Once weighted, median observed seat positions were calculated for each vehicle. In the normal distribution, the mean and median are the same. Although a weighted mean could have been used as the empirical measure of central tendency, the median was chosen, since it was felt that it would be less influenced by the effects of seat-track detents or other unusual characteristics of the distributions.

Using the expanded database, median seat position (for a 50%-male U.S. stature population) was regressed on H30, seat-cushion angle, wheel-to-BOF distance, and transmission type. Care was taken to identify any effects that differed between static and dynamic test conditions. In particular, seat-cushion angle had previously been found to have an accentuated effect on seat position in the laboratory buck as compared to dynamic test conditions. In addition, manual transmission testing was done only under dynamic conditions on the assumption that driving is necessary to adjust seat position in response to the clutch pedal. In such cases, only dynamically tested vehicles were used.

The resulting equation ($R^2 = 0.90$) is:

$$\hat{\mu} = 746 - 0.24h - 2.19p + 0.41w - 18.2t \quad (7)$$

where,

$\hat{\mu}$ = predicted mean seat position (mm aft of BOF)
h = H30 (mm)
p = seat-cushion angle (degrees)
w = wheel-to-BOF distance (mm), and
t = transmission type (0 if automatic; 1 if manual).

Figure 4 shows the observed versus predicted median seat positions for the vehicles tested dynamically. The regression equation predicts median seat position well, at least for a 1:1 male-female U.S. population distribution.

It is important to note that, although Equation 7 was generated from medians of the male-female combined distribution, it can be used to predict means for male-only and female-only seat-position distributions because the effects of vehicle variables and stature on seat-position are the same for males and females.

Modeling the Effects of Stature

The next step was to determine the relationship between seat position and stature, and to incorporate stature effects into the model. A subset of seat-position data from twenty one vehicles was chosen for this analysis. Fifteen vehicles were excluded because the sample size was too small after censored observations were excluded. Censored data result from too little rearward or forward seat-track travel to accommodate the tallest or shortest subjects, such that there is a "piling up" of subjects at the forward and rearward track positions. In such

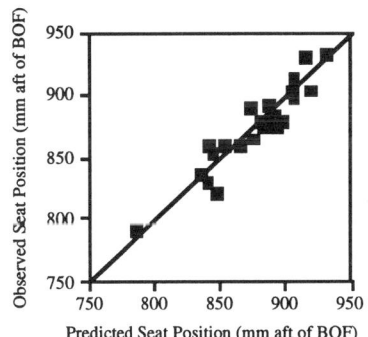

Figure 4. Observed versus SAM-predicted 50th-percentile seat position.

cases, all observations falling at the front (when forward censoring occurred) or rear (when rearward censoring occurred) of the track were eliminated from the analysis, even though some of those subjects may have been accommodated at the front or rear position.

For each vehicle, seat position was regressed on stature, and the slope, intercept, R^2, and mean squared error of the linear fit were recorded. The mean squared error is the estimate of the unexplained variance in the data, corresponding to σ_ε^2 in Equation 4.

The mean of the stature coefficients of the 21 vehicles, weighted by the number of subjects in each vehicle after excluding censored observations, is 0.433. This becomes the model coefficient for mean stature. To incorporate the stature component into Equation 7, the intercept was recalculated so that predictions of mean seat position remain the same for 1:1 male-female U.S. stature distribution. The new prediction equation for mean seat position becomes:

$$\hat{\mu} = 16.83 + 0.433\mu_s - 0.24h - 2.19p \\ + 0.41w - 18.2t \quad (8)$$

where,

μ_s = mean stature of single-gender (male or female) driver population
$\hat{\mu}$ = predicted mean seat position (mm aft of BOF)
h = H30 (mm)
p = seat-cushion angle (degrees)
w = wheel-to-BOF distance (mm), and
t = transmission type (0 if automatic; 1 if manual).

The weighted mean of the mean squared error is 29.75. Inserting values for the slope and mean squared error into Equation 6, the prediction equation for the standard deviation of the seat-position distribution becomes:

$$\hat{\sigma} = \sqrt{(.433)^2 \sigma_s^2 + (29.75)^2} \\ = \sqrt{.187\sigma_s^2 + 885} \quad (9)$$

where,

σ_s = standard deviation of stature distribution of male or female driver population
$\hat{\sigma}$ = predicted standard deviation of single-gender seat-position distribution

Combining Male and Female Predicted Seat Position

Equations 8 and 9 define SAM. However, these equations are designed to be used on male and female driver population distributions separately. For most vehicles, the target population will be some mixture of males and females, so the two predicted seat-position distributions need to be combined to estimate population percentiles of seat position.

In most cases, using the two distributions to generate percentiles is simple. For the median seat position, the two predicted means can simply be averaged in proportion to the gender mix. That is:

$$\hat{\mu} = k\hat{\mu}_M + (1-k)\hat{\mu}_F \qquad (10)$$

where,
$\hat{\mu}$ = predicted mean of mixed-gender seat-position distribution
k = proportion of males in the target population
$\hat{\mu}_M$ = predicted mean of male seat-position distribution
$\hat{\mu}_F$ = predicted mean of female seat-position distribution

For population seat-position percentiles at or below the 5th, or at or above the 95th, the overlap between the single-gender distributions is so small that the appropriate tail percentile of either the male or female distribution alone is sufficiently accurate. For example, Table 3 gives the percent of females and the corresponding percent of males under the left tail of a hypothetical (but typical) seat-position distribution. As indicated, less than one-half percent of males are seated forward of the 10th-percentile female seat position. Therefore, the 10th percentile of the female seat-position distribution is a reasonable approximation of the 5th percentile of a 1:1 mixture of males and females. The general formula is given in Equations 10 and 11.

Table 3
Proportion of Females and Males under the Left Tail for a Seat-Position Distribution in a Typical Vehicle

Percent females in left tail	Percent males in same tail	Ratio Male/Female
2.5%	0.06%	0.024
5.0%	0.16%	0.032
10.0%	0.47%	0.047
25.0%	2.14%	0.086
40.0%	5.18%	0.130
50.0%	8.26%	0.165

To find the appropriate percentile at the lower tail, use:

$$p_f = \frac{p_t}{1-k} \qquad (11)$$

where,
p_t = target population percentile,
k = proportion of males in the target population, and
p_f = percentile of the female-only distribution that corresponds to the target percentile of the combined distribution.

At the upper tail, use:

$$p_m = 1 - \frac{1-p_t}{k} \qquad (12)$$

where,
p_t = target population percentile,
k = proportion of males in the target population, and
p_m = percentile of the male-only distribution that corresponds to the target percentile of the combined distribution.

For cases in which target seat-position percentiles are between 5% and 95% but not 50%, or the male-female ratio is very different from 1:1, the normal mixture function must be solved for the desired value. The function is given in Equation 13 and can be solved for x using various numerical methods.

$$p = k\Phi_M(x) + (1-k)\Phi_F(x) \qquad (13)$$

where,
$\Phi(x)$ is the cumulative normal distribution for random variable X with mean, μ, and variance, σ^2, and
p is the target percentile of the population seat-position distribution.

Model Summary

To summarize the model from the user's point of view:
1) Determine the values for vehicle measures of H30, wheel-to-BOF distance, seat-cushion angle, and transmission type, and also the values for mean and standard deviation of male stature, mean and standard deviation of female stature, and the proportion of males in the driver target population.
2) Estimate the mean of the male seat-position distribution by using H30, wheel-to-BOF distance, seat-cushion angle, transmission type, the mean of the male stature distribution, and Equation 8:

$$\hat{\mu} = 16.83 + 0.433\mu_s - 0.24h - 2.19p + 0.41w - 18.2t \qquad (8)$$

3) Calculate the standard deviation of the male seat-position distribution using the standard deviation of the male stature distribution and Equation 9:

$$\hat{\sigma} = \sqrt{.187\sigma_s^2 + 885} \qquad (9)$$

4) Repeat steps 2 and 3 for females.

5) Generate population percentiles of seat position as follows:
1) 50th percentile:

$$\hat{\mu}_{.5} = k\hat{\mu}_M + (1-k)\hat{\mu}_F \quad (10)$$

2) 5th percentile or less:

$$p_a = \frac{p_t}{1-k} \quad (11)$$

3) 95th percentile or greater:

$$p_m = 1 - \frac{1-p_t}{k} \quad (12)$$

4) all other percentiles, solve for x in:

$$p = k\Phi_M(x) + (1-k)\Phi_F(x) \quad (13)$$

Model Performance

To test the performance of the model, observed 2.5th, 50th, and 97.5th percentiles of the seat-position distribution were calculated for each of 26 vehicles. All of these vehicles were included in some part of the model development, but several had been excluded from the modeling of the effect of stature. In addition, the three seat-position percentiles were calculated for a 1:1 U.S. stature population, a 1:1 Japanese stature population, and a 1:1 Central European stature population (4,5).

The observed percentiles were calculated using the same approach embodied in SAM. For each vehicle, seat position was regressed on stature, and the slope, intercept, and mean squared error were recorded.

Three target distributions were defined for the purpose of testing the model. These were nominally Japanese stature distribution, U.S. stature distribution, and Central European stature distribution. The source for parameters of the U.S. stature distribution is Abraham et al. (4), and the source for Japanese and Central European stature distributions is Jurgens et al. (5). Table 4 gives the mean and standard deviation of male and female statures used.

The subjects in the samples were all U.S. drivers of different statures. To the extent that stature is the primary subject factor influencing seat position, non-U.S. seat-position distributions can be estimated from these data by using the appropriate parameters of the stature distributions as described below. It is possible that other factors common to drivers of a particular nationality could influence seat position. Such factors will not be reflected in these estimates, but they are unlikely to have a large effect on seat position when compared to the effect of population stature.

To calculate observed mean seat position for males, the mean male stature of the target stature distribution, and the slope and intercept of the seat-position/stature regression were entered into the formula for the seat-position-distribution mean in Equation 8. Similarly, male standard deviation of the target stature distribution, and the slope and mean squared error from the regression were entered into the formula for the seat-position-distribution variance in Equation 9. The process was repeated for the female mean and standard deviation of stature. Once the mean and standard deviation of the seat-position distributions were calculated for the target population, percentiles were calculated according to Equation 10, 11, 12, or 13, as appropriate.

Table 4
Means and Standard Deviations of Stature Distributions for Different Populations

	Male		Female	
	Mean (mm)	Std Dev (mm)	Mean (mm)	Std Dev (mm)
US	1753	71	1615	64
Japan	1720	58	1590	61
Central Europe	1770	58	1660	49

For each vehicle, SAM was used to predict percentiles of the seat-position distribution for each of the three target populations. Figure 5 shows the model performance at the 2.5th percentile across all target populations for all vehicles. On the whole, the model does very well at the 2.5th percentile, with an average residual of -2.4 mm. The largest residual was -34 mm, and 86% of residuals were less than 25 mm in absolute magnitude.

Figures 6 and 7 show model performance at the 50th and 97.5th percentile, respectively. At the 50th percentile, the average residual is -0.1 mm, and no prediction was in error by more than 25 mm. At the 97.5th percentile, the largest residual was 33 mm, and 78% of residuals were less than 25 mm in absolute magnitude.

Clearly, the model performance is quite good, particularly across vehicles. This is to be expected at the 50th percentile of the U.S. population, given that all of these vehicles were included in the modeling of the prediction of the mean seat position for U.S. population. However, not all of these vehicles contributed to the modeling of the effect of stature on mean or standard deviation of seat position, which influences both the estimates of outer percentiles and estimates of all percentiles of non-U.S. populations. At the 50th percentile, the average residual for the Japanese population is -0.92 mm, and -0.54 mm for the Central European population. Even at the outer percentiles, which are more difficult to predict accurately, the average residual for any population is less than 3 mm.

SUMMARY AND CONCLUSIONS

Early efforts to model seat position used a regression approach to produce separate models of seat position at seven fixed percentiles for a U.S. stature distribution (1). Flannagan et al. (3) improved on the earlier approach by incorporating effects of other vehicle variables and proposing a more flexible modeling approach that can be used for different stature distributions and gender proportions.

In the present paper, an innovative approach to modeling seat position is presented. The new approach predicts seat-position distributions for males and females separately. In addition, it makes use of basic properties of the normal distribution to model the effect of stature on seat position.

The new Seating Accommodation Model or SAM allows the user to adjust the prediction for driver populations of any composition. As an added benefit, the prediction equation for mean seat position, can be used to predict the mean seat position of individuals with a specified stature. In previous models of seat position (including J1517 and earlier versions of SAM), the model produced only the nth percentile of the seat-position distribution, which is generally not the same as the seat position of the nth percentile person (in stature).

The effect of vehicle factors on mean seat position has been demonstrated extensively under both laboratory and dynamic testing conditions. The addition of a model component reflecting population composition is an important advancement in seat-position modeling. At this time, the validity of the population stature component relies on the validity of the assumption that stature influences seat position in the same way, regardless of the nationality of the driver. It is likely that stature is far more influential than any other variable. However, testing with other populations is an important future step in validating SAM.

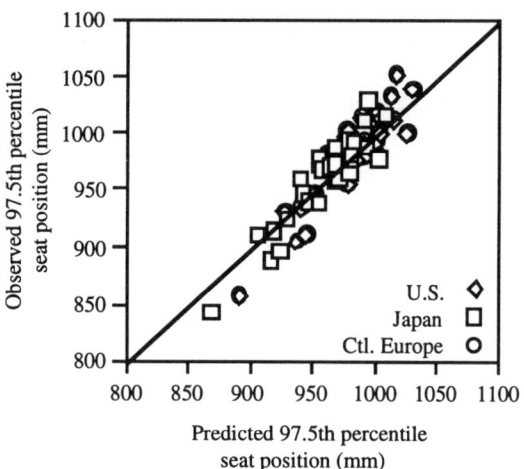

Figure 7. Observed versus SAM-predicted 97.5th percentile seat position.

ACKNOWLEDGMENTS

The authors would like to thank members of the UMTRI Biosciences Division staff who contributed to this work. They include: Bethany Eby, Jeff Lehman, Brian Eby, and Cathy Harden. The authors would like to acknowledge the sources of financial support for this research, including the American Automobile Manufacturers Association (AAMA), General Motors Corporation, Ford Motor Company, and Chrysler Corporation. The authors express sincere gratitude to the members of the project task group who provided valuable input to the experimental methods, the development of the test facility, and interpretation of the data. These include Ron Roe of General Motors Corporation, Gary Rupp of Ford Motor Company, and Howard Estes of Chrysler Corporation.

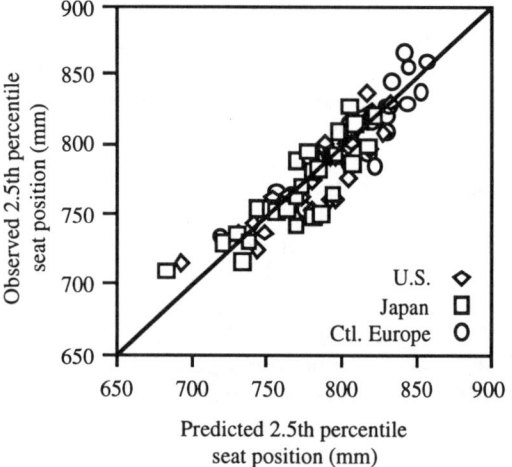

Figure 5. Observed versus SAM-predicted 2.5th percentile seat position.

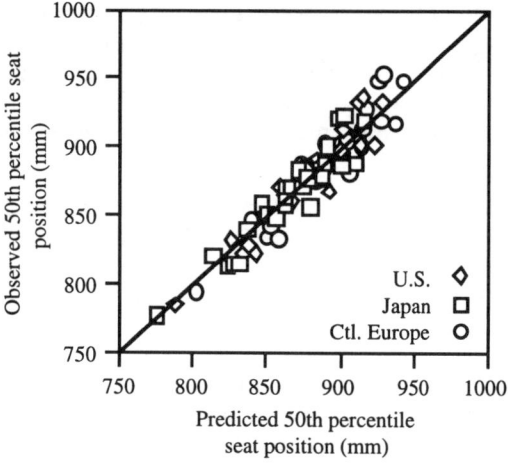

Figure 6. Observed versus SAM-predicted 50th percentile seat position.

REFERENCES

1. SAE Handbook (1991). Volume 4, *Driver selected seat position-SAE J1517*, pp. 34.175-176. Society of Automotive Engineers, Inc., Warrendale, PA.
2. Philippart, N.L., Roe, R.W., Arnold, A.J., and Kuechenmeister, T.J. (1984). *Driver selected seat position model.* SAE Technical Paper 840508. Society of Automotive Engineers, Inc., Warrendale, PA.
3. Flannagan, C.C., Schneider, L.W., and Manary, M.A. (1996). *Development of a Seating Accommodation Model.* SAE Technical Paper 960479. Society of Automotive Engineers, Inc., Warrendale, PA.
4. Ross, S. (1984). *A First Course in Probability.* MacMillan, New York.
5. Abraham, S., Johnson, C.L., and Najjar, M.F. (1979). Weight and height of adults 18-74 years of age. *Vital and Health Statistics.* Series 11, No. 211.
6. Jurgens, H.W., Aune, I.A., and Pieper, U. (1990). *International Data on Anthropometry.* International Labour Office, Geneva.

980653

Seating Physical Characteristics and Subjective Comfort: Design Considerations

Se-Jin Park
Ergonomics Research Group
Korea Research Institute of Standards and Science

Young-Shin Lee and Yoon-Eui Nahm
Department of Mechanical Design Engineering
Chung-Nam National University

Jung-Woo Lee and Jin-Sun Kim
Department of Industrial Engineering
Jeon-Ju University

Copyright © 1988 Society of Automotive Engineers, Inc.

ABSTRACT

The seating comfort is one of the most important indices that determine the performance of automotive seat. A subjective evaluation has been the general method for seating comfort evaluation of automotive seat. In our study, we have attempted to determine a method for objectively evaluating seating comfort. This paper discusses design parameters for automotive seats intended to meet the objectives: comfort. The body pressure distribution for automotive seat, objective technique, is investigated and compared with the subjective evaluation. In addition, this paper reports the results of a study comparing the subjective assessment with physical characteristics. The results show that the body pressure distribution and physical characteristics can be used as the objective technique for the seating comfort evaluation.

INTRODUCTION AND BACKGROUND

The improvement of automotive seating system, particularly for the driver, has been the subject of intense interest. Generally, there were two types of the published research in this area: studies of the positional and anthropometric requirements for automotive seat and studies of the comfort performance for the seat. The first type examines the anthropometric and biomechanical aspects of vehicle seating packages, and the second type rates the comfort of vehicle seat generally through a subjective evaluation procedure.

Vehicle seat is the most important part for the driver and passenger, and must accommodate the riders who have diverse physique. Vehicle seat plays a part in eliminating the shock and vibration in the car. Seat cushion should be designed to distribute the body weight and absorb the shock and vibration in the car as much as possible [1, 2]. As Dempsey indicated, the hips mainly support body weight. Specially, high body pressure is concentrated upon the ischial tuberosities [3]. In addition, 18% of the body weight are distributed in each ischial tuberosity [4]. That may interrupt the circulation of blood through the artery. This interruption may cause ache, palsy and pain [5, 6].

Therefore, the body pressure distribution between body and seat surface has been regarded as one of the most important factors that have an effect on the seating comfort [7, 8, 9, 10, 11, 12, and 15]. The design of automotive seating system for the improved occupant comfort is one of the primary goals for seat system engineering teams. Because of such factors as user subjectivity, occupant physique, seat geometry, and amount of time spent sitting, comfort measurement is difficult.

Recently, the automotive seat industry has begun to examine and use seating evaluation methodologies that provide objective, quantitative datum [17]. The predominating philosophy is to combine the subjective and objective techniques for the seat evaluation process. For the practical assessment for seat products, more attention has been paid to seat system and component tests than sensory and posture measurements. Indeed, from an engineer's point of view, the physical performance and comfort properties of seating system are the major concern [13].

Therefore, this study investigated some of these concerns about the possibility of using the body pressure distribution and physical characteristics in today's production seats. The primary goal was to evaluate the comfort of the automobile seats by comparing subjective technique with objective techniques like the body pressure distribution and physical characteristics for a sample of seventy-five drivers in fourteen seats: Thirty drivers were chosen and seven seats were selected in the first session, while thirty-five drivers were chosen and seven seats were selected in the second session.

The important factors for the overall seating comfort as well as the body pressure ratio for the comfortable seat were presented. Additionally, the relationship between the subjective technique and physical characteristics was shown. The differences of the physical characteristics between comfortable and uncomfortable seats were statistically significant. All of the statistical analyses were conducted using SAS statistics' package.

METHODS

SUBJECTS AND SEATS

For the subjective evaluation of seat comfort, the study was performed with seventy-five healthy subjects. Their characteristics appear in Table 1. Two different sets of tests were performed: one set involving seven seats and thirty drivers was to evaluate the relationship between the subjective technique and the body pressure ratio while the second set involved seven seats and thirty-five drivers was to examine the relationship between the subjective technique and physical characteristics (four domestic products and ten foreign products).

For detailed subjective sensing evaluation, subjects were selected from people who were engaged in the development of seats in the automobile company. The subjects selected for participation in the study were within normal body weight levels for their stature. One subject representing Korean standard physique was chosen for the measurement of the body pressure distribution (height; 174cm, weight; 72kg). Each seat was placed in the design position as shown in Figure 1, and the subjects were allowed to move the seat forward and aft to accommodate placement of the feet on the footplate.

Table 1 Subjects Characteristics

Classification	Age (year)	Height (cm)	Sitting height (cm)	Weight (kg)
Mean	29.49	171.36	91.01	69.09
S.D	3.63	6.28	3.15	13.30

Figure 1. Experimental set up.

TEST PROTOCOL

Two different sets of tests were performed; one set is to investigate the correlation between the subjective technique and body pressure distribution by including thirty subjects and seven seats, while the second set is to examine relationship between the subjective evaluation and physical characteristics by including thirty-five subjects and seven seats. The order of testing in each seat was randomized between subjects.

For the measurement of body pressure distribution, one subject was chosen. In order to minimize the difference of set-up about the chosen fourteen seats, the subject took the posture such as 110-degree hip angle, 15-degree thigh angle, 150-degree knee angle, 100-degree ankle angle and 25-degree torso angle documented by Yoshiyuki Matsuoka.

In the second set, seat tests were conducted to find the static and dynamic spring constant, damping ratio, force-deflection curve, the hardness of the foam padding, and the hardness of the seat cover.

SUBJECTIVE TECHNIQUES

One questionnaire was employed to record each subject's short-term evaluation of the test seats. The questionnaire was composed of twenty-three seating comfort items. For the statistics' analysis, each item in the evaluation sheet was defined as $X1, X2, X3, .., X22$ for the comfort type and $Y1, Y2, Y3, .., Y23$ for the satisfaction rate (Table 2).

The subjects were asked to rate each seat characteristic. For example, the subjects were asked to describe the firmness of the cushion by making a mark between the words "soft" and "firm". Additionally, the subjects expressed the satisfaction of the item by circling a number from one to five, with five indicating the highest satisfaction. These tests were completed for thirty-minute intervals. The subjects were outfitted in fitting clothing. In order to evaluate the overall seat comfort, evaluators who engaged in the development of seat in the automobile company were chosen.

Table 2 Evaluation Sheet of Seat Characteristics

DATE:　　　　SEAT:　　　　SUBJECT NO:

	type	satisfaction
1. firmness of cushion padding	hard 1 2 3 4 5 soft	discomfort 1 2 3 4 5 comfort
2. sense of hip sinking	large — small	discomfort — comfort
3. fitness of cushion (horizontal direction)	bad — good	discomfort — comfort
4. fitness of cushion (vertical derection)	bad — good	discomfort — comfort
5. sense of thigh oppression	large — small	discomfort — comfort
6. sense of thigh by side support	small — large	discomfort — comfort
7. sense of hip by side support	small — large	discomfort — comfort
8. sense of hip sliding	large — small	discomfort — comfort
9. sense of frame touching	a lot — none	discomfort — comfort
10. firmness of back padding	hard — soft	discomfort — comfort
11. fitness of back (horizontal direction)	bad — good	discomfort — comfort
12. fitness of back (vertical direction)	bad — good	discomfort — comfort
13. position of lumbar support	high — low	discomfort — comfort
14. sense of lumbar oppression	large — small	discomfort — comfort
15. discomfort feeling of armpit	a lot — none	discomfort — comfort
16. sense of bending front	a lot — none	discomfort — comfort
17. sense of bending back	a lot — none	discomfort — comfort
18. constricted feeling on stomach	large — small	discomfort — comfort
19. sense of touching frame	a lot — none	discomfort — comfort
20. firmness of head rest	hard — soft	discomfort — comfort
21. distance between head and head rest	far — near	discomfort — comfort
22. fitness of head rest	good — bad	discomfort — comfort
23. overall seating comfort		discomfort — comfort

OBJECTIVE TECHNIQUES

Body pressure distribution

To measure the body pressure distribution, we used the measurement system made by Novel Company. The thin polymer-type films were arranged in form of a matrix and sandwiched with cloth to form a mat. The thin polymer-type films were as follows: for the seat cushion and back, 65 rows' lengthwise and 32 rows' width.

To obtain the body pressure ratio, we divided each seat into 5 areas as shown in Figure 2. In case of the seat back, the area was limited to only lumbar region, because back pressure was mainly concentrated upon the lumbar area. The height of area (Rl) was selected from cushion surface to 20cm. Hip area (Rh) was determined by considering the hip point and the distance between ischial tuberosities (about 12cm). Each pressure ratio was calculated as shown Table 3.

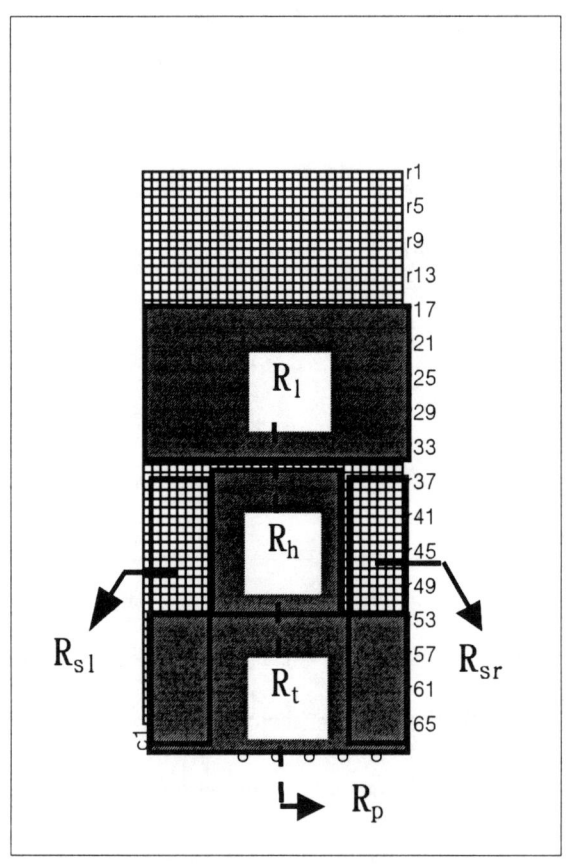

Figure 2. Division of seat areas.

Table 3 Calculation of Pressure Ratios

Classification	Calculation method
Ratio of hip pressure	Rh = pressure of Rh area / pressure of cushion
Ratio of side pressure(right, left)	Rsl, Rsr = pressure of Rsl or Rsr area / pressure of cushion
Ratio of thigh pressure	Rt = pressure of Rt area / pressure of cushion
Ratio of lumbar pressure	Rl = pressure of Rl area / pressure of back
Ratio of symmetric pressure	Rp = pressure of Rsl (or Rsr) area / pressure of Rsr (or Rsl) area

Physical characteristics tests

From an engineer's point of view, the physical performance and comfort properties of seating system are the major concerns. Therefore, we performed the seat tests to find the static spring constant, dynamic spring constant, damping ratio, force-deflection curve, the hardness of the foam padding, and the hardness of the seat cover.

To find the static spring constant, we gave a static load to the seat: gradually in the order of 45, 100 and 15kg for the cushion and in the order of 15, 50 and 10kg for the back. In case of the damping ratio test, we dropped the 60kg weight from the height of 100mm, and then recorded that curve. Additionally, to examine the hardness of the foam padding and the hardness of the seat cover, we gave a static load to the seat: for the hardness of the foam padding, the 100mm diametric weight in the order of 0, 6 and 0kg, and for the hardness of the seat cover, the 50mm diametric weight in the order of 0, 6 and 0kg. Of those seat tests, only the hardness test of the seat cover was wrapped up in the seat cover because the test was to find the properties about seat cover. The test procedures were shown in Figure 3. For the statistics' analyses, each item was defined as variables as shown in Table 4.

Table 4 The Definition of Variables

Variable	Definition		Variable	Definition	
CKS	Cushion	Static spring constant	CRF15	Cushion	Hardness of the foam padding at the 150, 300mm position from the rear edge respectively
CD3		$\delta 3$ for the 15kg weight	CRF30		
BKS	Back	Static spring constant	CR10	Back	Hardness of the foam padding at the 100mm position from the right edge
BD1		$\delta 1$ for the 15kg weight	BBT20		Hardness of the foam padding at the 200mm position from the bottom edge
XX1	Displacement X1		BL5		Hardness of the foam padding at the 50mm position from the left edge
XX2	Displacement X2		CD15	Cushion	Hardness of the seat cover at the 150mm position from the rear edge
DAMP	Natural damping ratio		CDL5		Hardness of the seat cover at the 50mm from position the left edge
CKD	Dynamic spring constant		BDL5	Back	Hardness of the seat cover at the 150mm position from the left edge
CRF5	Cushion	Hardness of the foam padding at the 50mm position from the rear edge			

RESULTS

OVERALL SEATING COMFORT

To examine relationship between each evaluation item (Y1, Y2,.. , Y22) and the overall seating comfort (Y23), analysis of variance, correlation analysis and regression analysis were performed with the summation values, which were obtained by the questionnaire. First, in order to classify seats as comfortable or uncomfortable, ANOVA (analysis of variance) was performed. Of the two different types of sets, the results in the first session revealed that there was not necessarily a significant difference between the seats ($p>0.05$), while the results in the second session revealed that there was a significant difference between the seats ($p<0.01$). Seat 6 in the first session and seat 5 in the second session received the highest value. On the other hand, seat 3 in the first session and seat 1 in the second session received the lowest value (Figure 4). Note that seat 6 and seat 5 were the domestic products. The result shows that the domestic seat fits well to the physical condition of the Korean. Next, in order to find relationship between overall seating comfort and each evaluation item, the correlation analysis was performed.

The correlation coefficients and significance were shown in Table 5.

The results revealed that the important factors about the seating comfort were the fitness of cushion (Y3), the fitness of back (Y12) and the fitness of head rest (Y22)($p<0.05$).

According to the results, the comfortable feeling for the seat seems to be based on the sense of the fitness of the seat shape. Last, the evaluation datum of each satisfaction item (Y1, Y2,.. , Y22) were regressed against the satisfaction datum for the overall seating comfort (Y23) to determine subjects' preferences on each seat. By the stepwise regression analysis, following regression equation was estimated and the result of the analysis was shown Table 6.

$$Y23 = -25.885 + 2.301\ Y1 - 0.275\ Y3 - 0.367\ Y10 + 1.099\ Y12 - 1.249\ Y19 \quad \text{-----------------------}(1)$$

Table 6 shows that the estimated model is regarded as the suitable equation (R-square=0.99). The significant variables associated with the overall seating comfort are well expressed by the firmness of cushion, the fitness of cushion and the fitness of back etc.

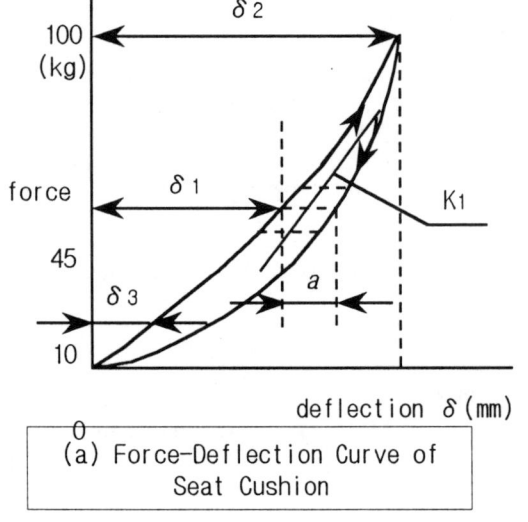
(a) Force-Deflection Curve of Seat Cushion

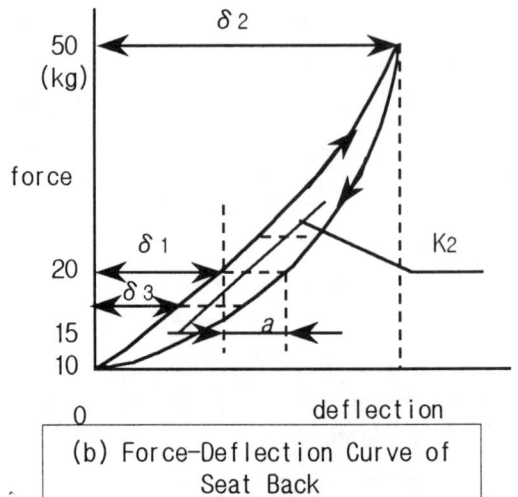
(b) Force-Deflection Curve of Seat Back

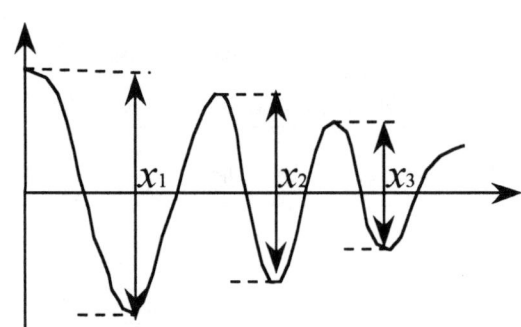
(c) Damping Curve

Damping Ratio: $\zeta = \dfrac{1}{\pi} \log_e \alpha$

Slope: $\alpha = \dfrac{1}{n}\left(\dfrac{x_1}{x_2} + \dfrac{x_2}{x_3} + \Lambda + \dfrac{x_n}{x_{n+1}}\right)$

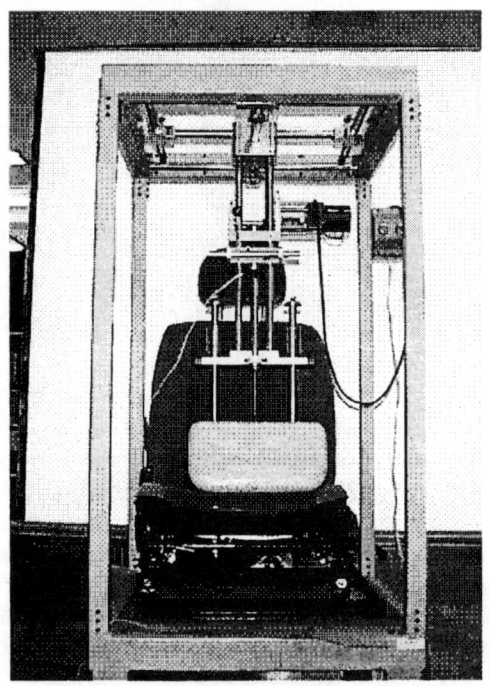
(d) Static Load Tester

Dynamic Spring Constant: $Kd\,(kgf/mm) = (2\pi f_0)^2 \dfrac{W}{G}$

Where, W(kg):weight G(9800mm/s2):gravity fo(Hz):resonance frequency

Figure 3. Physical Characteristics Tests

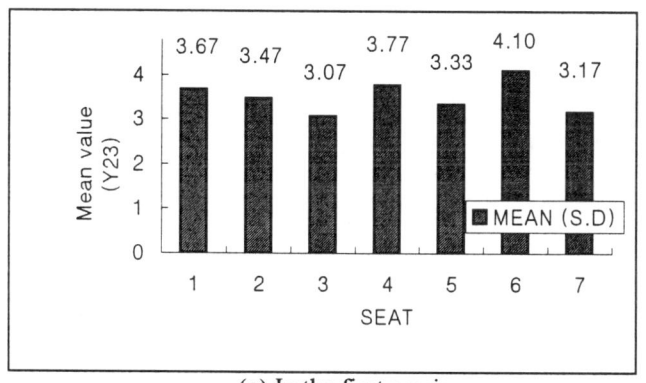

(a) In the first session

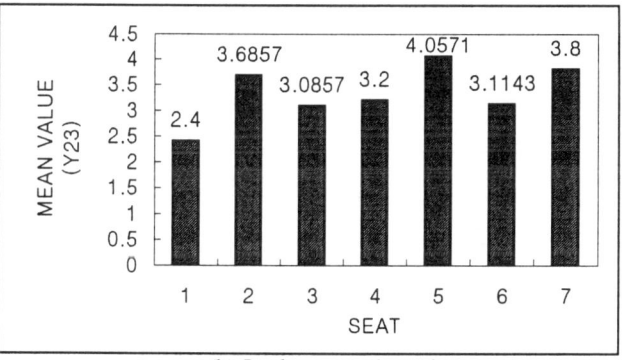

(b) In the second session

Figure 4. Mean values of overall seating comfort (Y23).

Table 5 The Result of the Correlation Analysis

Variable	Y1	Y2	Y3	Y4	Y5	Y6	Y7	Y8
R-square	0.6468	0.1285	0.7612	0.7468	0.1233	0.2942	0.1379	-0.0425
P-value	0.1164	0.7836	0.0468*	0.0538	0.7923	0.5219	0.7681	0.9279
Variable	Y9	Y10	Y11	Y12	Y13	Y14	Y15	Y16
R-square	0.2567	0.7094	0.7325	0.8312	0.0638	0.2192	0.1779	0.6026
P-value	0.5784	0.0742	0.0612	0.0205*	0.8919	0.6368	0.7027	0.1521
Variable	Y17	Y18	Y19	Y20	Y21	Y22		
R-square	0.6049	0.7155	0.1824	0.1523	0.5356	0.8216	.	.
P-value	0.1502	0.0706	0.6954	0.7444	0.2153	0.0234*	.	.

*; Statistically significant $p<0.05$

Table 6 The Result of the Regression Analysis

Source	Sum of Squares	DF	Mean Square	F	Prob.>F
Model	707.427	5	141.485	100513	0.0024
Error	0.002	1	0.002		
Total	707.429	6	.	.	.

RELATIONSHIP BETWEEN SUBJECTIVE EVALUATION AND BODY PRESSURE DISTRIBUTION

The body pressure ratios obtained by the calculation processes as shown in Table 3 were demonstrated in Table 7. To examine the relationship between the pressure ratio and the subjective evaluation (the summation value of the overall seating comfort item: Y23), the correlation analysis was conducted. As shown in Figure 5, we can see that the comfortable levels increase with the increasing Rh, Rp and Rl, while the comfortable levels decrease with the increasing Rt.

This means that the pressure ratios are related to the important factors of subjective evaluation such as the fitness of the cushion and back (Y3, Y12).

Note that seat 6 evaluated as the comfortable seat in the subjective technique had the highest Rh, Rl and the lowest Rt. According to the results, it may be supposed that the comfortable seat affords the sufficient support to the hip and lumbar region while the uncomfortable seat does not afford the sufficient support to those regions.

Table 7 The Body Pressure Ratios

Seat	Rh	Rsl	Rsr	Rsm	Rt	Rl	Rp
1	0.4837	0.1537	0.1806	0.1672	0.3397	0.6984	0.85112
2	0.4578	0.1396	0.1714	0.1555	0.4022	0.5217	0.81436
3	0.4482	0.2270	0.1407	0.1838	0.4097	0.6680	0.61982
4	0.4883	0.1856	0.1926	0.1891	0.3088	0.6886	0.96350
5	0.4509	0.1289	0.2086	0.1688	0.3661	0.6655	0.61779
6	0.5097	0.1737	0.1685	0.1708	0.2861	0.7698	0.97330
7	0.4482	0.1880	0.1599	0.1739	0.3761	0.5533	0.85033

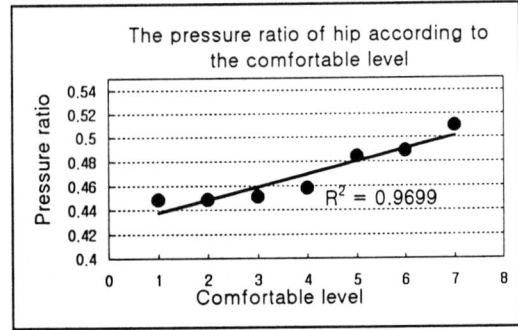
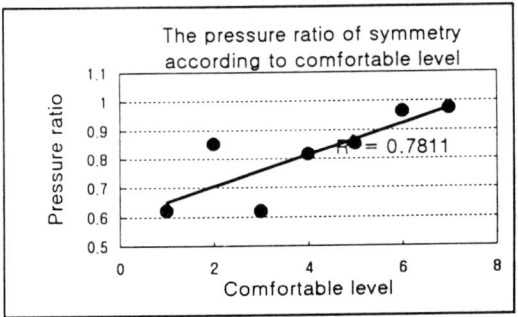
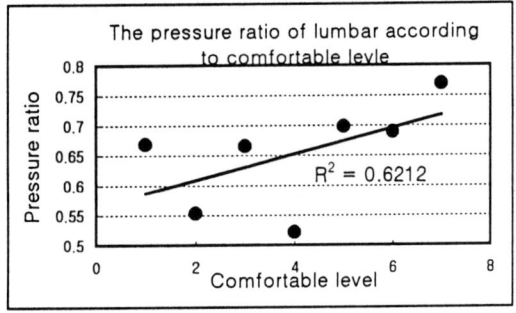
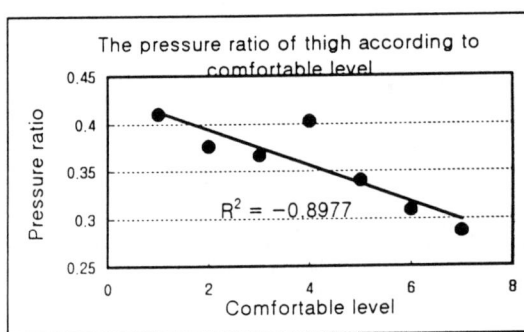

Figure 5. The result of correlation analysis.

Therefore, the pressure ratio upon those regions can be regarded as one of the standards for evaluating the quality of seats. To further clarify the difference of the pressure ratios between the comfortable and uncomfortable seats, t-test was conducted as shown in table 8 and figure 6. Rl, Rh and Rp were the greater for the comfortable seats than for the uncomfortable seats, while Rt was the less for the comfortable seats. The greatest difference of the pressure ratio was observed at Rp. The difference of the pressure ratio in the hip, lumbar, thigh and left-right symmetry was statistically significant ($p<0.1$).

Table 8 The Results of T-test between the Comfortable and Uncomfortable Seats

Classification		Rh	Rsl	Rsr	Rsm	Rt	Rl	Rp
Comfortable	Mean	0.494	0.171	0.181	0.176	0.316	0.719	0.929
	SD	0.014	0.016	0.012	0.012	0.027	0.044	0.068
Uncomfortable	Mean	0.451	0.171	0.170	0.171	0.389	0.602	0.726
	SD	0.005	0.045	0.029	0.012	0.021	0.076	0.124
Significance		p<0.01**	p>0.1	p>0.1	p>0.1	p<0.01**	p<0.1*	p<0.1*

*; Statistically significant difference ($p<0.1$), **; Statistically significant difference ($p<0.01$)

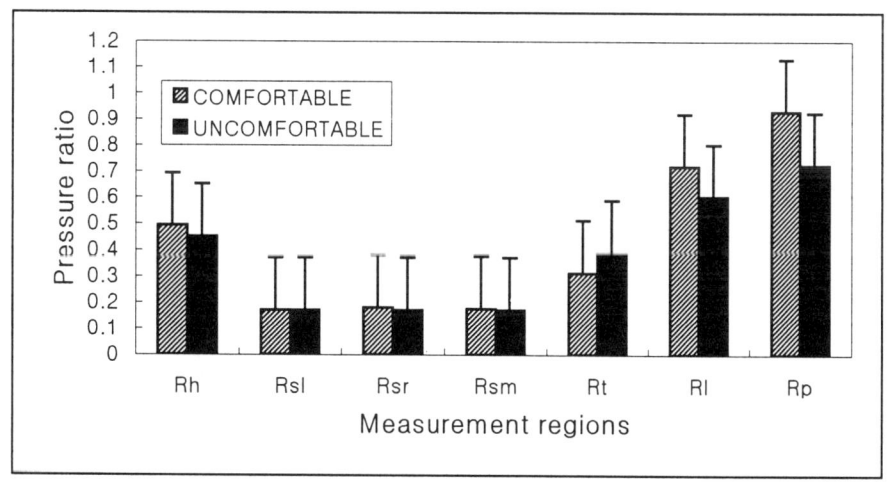

Figure 6. Comparison of mean pressure ratios between the comfortable and uncomfortable seats.

Figure 7 shows the graphs of body pressure distribution of seat 6 and 3 classified into the comfortable and uncomfortable seat respectively as a result of the subjective evaluation. Seat 6 can be regarded as a good pattern of body pressure distribution without any unnatural peak of pressure and asymmetrical pressure. The pressure pattern of this seat cushion represented the contour form centering symmetrically on the ischial tuberosities. In case of the seat back, the pattern showed a symmetrical form at the lumbar part.

This means that the lumbar and hip parts are well supported by the seat and the driver can make the good driving posture. Seat 3 can be regarded as a bad pattern of body pressure distribution. In this graph, the body pressure distribution of the seat extends in the lateral direction, as compared with a good seat. Further, asymmetrical pressure was detected in the cushion and back of the seat. Thus, the patterns as well as the ratios of the body pressure distribution can be correlated with the subjective evaluation.

RELATIONSHIP BETWEEN SUBJECTIVE EVALUATION AND PHYSICAL CHARACTERISTICS

For the substantial evaluation of seat products, more attention has been paid to seat system/component tests than sensory and posture measurements, because posture is more individual and undetermined whereas seat features are more general to the interest of product engineers. Indeed, from an engineer's point of view, the physical performance and comfort properties of a seat system are the major concerns and they can only be improved through work on seat features and components.

As mentioned earlier, the difference between two groups was statistically significant through ANOVA (Analysis of Variance Procedure)($p<0.01$). As shown in Table 9, to examine the relationship between the subjective technique and physical characteristics, the correlation analysis was performed. CD3, CD15, CDL5, CR10 and BDL5 were well correlated with the overall seating comfort: Y23 ($p<0.1$). This shows that the main factors for the seating comfort are the deformation level, the hardness of the foam padding and the hardness of the seat cover.

Table 9 Correlation Analysis between Overall Seating Comfort (Y23) and Physical Characteristics

Variable	CKS	CD3	BKS	BD1	XX1	XX2
R-square	-0.1290	0.8918	-0.5688	0.6274	0.5682	0.5875
P-value	0.7828	0.0070**	0.1826	0.1315	0.1833	0.1654
Variable	DAMP	CKD	CRF5	CRF15	CRF30	CR10
R-square	-0.4413	0.0168	0.4116	0.7117	0.5307	0.7263
P-value	0.3216	0.9714	0.3589	0.0729	0.2204	0.0645*
Variable	BBT20	BL5	CD15	CDL5	BDL5	
R-square	0.6550	0.7054	0.7959	0.8036	0.8535	
P-value	0.1103	0.0766	0.0323*	0.0295*	0.0146*	

*; Statistically significant $p<0.1$, **; Statistically significant $p<0.01$

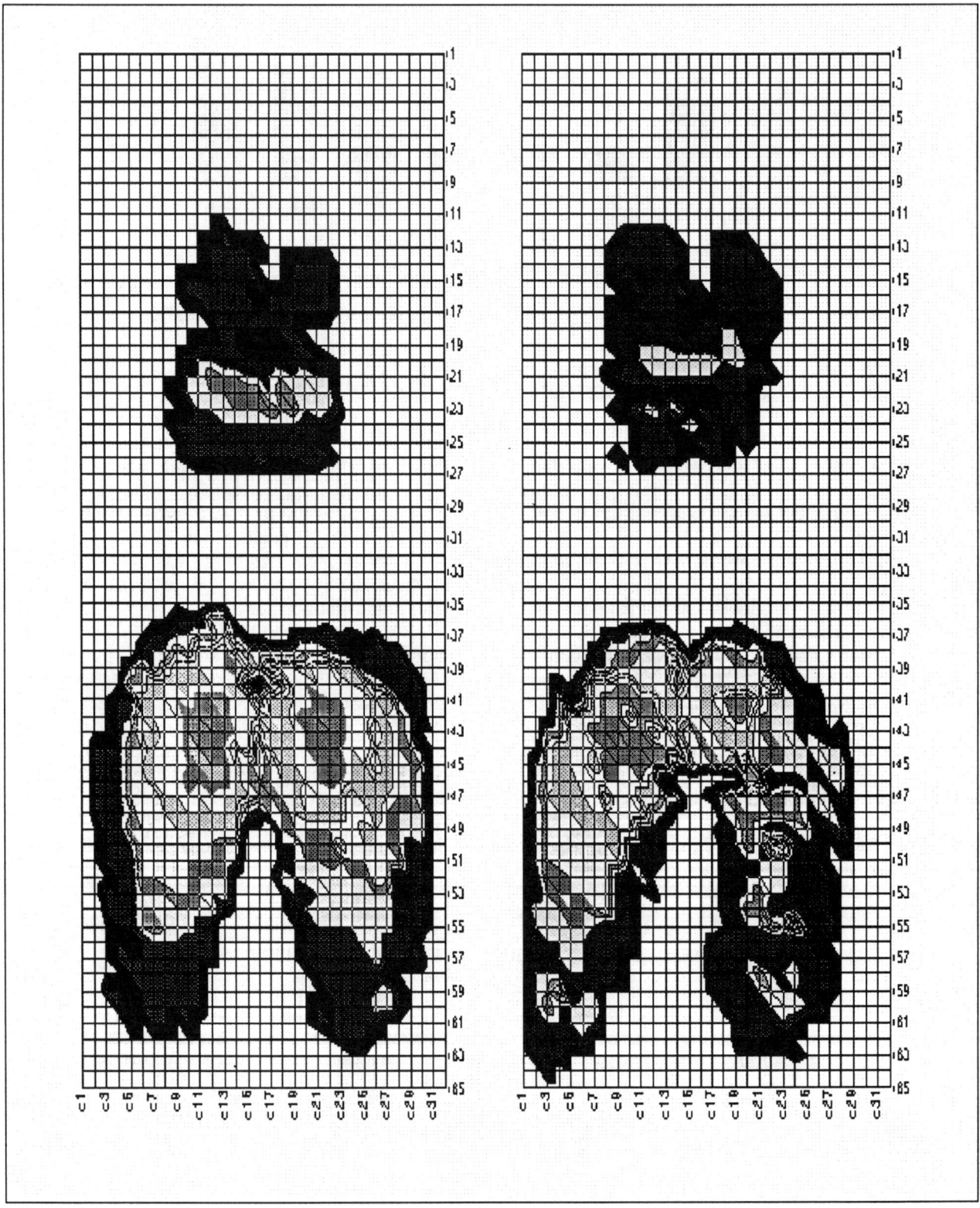

Figure 7. Body pressure distribution of a comfortable seat and an uncomfortable seat.

By the stepwise regression analysis, following regression equation was estimated as shown in Table 10.

$$Y23 = 16.041 + 2.435\ CD3 + 0.447\ BD1 + 5.335\ CKD - 2.211\ BL5 + 5.271\ BDL5 \quad \text{------ (2)}$$

Table 10 shows that the estimated model is regarded as the suitable equation (R-square: 0.99). The main variables significantly associated with the overall seating comfort are well explained by the deformation level of the cushion and back, dynamic spring constant, and the hardness of the foam padding and seat cover in the back.

Table 10 Multiple Regression Analysis between Dependent Variable Y23 and Physical Characteristic Variables

Source	Sum of Squares	DF	Mean Square	F	Prob.>F
Model	2283.3925	5	456.678	12670.4	0.0067
Error	0.036	1	0.036		
Total	2283.428	6		R-square	0.99

T-test results about the physical characteristics between two groups showed significance (Table 11 and Figure 8). In preliminary tests involving 35 subjects, utilizing seven different seats, findings support the use of physical characteristics as a viable means of objectively measuring seating comfort.

Table 11 T-test Results about Physical Characteristics between Two Groups (Comfortable/Uncomfortable) Determined by ANOVA

variable	CKS	CD3	BKS	BD1	XX1	XX2	DAMP	CKD	CRF5
p-value	0.6859	0.0075**	0.0360*	0.0080**	0.3396	0.3898	0.2826	0.9123	0.2100
variable	CRF15	CRF30	CR10	BBT20	BL5	CD15	CDL5	BL5	
p-value	0.0435*	0.4865	0.0050**	0.0116*	0.0153*	0.0503	0.0375*	0.0355*	

*; Statistically significant difference ($p<0.05$), **; Statistically significant difference ($p<0.01$)

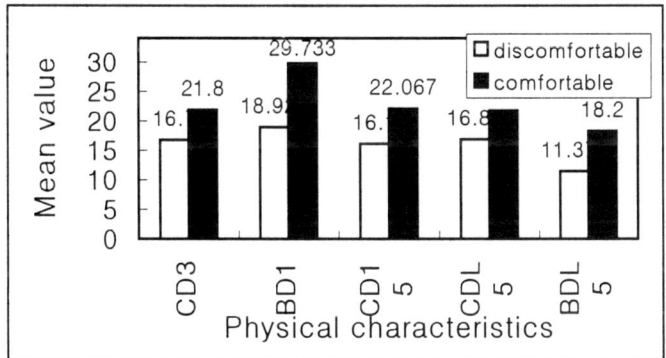
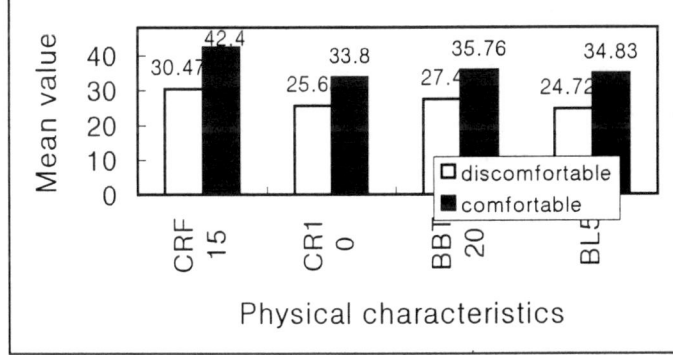

Figure 8. T-test results about physical characteristics between two groups (comfortable/uncomfortable) determined by ANOVA.

CONCLUSIONS

To examine a method for objectively evaluating seating comfort, we conducted the measurement of the body pressure distribution and physical characteristics. In the subjective evaluation, the fitness of cushion, back and head rest were the important factors for the overall seating comfort.

Of two sets of tests, in the first session, the body pressure ratio and pattern were compared with the subjective technique. The body pressure ratios on the hip, lumbar and thigh regions were well connected with the overall seating comfort of the subjective technique: the fitness of cushion and back regarded as the important factor were well associated with the body pressure ratios such as Rh and Rl. In addition, there was a clear difference between comfortable and uncomfortable seats. First, the pressure ratios such as Rl, Rh and Rp of the comfortable seats were the greater than that of the uncomfortable seats while Rt was the smaller. This means that the comfortable seat affords the sufficient support to the hip and lumbar region. Second, in the comfortable seat, the body pressure was distributed well and symmetrically centering around the ischial tuberosities. On the other hand, in the uncomfortable seat, there was the asymmetrical form of the pressure distribution.

In the second session, the physical characteristics were compared with the subjective technique. The deformation level, the hardness of the foam padding and the hardness of the seat cover were well correlated with the overall seating comfort. Also, the physical characteristics were regressed with the subjective technique. According to these results, it was statistically significant that the overall seating comfort was related to the deformation level of the cushion and back, dynamic spring constant and the hardness of the foam padding and seat cover in the back.

Therefore, the body pressure ratio and physical characteristic can be recommended as the objective technique for the seating comfort evaluation. Further, to obtain the quantitative data for the subjective technique, we will attempt to establish the ergonomic parameters using the electromyography and vibration analysis etc.

REFERENCES

[1] S.J. Park, N.S. Lee, C.J. Kim, S.Y. Lee, "A Study on the development of body pressure measurement system, 10th Annual Conference Proceedings of The Ergonomics Society of Korea, pp.187-192, 1992.

[2] J.M. Lee, Y.H. Youm, M.H. Seong, and S.H. Shin, "A Study on the safety and human engineering to enhance design quality of vehicle seat" Report of DAEWON Technique, No.6, pp.28-40, 1987.

[3] Dempsey, C.A., Chapter9: Posture and sitting, pp.165-180, McGraw-Hill, New York, 1963.

[4] Drummond D.S., et al., "A Study of pressure distributions measured during balanced and unbalanced Sitting", J. Bone and Joint Surgery, Vol.64 (A), No.7, pp.1034-1039, 1982.

[5] Bader, D.L., Barnhill, R.L., and Ryan, T.J., "Effect of externally applied skin surface forces on tissue vasculature", Archives of Physical Medicine and Rehabilitation, 67:11, pp.807-811, 1986.

[6] Chow, W.W., and Odell, E.I., "Deformations and stresses in soft body tissues of a sitting person", J. of Biomech. Eng., Vol.100, pp.79-87, 1978.

[7] Diebschalg, W., Heidinger, F., Kuurz, B., and Heiberger, R., "Recommendation for ergonomic and climatic physiological vehicle seat design", SAE paper No.880055, 1988.

[8] Diebschlag, W., and Muller-Limmeroth, W., Physiological requirements on car seats: Some results of experimental studies, Human Factors in Transport Research, (Oborne, D.J., and Levis, J.A., ed), pp.223-230, 1980.

[9] Hertzberg, H.T.E., "The human buttocks in sitting: Pressure, patterns, and palliatives", SAE paper no. 720005, 1972.

[10] Kamijo,K., et al., "Evaluation of seating comfort", SAE paper no.820761, 1982.

[11] Kohara, J., and Sugi, T., "Development of biomechanical manikins for measuring seat comfort", SAE paper no. 720006, 1972.

[12] Their, R., "Measurement of seat comfort", Automobile Engineer, Feb, 1963.

[13] Wenqi Shen and Alicia M. Vertiz., "Redefining seat comfort", SAE paper no. 970597, 1997.

[14] Ronald L. Huston and Ashraf M. Genaidy, "Design parameters for comfortable and safe vehicle seats", SAE paper no. 971132, 1997.

[15] S.J. Park, Chae-Bogk Kim, "The evaluation of seating comfort by the objective measures", SAE paper no. , 1997.

[16] Kuntal, Thakurta., Daniel, Koester., Neil, Bush., and Susan, Bachle, "Evaluation short and long term seating comfort", SAE paper no. 950144, 1995.

[17] Tamara Reid, Bush., Frank T, Mills., Kuntal, Thakurta., Robert P, Hubbard and Joseph Vorro, "The use of electromyography for seat assessment and comfort evaluation", SAE paper no. 950143, 1995.

980654

Evaluation of Comfort Properties with Covering Textiles of Car Seats

Yoon-Sook Hur and Se-Jin Park
Korean Research Institute of Standards and Science

Copyright © 1988 Society of Automotive Engineers, Inc.

ABSTRACT

Material and properties of seat cover were investigated in order to help develop desired seat covers. The results showed that woven seat cover was used most widely, but the luxurious and soft textured leather seat cover was preferred. Low electrostatic propensity, high moisture transport, high resistance to soil, good cushion, and comfortable warm-cool sensation, mainly comfort properties were demanded for seat covers. The above properties were evaluated for three types of currently used seat covers, leather, woven, and pile knit. New apparatus were manufactured to measure moisture transport and warm-cool sensation. There were differences between seat covers. For leather seat cover, water vapor transmission was the lowest and warm-cool sensation was the coolest. Electrostatic propensity was the highest for woven seat cover. Woven and pile knit seat covers would be easily soiled by water-soluble soil.

INTRODUCTION

At the present time, many changes are being made to automotive textiles, particularly seating and headlining fabrics, which are required by consumer demands for new and more luxurious interiors. This demand has been emphasized since exterior styling changes were diminished.

Composition of automotive fabrics has been changed dramatically for the past several decades, primarily due to the use of synthetic fibers. Increased man-made fiber availability and the continuous improvement in finishing have provided superior performance and lower cost compared to natural fibers [1]. Especially, polyester has been mainly used due to its lower cost and higher tensile strength. Even greater efforts are being made to reduce the weight of automotive fabrics to take the advantage of the polyester's higher physical properties and relatively stable price.

Each year many new automotive fabrics are introduced. Evaluation of these new fabrics for automotive applications requires a great variety of tests to assure their durability, because automotive textiles are exposed to extreme environments than other commercial textile applications [2].

As a result, many automotive textiles with superior physical properties are being used. However, consumers might want to have other properties of automotive fabrics in addition to these physical properties. Therefore, this study primarily surveyed the types of seat covering materials that consumers use and prefer and the problems while they use seat covers. On the basis of the results, samples of seat covers of leather, woven, and pile knit were selected. They were quantitatively evaluated on the properties, in order to help develop more comfortable seat covers.

SURVEY

THE CHARACTERISTICS OF SUBJECTS

Four hundred and eight subjects who lived in Taejeon participated in this survey in January 1997. Two hundred and ninety five subjects were male and one hundred and thirteen subjects were female.

Ages of subjects are shown in Table 1. The proportion of the subjects in their thirties were highest and that of the subjects over 50 years was very low.

The types of occupation and car categories of the subjects are shown in Table 2. For the types of occupation, professional was the most frequent, because there were many research institutes in the surveyed region.

For car categories that subjects own, the proportions of compact, intermediate, and full-sized cars were similar, except that of premium cars on the whole. The proportion of big cars, such as premium cars was very small.

Table 1. Number of Subjects

Age(years)	Male	Female	Total
20-29	76	30	106
30-39	147	54	201
40-49	59	23	82
50-59	11	6	17
over 60	2	0	2
Total	295	113	408

Table 2. The Frequency of Types of Occupation and Car Categories of Subjects

	Compact	Intermediate	Full size	Premium	Total
Student	21	9	2	0	32
House wife	12	11	28	3	54
Clerical worker	41	24	14	4	83
Self-employed	11	7	11	3	32
Business man	1	0	2	2	5
Professional	56	51	72	4	183
Others	7	4	3	1	15
Total	149	106	132	17	404

No answer: 4

Table 3. Characteristics of Samples

Sample	Fiber type	Weight* (g/m^2)	Thickness** (mm)	Air permeability*** ($cc/cm^2/sec$)
Leather	Cowhide	780	1.74	0.002
Woven	Polyester	423	1.36	23.1
Pile knit	Polyester	432	2.17	28.3

* ASTM D3776-96
** ASTM D5736-95
*** ASTM D737-96

MATERIAL

In the questionnaire, subjects were asked about the types of automotive seat cover materials to be used and preferred now, and the problems that they have, while using them.

EXPERIMENT

FABRICS

One sample from each type of seat covers - leather, woven, and pile knit - was used, respectively. The characteristics of the samples are shown in Table 3.

WATER VAPOR TRANSMISSION (WVT)

Apparatus

As shown in Figure 1, an apparatus was made to measure the water vapor transport property of a seat [3-5]. To simulate the skin, a copper plate (20.5 cm x 18.5 cm x 1.0 cm) was used. One side of the plate was attached to the panel heater and the other side was attached to high moisture absorbing nonwoven fabric (Sontara H847, Pulp/Polyester 55/45, Du Pont). Its temperature was controlled to 37 °C by inserting a RTD type stick temperature sensor into the copper plate and the sensor was connected to an automatic controlling system.

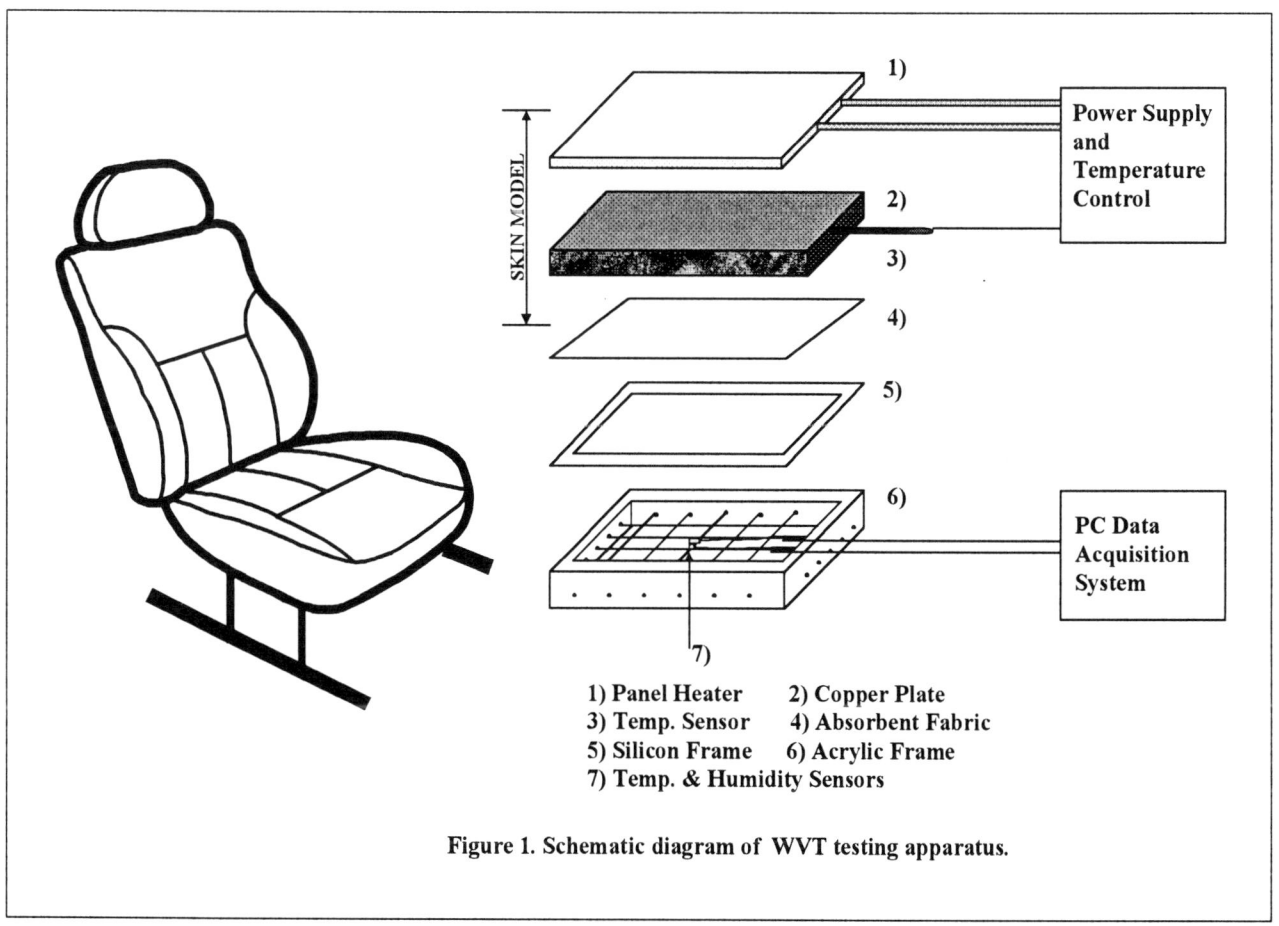

Figure 1. Schematic diagram of WVT testing apparatus.

1) Panel Heater 2) Copper Plate
3) Temp. Sensor 4) Absorbent Fabric
5) Silicon Frame 6) Acrylic Frame
7) Temp. & Humidity Sensors

Lucite acrylic frame with a net was inserted to secure the space for sensors between the skin model and the seat, as shown in Figure 1. A film type humidity sensor covered with the plastic cap (0.8 cm × 0.8 cm × 0.4 cm, Humicor 6100, Coreci Co., France) and a temperature sensor of RTD type were inserted in the middle of the net of the acrylic frame. These sensors were connected to the transmitter (4/20-mA output, Transmicor Model 131.2, Coreci Co., France).

To simulate the sweating of the human body, water was dropped onto the nonwoven fabric through the net, using a syringe. To prevent the leak of the vapor, a silicon frame was inserted between the nonwoven fabric and the acrylic frame.

Test procedure

The apparatus was switched on and the temperature of the skin model was set at 37 °C after 1g water was dropped onto the nonwoven fabric, the skin model was placed on the seat. This procedure took about 10 seconds. Data were collected for 1 hour using the C++ software program. The change of capacitance was converted to the change of voltage and it was recorded as relative humidity and temperature.

ELECTROSTATIC PROPENSITY

The test method of KS K 0555 [6] electrostatic propensity was used to measure the capacity of a nonconducting material that acquires and holds an electrical charge by rubbing with another material. Three test specimens, 4 cm × 8 cm, were prepared in both warp and fill directions, respectively. White test cotton and nylon clothes, 2.5 cm × 16 cm, were used as rubbing materials. All test specimens and rubbing clothes were conditioned at atmospheric condition of 65% R.H. and 20 °C for 24 hours.

Rotary Static Tester (Model RST-201, KOA SHOKAI Co.) was employed to measure an electrostatic voltage. A test specimen was clamped to a grounded metal disc, which was rotated at 450 rpm. The distance of measuring electrode to the specimen was adjusted to 0.15 cm. The rubbing cloth was clamped and was exposed by the load of 500 g. The motor drive was turned on and after 1 min and 5 min, the electrostatic capacity (Voltage) was read on the oscilloscope and on the recorder. The results of all measurements of both directions were averaged.

RESISTANCE TO SOIL

The test method TES K 311-79 (Test standards of automotive textiles of Mazda Co.) [7] was used to measure the amount of soiling of automotive textiles. A specimen, 5 cm x 5 cm, was cut and was placed on the base of the Crocmeter, and the artificial soil agent (Table 4) of 20 mg/cm^2 was added on the surface of the specimen. The meter handle rubbed the surface of the specimen at the rate of 65-80 turns per min for 10 minutes. Then, the front and rear surface of the specimen were brushed 20 times to remove the extraneous soil.

The soiled specimen was compared with the unsoiled original test specimen. Class 5 represents no soiling and Class 1 complete soiling.

Table 4. Composite Percentage for the Artificial Soil Agent

Component	Percentage
Brown coal or Charcoal	38
Cement	17
Soil or Chalk	17
Silica(200)	17
Carbon black	1.75
Ferric oxide	0.5
Nujol	8.75

WARM AND COOL SENSATION

To measure contact warm and cool sensation, a copper plate equipped with RTD type temperature sensor (0.15 cm x 0.2 cm, 4/20 mA output, Coreci Co., France) was used as a skin model [8-9]. The copper plate was heated to the skin temperature of the human body, 36.5 °C and the specimens, 5 cm x 5 cm, were kept in an environmental chamber of 45% R.H. and 25 °C for 3 hours. After a test specimen was mounted on the copper plate, the temperature changes of the copper plate were measured for about 2 mins. Each fabric was measured 3 times and the results were averaged.

WATER REPELLENCY

The test method of AATCC 22-89 Spray test [10] was used to measure the resistance of fabrics to wetting by water.

RESULTS AND DISCUSSION

THE PREFERRED TYPES OF SEAT COVER MATERIALS

The types of seat cover materials being used and preferred were surveyed (Table 5). The most frequently used material was woven and then followed by synthetic leather, pile knit, and leather.

Table 5. Frequency of Types of Seat Cover Materials Being Used and Preferred (N=408)

Sample	To be used	To be preferred
Leather	33	145
Synthetic leather	119	86
Woven	202	135
Pile knit	43	33
Others	11	9

No answer: 11

The most preferred material was leather, and then was followed by woven, synthetic leather, and pile knit. There was discrepancy between the types of seat cover materials being used and being preferred. The discrepancy comes from the fact that consumers could not choose the type of material when they buy a car. The expensive leather seat covers are being mainly used for the premium and luxury cars, but many below full-sized cars were sampled in this study.

The Reasons for the preference were surveyed (Table 6). Leather was preferred mainly due to luxury and soft texture and then due to easy soil removal and good seating comfort. Synthetic leather was chosen due to economy and easy soil removal. Woven was preferred due to economy and soft texture. Finally, pile knit was chosen primarily because of soft texture. Many people seem to think leather seat cover is more luxurious than the other seat covers and synthetic leather, woven, and pile knit seat covers do not give luxurious feeling to them.

Table 6. Reasons for the Preference of each Material as a Seat Cover (N=408)

Reason	Leather	Synthetic Leather	Woven	Pile knit
Luxurious	51	5	1	1
Economic	7	38	56	4
Soft texture	46	6	54	20
Easy soil removal	19	32	1	1
Various colors	1	1	8	3
Good seating Comfort	16	3	8	3
Others	5	1	6	0

THE PROBLEM OF CONSUMERS WHILE USING SEAT COVERS

As shown in Table 7, the problem areas while using seat covers were summarized. The subjective discomfort was not related to the durability or discoloration of the material, but was mainly related to the comfort properties, such as ventilation, static electricity, resistance to soil, etc. Therefore, three types of seat covers such as leather, woven, and pile knit, were evaluated in terms of the above comfort properties to find out the degree to which each type has these problems.

Table 7. Problem Areas (N=408)

Problem	Frequency
1. It is uncomfortable because of the static electricity.	83
2. It is uncomfortable because of the poor ventilation.	88
3. The seat is slippery.	19
4. Cushion is not good.	32
5. Soiling is easy and soil removal is difficult.	82
6. The seat cover is easily worn and torn.	11
7. The pilling due to rubbing between seat cover and clothes is easily formed.	6
8. The color of seat cover is easily discolored.	13
9. Cool contact sensation on seating is significant.	31
10. The texture of seat cover is not soft.	6
11. Color is dark.	8
12. Others	8
13. No answer	21

EVALUATIONS OF THE COMFORT PROPERTIES

Water vapor transmission

At atmospheric condition of 25 °C, 55±2% R.H., sweat pulse was manipulated by adding 1 g of distilled water on the absorbing nonwoven fabric of the skin model. Three seats with the different seat cover and the same sponge were used. The changes of humidity were measured at the space between the seat and the skin model.

Figure 2 revealed that the leather seat showed the highest relative humidity, indicating the worst water vapor transmission. The woven seat revealed the fastest water vapor diffusion.

Natural leather is able to transport the water vapor because they have many micropores on the surface and absorb water. However, the leather used as a seat cover usually goes through the process of dying and surface finishing, and thus the micropores of leather could be clogged with dyes and finishing agents [11]. As a result, the finished leather hardly seemed to transport water vapor. In order to verify the idea, the air permeability of seat covers was checked (Table 3). The low air permeability of leather seat cover could be attributed to the poor ventilation. According to Tomming [4], when a piece of synthetic film is placed on a seat, it acts as a vapor barrier and results in high humidity build up. Under this condition, a ventilated seat showed the lowest humidity.

For the pile knit seat and the woven seat, the relative humidity of the pile knit seat was higher than for woven. Because pile knit seat cover was slightly heavier and thicker than woven seat cover and this might have operated as a barrier to the water vapor transmission. Unlike water vapor transmission, the air permeability of pile knit seat cover was a little higher than that of woven seat cover. The air permeability was measured by blowing the airflow with high pressure. The airflow could be transported through the loose weave, knit structure.

the pile knit seat and the woven seat to reach the maximum temperature. On the other hand, the leather seat reached to it after passing about 27 minutes. The slow temperature change of leather was attributed to the lower surface temperature of leather seat covers than the other fabric seat covers.

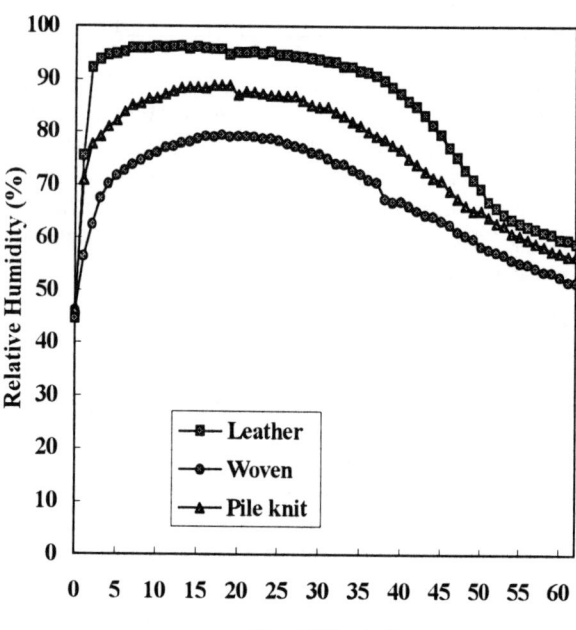

Figure 2. Relative humidity changes in the space between seats and the skin model.

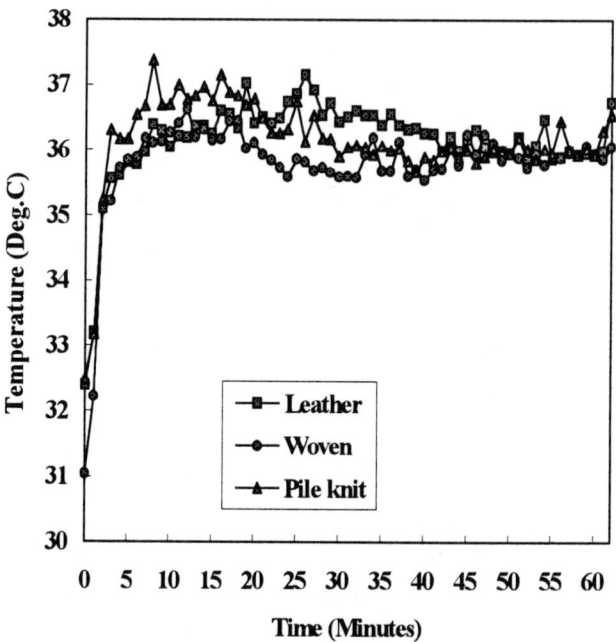

Figure 3. Temperature changes in the space between seats and the skin model.

Thus, woven seat cover would be the most comfortable fabric in water vapor transmission. Although leather seat cover was preferred the most due to its luxury and soft texture, its water vapor transmission was worse than the other seat covers. Especially, in the hot and humid environment, leather seat cover could give bigger discomfort to drivers and passengers due to the lack of ventilation.

Figure 3 showed the changes of temperature in the space between the seat and the skin model during transport of water vapor.

Overall temperature increased drastically at the very beginning due to the hot vapor produced by the skin model. After reaching the maximum temperature, it slightly decreased due to heat conduction while the absorbing nonwoven was slowly drying. The decreased temperature slowly increased again as dry heat was accumulated between the seat and the skin model.

After the sweating, it took about 20 minutes for

Electrostatic propensity

Electrostatic voltages of seat covers were measured (Table 8). Cotton and nylon fabrics were used as rubbing cloth, and rubbing time was 1min and 5min, respectively. When hydrophilic cotton fabric was used, electrostatic voltages were lower except for woven seat cover and when rubbing time was 5 min, electrostatic voltages were higher. As the moisture regain of fabric and the relative humidity get larger, electrostatic charge becomes smaller. In addition, when the rubbing time is longer, voltage is higher [12].

For leather seat cover, voltage was the lowest. According, leather seat covers are expected to low in producing discomfort related to the static charge than the other seat covers. The voltage of woven seat cover was much larger than that of pile knit cover, even though the compositions of pile knit and woven seat covers were the same, polyester fiber. The difference in voltage could be attributed to the difference of fabric structure. However, it needs further research.

This electrostatic charge causes the problems, such as soiling of seat covers and clothes due to attached dusts cling clothes, and body shock [13-14]. If above 3 kV is charged and then discharged in the human body, the human body is shocked. For woven seat cover, at atmospheric condition 20 °C, 65% R.H., electrostatic voltage was above 2 kV. Under the atmospheric condition of low temperature and relative humidity, woven seat covers could generate high voltage and thus produce discomfort to the human body. Therefore, the woven seat covers made of polyester fiber need to be improved through the blending with hydrophilic natural fibers, the permanent anti-static finishing, and the improvement of yarn and fabric structure to be able to reduce the frictional force on the surface of seat cover.

Table 8. Electrostatic Voltages of Seat Covers (volt)

Sample	Cotton		Nylon	
	1 min	5 min	1 min	5 min
Leather	70	80	130	160
Woven	2110	2730	1230	1720
Pile knit	190	420	580	890

Resistance to soil

Resistance to soil of seat covers was measured by using soiling agent made by TES K 311-79 (Table 9). A solid type of soiling agent was used.

Although many consumers reported the problem of soiling of seat cover in the survey, all seat covers were rated Class 4, which represented almost no soiled. However, water repellency scores showed that leather seat cover was not wetted whereas woven seat cover was partially wetted and pile knit seat cover was almost completely wetted. Actually, in the survey easy soil removal was the reason for the preference of seat cover in synthetic leather and natural leather seat covers, but was scarcely selected in woven and pile knit seat covers. Therefore, woven and pile knit seat covers can be easily exposed to water-soluble soil and need to be improved.

Table 9. Resistance to Soil and Water Repellency of Seat Covers (rating)

Sample	Resistance to soil	Water repellency
Leather	4	100
Woven	4	70
Pile knit	4	03

Warm and cool sensation

To simulate the warm and cool sensation at the moment of contact with the seat, temperature changes of the copper plate, which was set at 36.5 °C, the skin temperature, were measured when fabrics of 25 °C were placed on it. Figure 4 shows the results.

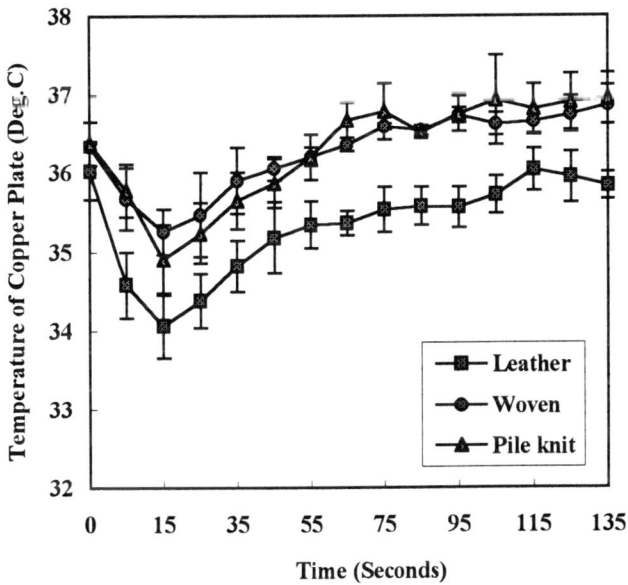

Figure 4. Temperature changes of the copper plate covered with seat covers.

The leather seat cover had the lowest temperature and the temperature changes for the leather seat cover were significantly different with those for the other seat covers. However, woven and pile knit seat covers showed similar temperature changes. There was no big difference in the contact warm and cool sensation regardless of the types of fabric. Therefore, the leather seat could give the cool sensation to the human body and also provide lower thermal insulation than fabric seats in winter.

CONCLUSIONS

The present study was carried out to help develop desired seat covers.

First, the luxurious and soft textured leather seat cover was preferred the most and then the economic and soft textured woven seat cover was preferred. If woven, synthetic leather, and pile knit seat covers additionally have the luxury, the preference of seat covers could be different.

Second, consumers demanded that seat covers have both durability and superior comfort properties, such as high moisture transport, resistance to soil, low electrostatic propensity, good cushion, and comfortable warm-cool sensation.

Third, the results of the evaluation of the above properties for three types of currently used seat covers (leather, woven, and pile knit) showed that for leather seat cover, water vapor transmission was the lowest and warm-cool sensation was the coolest. But electrostatic propensity and resistance to soil was the lowest. For woven seat cover, water vapor transmission was the fastest and cool sensation was small, but electrostatic propensity was highest and resistance to soil was low. Water vapor transmission and electrostatic propensity of pile knit seat cover was in the middle of the other seat covers. Its cool sensation was small, but it can be easily soiled by water-soluble soil.

Therefore, there was no seat cover that satisfies all consumer demands at the same time, and thus the consumer's need should be considered in the future development of seat covers.

The subjective evaluation of comfort properties will be carried out for the three types of seat covers and the relationship between the objective and subjective evaluations of comfort properties will be investigated in the future study.

REFERENCES

1. McCallum, J. B., Engineering Requirements for Automotive Textiles, *SAE Paper No.750340*, 1-6, 1975.
2. Bard, L. A., USA Automobile Interior Material Testing Yesterday, Today and Tomorrow, *SAE Paper No.931959*, 5-52, 1993.
3. Kim, E. A. and Barker, R. L., Evaluation Method for the Water Transport Properties of Sweat Absorbent Fabrics -Water Vapor Transport in the Human-Clothing-Environment System-, *Journal of the Korean Society of Clothing and Textiles, 17(2)*, 329-338, 1993.
4. Temming, J., A New Method to Assess the Summer Suitability of Car Seats, *SAE Paper No.930106*, 7-12, 1993.
5. Hur, Y. S., Yoo, H. S. and Kim, E. A., Measurement of Buffering Capacity against Water Vapor of Fabrics Using a Human-Clothing-Environment Modeling System, *Journal of the Korean Fiber Society, 33(2)*, 183-191, 1996.
6. KS K 0555, Standard name: Testing Method for Electrostatic Propensity of Woven and Knitted Fabrics, Korean Industrial Standard, 1983.
7. TES K 311-79, Standard name: Resistance of Soil, A Standard of Automotive Textiles of Mazda Co., 1979.
8. Rees, W. H., Physical Factors Determining the Comfort Performance of Textile, *In Papers presented at the 3rd Shirley International Seminar: Textile for Comfort*, Shirley Institute, 1971.
9. Kim, E. A. and Park, S. J., *Basic Clothing Comfort*, Gung Chun Co., 99-100, 1994.
10. AATCC Test Method 22-89, Standard name: Water Repellency: Spray Test, An American National Standard, 1989.
11. Trotman, E. R., *Dyeing and Chemical Technology of Textile Fibers, Chapter 12: Introduction to chemical constitution and color, theory of dyeing, and classification of dyes*, Charles Griffin & Co. Ltd., 318-354, 1975.
12. Lyle, D. S., *Performance of Textiles, Chapter 5: Testing for product performance part 3 comfort*, John Wiley & Sons, Inc., 158-177, 1977.
13. Morton, W. E. and Hearle, J. W. S., *Physical Properties of Textile Fibers, Chapter 21: Static electricity*, The Textile Institute, 529-563, 1975.
14. Choi, S. C., Jo, G. L. and Jang, J. J., *Clothing Comfort*, Hyung Sul Co., 281-288, 1989.

980655

A Comparison Test of Transmissibility Response from Human Occupant and Anthropodynamic Dummy

Yi Gu
Lear Corporation

Copyright © 1988 Society of Automotive Engineers, Inc.

ABSTRACT

In order to specify the human dynamic comfort of seat vibration, a lot of work has been conducted in laboratory research. From the seating manufacturer's perspective, the author proposed a test method to measure the seat ride comfort by using a spring-mass dummy which was designed to match the human response in low frequency in vertical direction. A hydraulic table was employed as the excitation source to the occupied seat. Two seat samples, both measured with human occupants before, were used for this study. For simplicity and comparison, a sweep sine signal in vertical direction was used as the excitation signal. The transmissibility results measured of the dummy loaded seat were compared to those of human occupants. In this paper, a continuing effort focused on correlating the vibration response from dummy occupied seat to that from human occupied seat. A consistent relation was shown between the two measurements. The results provide a promising analysis tool that can be very helpful in seat prototype design to improve ride comfort.

INTRODUCTION

The human vibration comfort has attracted more attention in recent development effort by OEM's and suppliers. This is due to the new phase of NVH improvement and competition in vehicle design. Improving human ride comfort has always been an issue for vehicle component designers and manufacturers. For seating and interior integrators, a great deal of effort has been focused on improving the ride comfort.

Vibration transmissibility has been used as an indicator to measure the comfort of ride by researchers [1,2] for many years. The ratio of acceleration of the seat cushion or seat back to that of the floorpan is used to measure the transmission of vibration from vehicle to the human occupant. The difficulty arises when a large number of human occupants are needed for a valid test or a component quality evaluation. Component suppliers usually can not afford such expensive tests during prototype development stage. The number of tests involved to improve the quality of seating system requires a quicker and more repeatable way to measure the seat vibration performance.

In the last development in Lear [3], the author used a rigid mass dummy to set a "benchmark" for the seat vibration performance measurement. The rigid dummy could not provide a similar transmissibility measurement to that of human so its application is largely limited. In this test, however, the stage 4 anthropodynamic dummy provides a better match between the results obtained from a number of human occupants [4] and that of the dummy. Based on this result, the dummy will be used as a measurement tool for vibration ride comfort.

TEST SETUP

The stage 4 dummy was mounted on a hydraulic table in Lear Corporation. The test sample seats were the same ones tested with human occupants by Rakheja et al[4]. The weight of the dummy was about 52.4 kg, 66% of the weight of a 50th percentile male. A B&K acceleration pad was placed under the dummy for the transmissibility measurement. Only vertical exaction was used for the measurement. Because the human occupants were driven with sweep sine signals, the sweep sine was also used for the dummy test.

An HP3566A analyzer was used for data acquisition and the analysis was further performed on a PC with Matlab.

Both the sweep sine signals for human and dummy tests took a long time in order to let the seat cushion settle. In the dummy test, the sweep signal took about 120 seconds with a sampling rate of 512 Hz. Because of the hydraulic table stroke limit, stroke control (2") was used for signal below 2 Hz.

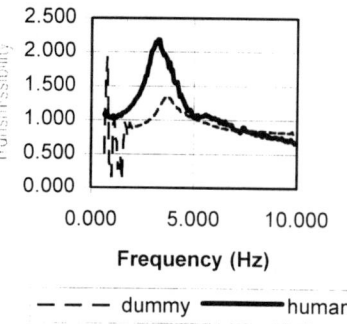

Fig.1. Comparison between human and stage 4 anthropodynamic dummy - sample seat #1

RESULTS AND DISCUSSION

The human occupant and dummy test results in Fig.1 to 3[4] show good correlation between the two measurements, which is an improvement compared to the results of rigid dummy [3]. In general, the transmissibility level of dummy is lower than that of human occupants, the err could be caused by poor friction in the dummy damper. The resonant frequencies of seat match quite well in sample #1(Fig.1) but not so well in sample #2.

The dummy was modified with a plywood base in order to solve the posture stability problem when it came. The back support was also strengthened to prevent losing posture. However, the case of the dummy turned out to be too weak to sustain such strengthening, so that the case got slightly deformed over time. Another issue of dummy performance is the damper weariness. The functionality of damper will significantly affect the dummy performance. This factor is also under examination.

Despite of the problems, the dummy provides a promising test tool for transmissibility measurement, which is a critical measurement of vibration ride comfort.

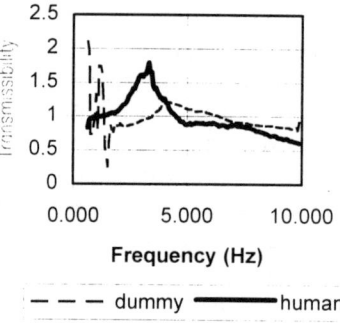

Fig.2. Comparison between human and stage 4 anthropodynamic dummy - sample seat #2

CONCLUSION

A correlation test to relate the test results from human occupants to dummy was performed. The test showed potentials to reduce the risk and expense of using human occupants by using the dummy to replace human occupants to measure the seat transmissibility in vertical direction.

More tests are needed to include different road profiles and vibration levels. A small number of human tests is still expected to validate the test results.

REFERENCE

[1] M.J. Griffin, Hand of Human Vibration, Academic Press, 1990

[2] C. Vail, Fundamentals of Seat Ride Dynamics, SAE Seminar, 1995

[3] Y. Gu, A Vibration Transmissibility Measurement Using a Mass Dummy, SAE Technical Paper 960475

[4] S. Rakheja et al, Interim Report on Static and Dynamic Characterization of an Automotive Seat, Concave Research Center, 1996

980656

Automotive Seating Comfort:
Investigating the Polyurethane Foam Contribution-Phase 1

G.R. Blair*, R. So, A. Milivojevich, J.D. van Heumen
THE WOODBRIDGE GROUP

Copyright © 1998 Society of Automotive Engineers, Inc.

ABSTRACT

In this investigation, we have selected a number of PUF driver seat cushions from a wide range of vehicles assembled in the U.S.A. and Canada. These driver seat cushions have been characterized with respect to both static and dynamic foam comfort. It has been determined that cushions considered good with respect to static comfort may not have the same rating when analyzed via vibrational transmissivity as a measure of dynamic comfort. Furthermore, we have also examined the vibrational transmissivity response of various PUF chemical systems in an attempt to measure the impact of the PUF chemistry on cushion design. This was accomplished by selecting two driver seat cushion molds and pouring four different chemistries into each mold. In addition to the observed effect of the PUF chemistry, the cushion design was also found to play a role in the dynamic comfort as indicated by the vibrational response. In the future, we plan to further examine the role of the cushion chemical design parameters on both static and dynamic seat comfort. The goal is to fully understand the key chemical factors such as type and level of isocyanate, base polyol, polymer solids, and additives and how these PUF ingredients affect the overall cushion comfort.

INTRODUCTION

For some time, many of the automotive manufacturers have focused on understanding the complex topic of seating comfort. Due to the subjective nature of seating comfort it has been difficult to identify those key elements of the seat construction which affect comfort most significantly. Recently, seat comfort surveys have been included as part of the overall vehicle satisfaction index. These independent surveys requested new vehicle owners to rate front and rear seats for comfort, support and ease of use. New owners indicated a wide range of comfort ratings. We have selected a number of front seat driver cushions from this survey. Cushions were chosen from the top and bottom of the comfort rating scale from each vehicle segment.

Various scientific techniques have been applied to understand the role of the seating components in comfort. One of the major contributors to seating comfort is the polyurethane foam (PUF) portion of the seat construction. This is evident from the recent increase in the number of publications dealing primarily with the role of the PUF in seating comfort (Casati, Herrington, Broos, Miyazaki, 1997; Shears, Bastin and Banner, 1997; Leenslag, Huygens and Tan, 1997; Utsumi, Isobe, Hiraide, Obata, Ohkubo, Sakai, 1997).

Comfort has been characterized from both the static and dynamic comfort perspectives. Static or showroom comfort is considered largely when the user first makes contact with the seat construction. Dynamic comfort, as the name implies, considers long term, in-vehicle performance of the seating components (Blair & Horn, 1996).

Static comfort is measured, in some cases, by more traditional foam test methods such as foam hardness, density, hysteresis, ball rebound, airflow, etc. as well as Body Pressure Distribution analyses (Runkle, 1994). Dynamic comfort is measured by methods such as vibrational dampening (Kinkelaar, Cavender & Crocco, 1996), dynamic hysteresis (Hilyard, Lee, and Cunnigham, 1991), dynamic fatigue (Cavender, 1990), S.E.A.T. dynamic testing (Crocco, Kinkelaar, and Neal, 1997), and long term in-vehicle service (Blair and Horn, 1996; Blair and Wilson, 1993).

Some published literature has focused on controlled changes in polyurethane chemistry to illustrate the potential contribution of PUF on the overall seat comfort performance. In this investigation we have considered the impact of PUF cushion chemistry on the static and dynamic comfort performance of the selected cushions. Woodbridge Foam Corp. is a global manufacturer of molded PUF. With this opportunity, Woodbridge Foam Corp. is embarking on an extensive long term program that is intended to illustrate and better understand the contribution of PUF cushion chemistry to automotive seating comfort. Through controlled PUF chemistry and processing, we have the ability to modify foam properties including those that may impact cushion comfort. This program will examine the influence of PUF cushion shape and design, but more importantly will include the effect of controlled formulation changes such as polyol and isocyanate types and additives.

In this preliminary study we have selected a number of PUF driver seat cushions from a wide range of vehicles assembled in North America. These cushions are from vehicles assembled by both domestic and transplant OEMs. Here we report on the characterization of these PUF cushions from both static (e.g. hardness and physical properties) and dynamic (e.g. vibrational characteristics) comfort perspectives.

This program is long term but our measurements and development work-to-date is reported here. We hope to be able to correlate the controllable and measurable foam properties to an acceptable comfort rating and then to recommend to the OEMs and seat assemblers those seat design characteristics that can further improve seat comfort.

BACKGROUND

SEAT USE

During extended periods of use (hundreds of kilometers), seats may exhibit some temporary fatiguing and the dynamic comfort may change. These changes are transient and after a short recovery period, the seat will return to its original state and the original degree of comfort will be experienced. That is, the degree of seat support (thigh, buttocks, lumbar) will offer the rider a comfortable environment without excessive pressure points in critical areas, eg. buttocks, lumbar. Also, the ability of the PUF components to absorb road vibrations by dampening or attenuation must be maintained during both short and long haul journeys.

SEAT DESIGN VERSUS COMFORT

Initially PUF components were used as topper pads on seat spring assemblies. Gradually as the benefits of PUF were established, PUF components especially cushions expanded to fill the entire seat interior. Later combinations of PUF components and various suspension mechanisms were developed and used to optimize seat comfort.

Some OEMs have indicated that changes in seat design are being considered, (e.g deep foam, thinner foam and multi-hardness foam). These changes in design are partially due to the excellent comfort and durability of PUF components and are influenced by the desire to reduce both seat mass, overall cost and to increase usable vehicle space. Thus, the chemistry of the PUF component will have to selected to provide the necessary seating support and to remain comfortable over a range of driving conditions such as short and long distances, low and high temperatures and relative humidities, and smooth and rough road. Automotive seating comfort is also influenced, in part, by the support frame type, adjusting and suspension mechanisms.

Our focus here is on the major contribution made by the PUF to seating comfort. Woodbridge Foam Corp. has the ability through proprietary processing and formulating technologies to deliver a wide range of PUF types, e.g. high resiliency foam based on TDI, MDI and isocyanate blends, multi-hardness parts and hot foam. Using these chemical and processing flexibilities, we can manufacture, evaluate, and deliver custom formulated parts to seat designers. These parts can then be used to explore various seat foam options and their contribution to overall seat comfort.

TESTING METHODOLOGY

The polyurethane foam component contribution to Seat Comfort Quality can be summarized as in scheme 1.

Experimental

This investigation has two parts. Part 1 in which production cushions have been analyzed from a static and dynamic comfort perspective and Part 2 in which chemistries evaluated in Part 1 have been examined in two specific cushion designs.

1. Cushion Selection

a) Part 1: Production Cushions

Based on independent surveys, production cushions were selected from both domestic and transplant vehicles having either suspension or dead pan support mechanisms. Three driver front seat cushions from each of the selected vehicle classes shown below were sampled for thorough evaluation.

Vehicle Class:

A1	Small car	(dead pan support)
A2	Small car	(suspension support)
B1	Small sport utility	
B2	Large sport utility (large metal frame moulded in place)	
C1	Sporty car	
C2	Sporty car	
D1	Midsize car	
D2	Midsize car	

E Pick up

F Minivan

G1 Luxury car

G2 Luxury car

Part 2: Modified Production Cushions

In this phase, our objective was to examine the results that would be obtained if a cushion was made with various polyurethane formulations. We selected two moulds and into each poured four different chemistries. This was done by using the molds for the A2 and F cushions and pouring four chemistries into each of the two molds as shown in the table below. The chemistries selected for evaluation were those used in cushions A2, A1, B1 and F, as illustrated in part 1. The molds chosen were for cushions A2 (small car with suspension support) and F (a minivan with suspension support). The following combinations were poured in triplicate:

VEHICLE CUSHION	FORMULATION POURED
A2	A2
	A1
	B1
	F
F	F
	A1
	A2
	B1

2. Cushion Evaluation

All cushions were tested in our Corporate Testing Laboratory in Woodbridge, Ontario after at least twenty-four hours of conditioning at $23 \pm 2°C$ and $50 \pm \%$% R.H. Cushions evaluated for vibrational tranmissivity and dynamic fatigue were tested at ARCO Chemical Company's South Charleston, West Virginia test laboratory.

a) Indentation Force Deflection (IFD), Hysteresis

In all cases the cushion hardness was determined in the same manner irrespective of the automotive manufacturers specifications. Cushion hardness was determined using a 4200 series Instron equipped with series XII control software. The test was conducted at a crosshead speed of 200mm/min with two 75% preflexes prior to a final compression of 65% in which the IFD at 5, 20, 25, 50 and 65% strain were recorded. The second 75% preflex were recorded in order to generate the hysteresis response. The comfort ratio was calculated from the IFD at 20 and 5% strain, while the support (sag) ratio was calculated from the IFD at 65 and 25% strain.

b) Physical Properties

Physical properties of all cushions were evaluated as in compliance with the ASTM D 3574-95 specification. Flammability was evaluated as indicated in the FMVSS302 specification. In addition, ball rebound tests were performed on core foam samples.

c) Foam Cell Structure

Foam cell structure was examined at both the foam surface and the pad interior by use of an optical microscope equipped with a photographic attachment. Cell structure measurements were made at a constant magnification of 20x.

d) Constant Force Pounding (ASTM D3574-95, Test I3)

The constant force pounding (CFP) test subjects a foam sample to 80,000 cycles of pounding with a $750 \pm 20N$ force under standard temperature and humidity conditions. Initial and final sample thickness and hardness at 40% deflection are measured and results are presented as a percentage of the initial foam thickness and hardness.

e) Vibrational Transmissivity

In the transmissivity test, complete cushions are subjected to a load of 22.7 kg over an indenter foot area of 314 m^2 the load is then oscillated over a range of frequencies (1-16 Hz) in 150 seconds. The natural or resonance frequency (v_r) is thus obtained as well as the peak maximum at resonance A_p. It is also possible to obtain the attenuation frequency (v_A), i.e. the frequency at which the output vibrational amplitude, A, is equal to the input amplitude, Ao. The transmissibility at 6 Hz (A_{6Hz}) is also recorded, as this is the frequency at which the internal organs of the human body resonate.

f) Dynamic Fatigue

The dynamic fatigue test was conducted as previously outlined by Cavender (Cavender, 1990, 1992). The dynamic fatigue test exposes the foam cushion to a 5Hz perturbation having an amplitude of ±5% deflection after a pre-compression of 50%. Changes in load, hysteresis, and height are monitored as a function of time and used to generate a composite number (fatigue number) depicting the potential durability of a particular PUF cushion.

Results and Discussion

As discussed, comfort has traditionally been evaluated from both a static and dynamic perspective. We will consider the static and dynamic comfort properties of both the production cushions (Part 1) and the modified production cushions (Part 2).

Part 1: Production Cushion Evaluation

Static Comfort Analysis

Table 1 illustrates the key static comfort elements. Hilyard (Hilyard, 1994) has indicated that hysteresis and ball rebound are inversely related and are strongly affected by the overall polyurethane polymer structure (i.e. soft segment functionality and hard domain concentration and structure). The trend has traditionally been towards higher resiliency foam (i.e. high ball rebound, low hysteresis loss) in the driver seat cushion, Table 1 illustrates the effect of foam hardness, density, and cushion thickness on ball rebound and hysteresis loss. Given the cushion specifications (hardness, mass), ball rebound values ranging from 32 to 61% were measured. We have the capabilities of producing cushions with higher ball rebound values and these will be reported on in future phases of this seat cushion comfort program. Considering ball rebound, cushions A1, B1 and E have the highest observed ball rebound. Both A1 and B1 also have the highest core densities. Conversely, E has both the lowest density and measured hardness value. Furthermore, A1 and E have a relatively large cross-sectional thickness of 89mm and 80.9mm respectively. From the hysteresis loss perspective A1 and B1 have relatively low hysteresis loss values, whereas E has a measured hysteresis loss of 28.

Cushions C1, D2, and G1 all have the lowest measured ball rebound and as well the highest hysteresis loss. From these cushions, C1, having a core density of 38kg/m^3, has the lowest cushion thickness, 31.7mm, and the highest measured load, 426N. D2 also has a high measured hardness, 408N, as well as a low core density of 36kg/m^3. The luxury car cushion G1, with the highest thickness of the cushions tested, a moderate hardness, and a relatively high core density has only a measured ball rebound of 48% and hysteresis loss of 25%. With the above cushion characteristics of thickness, hardness and density, we would have expected both a higher ball rebound and lower hysteresis loss. These observations will be discussed in relation to their effects on the vibrational characteristics of this cushion.

These observations indicated that some relationship may exist between the static comfort factors of hysteresis loss and ball rebound and the seat design parameters of cushion hardness, core density and cushion thickness. To further establish the relationship between static comfort factors and seat design parameters a correlation analysis was conducted.

Ball Rebound versus Hysteresis Loss

We have found that there is a reasonably good correlation between ball rebound and hysteresis loss, as shown in Figure 1. As the ball rebound capacity of foam increases in value, the hysteresis loss value decreases. This correlates well with the inverse relationship between ball rebound and hysteresis illustrated by Hilyard. (Hilyard, 1994).

Normally, cushions having a low hardness, high core density and high thickness exhibit high ball rebound and low hysteresis loss. However, cushion B1 with a moderate hardness, and core density of 47 kg/m^3 and a relatively low thickness of 49mm had a ball rebound of 60% and a relatively low hysteresis loss of 25%.

Static comfort was also considered in relation to foam cell structure. In the cushions evaluated the foam cell structure was found to vary at the foam surface between 28 and 40 cells/cm whereas the interior (core) foam has a coarser cell structure with 15 to 19 cells/cm. We believe that the cell morphology has an effect on the cushion comfort, but at this time the relationship has not been established. In addition, one cushion of each vehicle type was tested for physical properties according to the ASTM D 3574-95 specification and the results are shown in Table 2. Also, all cushions pass the FMVSS 302 flammability requirements with ratings of SE or SE/NBR.

Dynamic Comfort Analysis

Vibrational Transmissivity

The vibrational response of a PUF cushion is one of the important measurables required when assessing the dynamic comfort of a particular cushion. One of the functions of the PUF should be to dampen or attenuate incident vibrations encountered during use. More importantly the PUF should minimize vibrations in the 6 Hz region as this is the resonate frequency of a human's internal organs. The vibrational response is also critical when considering the type of support, suspension versus dead pan. Dead pan support requires that the foam assume the total vibrational dampening role. To minimize transmitted vibrations the literature indicates that a foam cushion should have a low natural frequency (v_r) as well as a low peak height at resonance (A_p) and a low attenuation frequency (v_a) (Griffin, 1997).

The results of the vibrational analysis are illustrated in Table 3 and examples of the transmissibility curves are shown in Figure 2. From the vibrational analysis, cushion A1 was found to have the lowest v_r and v_a. However, the A_p found for this cushion was the highest among the cushions tested. Recall that cushion A1 was found to be the most acceptable cushion with respect to static foam comfort (i.e. highest ball rebound and lowest hysteresis loss). In contrast, the luxury car cushion G1 was found also to have a low v_r and v_a as well as a low A_p. These vibrational characteristics are considered to contribute to seat comfort (Griffin, 1997). Static comfort analysis indicates that G1 has both a low ball rebound and a relatively high hysteresis loss suggesting moderate to low static comfort. Cushions A1 and G1 have approximately the same v_r, but since A1 has a higher A_p than G1 it would indicate that the G1 cushion provides better vibrational dampening ability than A1. Recently, similar conclusions were also drawn by both Kinkelaar and Crocco (Kinkelaar, 1996; Crocco, 1997). Cushions C1, C2 and D2, all previously shown to have a low static comfort rating, were found to have a v_r of greater than 6

Hz and a high v_a. Considering vibrational damping characteristics as one of the measures of dynamic seating comfort, these cushions may have a lower comfort rating than either A1 or G1.

It should also be noted that the support mechanism does contribute to the vibrational performance of the full seat assembly and that this investigation has considered the performance of the PUF only.

In an attempt to further understand the relationship between static and dynamic comfort, we analyzed for potential correlations between static and dynamic comfort properties.

Natural and Attenuation Frequencies versus Ball Rebound
In Figure 3, both the v_r and v_a for all cushions are plotted versus their respective ball rebound values. In the figure trend lines have been drawn to indicate that the frequency values decrease as the core foam ball rebound values increase. Therefore, within the boundaries of this evaluation, if foam is formulated to have higher ball rebound values, the v_r and v_a will decrease.

Natural Frequency Peak Height versus Ball Rebound
The A_p for these cushions appears to decrease as their ball rebound values decrease, Figure 4, i.e. less resilient foam will give a lower A_p than higher resiliency foam as illustrated in the comparison between cushion A1 and G1.

Natural and Attenuation Frequencies versus Cushion Thickness
Both v_r and v_a decrease with increasing cushion thicknesses as are shown in Figure 5. Thus as cushion thickness is increased, the natural frequency and the attenuation frequency decrease and this may contribute to an increase in dynamic comfort, especially if these frequencies can be shifted below 6Hz.

A/Ao Peak Height versus Cushion Thickness
In Figure 6, it is clearly seen that A_p decreases with decreasing cushion thickness (i.e. the thinner the cushion, the lower the A_p).

Natural and Attenuation Frequencies versus Hysteresis Loss
As the hysteresis loss decreases both v_r and v_a decrease in value, Figure 7. By formulating the foam in a cushion to have lower hysteresis loss in an force versus deflection test, v_r and v_a should be lowered. A_p is not well correlated with hysteresis loss.

Constant force pounding
A second cushion was sectioned and a 50mm thick sample cut from the IFD area was prepared and subjected to the CFP test. The results are found in Table 4. Only one cushion, D2, shows a significant loss in thickness (11.6%). All cushions exhibited less than 30% loss in hardness with two-thirds of those tested having less than 20% loss in hardness. Based on our previous long-term testing experience (Blair, Wilson & Horn, 1996), these cushions should also perform well in vehicles, even high-use vehicles such as police cruisers or taxi cabs.

Dynamic Fatigue
In the Dynamic Fatigue test, large differences were found, Table 5. Cushion E exhibited an excellent value of 69 (the lower the number the better the fatigue value) to 154 for cushion C2. This latter value is extremely high and may be due to the low thickness of this sporty vehicle cushion. Normally a Dynamic Fatigue Number of more than 100 is considered poor and potentially such cushions may exhibit in-vehicle durability problems unless the seat construction, i.e. support mechanism, has been designed to alleviate this problem.

Dynamic Fatigue Number versus Ball Rebound
With the exception of one cushion (D1), there is a good correlation between these two properties. The cushion property (dynamic fatigue number) appears to be dependent on the foam static property, ball rebound, Figure 8. Similarly, the CFP IFD loss may correlate with ball rebound but the extent of correlation is not as good as it is for dynamic fatigue number.

Part 2: Modified Production Cushions

Dynamic Comfort Analysis
In this part of the investigation we examined the effect of cushion design on both the static and dynamic foam comfort properties. This was done by using molds A2 and F and pouring four chemistries into each of the two molds. The chemistries selected for evaluation were those used in cushion A2, F, A1 and B1 illustrated in part 1.

Vibrational Transmissivity
The effect of seat design on the measured vibrational parameters is shown in Table 6. In addition the vibrational response plots for all cushions are shown in Figures 9A and 9B for cushions A2 and F respectively. Cushion A2, having a suspension support mechanism, was made using its regular chemistry and was found to have a v_r of 5.82 Hz and a A_p value of 3.34. The cushion chemistry, as expected, has an influence on the vibrational performance of the PUF cushion. This is illustrated by substituting the B1 chemistry for A2 chemistry the v_r can be reduced from 5.82Hz to 5.33Hz. Similarly, a substitution of the F chemistry for the A2 chemistry yields an increase in the v_r to 6.80Hz. This effect of change in chemistry is most clearly evidenced in Figure 9A where the vibrational response plot is illustrated for cushion A2. Clearly pouring the A1, B1, and F chemistries into a cushion A2 mold creates A2 type cushions having distinctly different vibrational responses.

This is contrasted with a similar experiment conducted using the cushion F mold and illustrated in Figure 9B. The use of A1, A2, and B1 chemistries has resulted in F type cushions with similar vibrational characteristics. These three chemistries have similar v_r and A_p. All of these formulations yield lower v_r and A_p values than the original F cushion.

By considering the observations made on cushions prepared from the A2 and F molds it is also evident that the cushion geometry also plays a significant role in the PUF cushion vibrational response.

Conclusions

A range of commercial front seat cushions evaluated by static and dynamic tests exhibit a wide range of responses.

Some cushions with good static properties do not have acceptable vibrational characteristics.

There is a good correlation between the static comfort properties (e.g. ball rebound, hysteresis loss) and cushion hardness, thickness and density.

The vibrational analysis indicates that, within the scope of this investigation, cushions with high thickness, moderate hardness and foam core density provide the best vibrational characteristics, i.e. low resonance frequency, and low peak maximum at resonance.

Cushion geometry influences seat comfort as measured by vibrational transmissivity.

Cushions with the same geometry but made with selected foam formulations have different vibrational transmissivities.

Woodbridge Foam Corp. has the capability to design and control foam chemistry and processing to achieve specific static and dynamic cushion responses and hence obtain improved seat comfort.

Future Work

We intend to study various contributions to seating comfort resulting from polyurethane formulation modifications. Such changes will include the type of isocyanate (i.e. TDI, MDI, and TDI/MDI blends), various base polyols (i.e. functionality, molecular weight), the type and level of polymeric solids (e.g SAN, AN, PHD, and PIPA), and other additives (e.g. surfactants, catalysts). In addition, foam characteristics such as foam surface breathability, core foam airflow and cell morphology will also be studied to ascertain their contributions to seating comfort.

Furthermore, other seat comfort analyses will be investigated. Some of these techniques will include dynamic modulus response, body pressure distribution (i.e. pressure mapping), seat complete dynamic testing (i.e. S.E.A.T. analysis), and other appropriate techniques.

ACKNOWLEDGMENTS

Special thanks to Tuan Le at our Corporate Testing Laboratory who supervised physical property testing and to Primo Tarantini at our Woodbridge, Ontario plant who produced the cushions for Part 2 of this study.

Thanks to Doug Cavender and Mark Kinkelaar of ARCO Chemical Company, West Virginia for their supervision and reporting of the Transmissivity and Dynamic Fatigue investigations.

*To whom correspondence should be addressed
8214 Kipling Ave. North
Woodbridge, Ontario
Canada
L4L 2A4

REFERENCES

Blair, G.R. and Horn, R.J., 1993, "OEM Seating Foam Specifications - an International Suppliers Viewpoint", PU Forum, Nagoya, Japan. pp. 221-229.

Blair, G.R. and Wilson, A.R., 1995, "Polyurethane Automotive Cushioning; Material Properties After in Car and Simulated Durability Testing", 36th Annual Polyurethane Technical/Marketing Conference, Sept. 26-29, Chicago, IL. pp. 413-417.

Blair, G.R. and Horn, R.J., 1996, "Fleet Durability Testing of Moulded Polyurethane Foam and Competitive Automotive Cushions", Utech '96, March 26-28, paper 5.

Blair, G.R. and Horn, R.J., 1996, "Long Term Evaluation of High Resiliency Moulded Foam Comfort and Durability", PUKOREA, pp. 21-37.

Casati, F.M., Herrington, R.M., Broos, R., Miyazaki, Y., 1997, " Tailoring the Performance of Molded Flexible Polyurethane Foams for Car Seats", Polyurethanes World Congress'97, Amsterdam, The Netherlands pp. 402-420.

Cavender, K.D., 1990, "New Dynamic Flex Durability Test" 33rd Annual Polyurethane Technical Marketing Conference, Sept. 30 - October 3, pp. 282-288.

Cavender, K.D., 1992 "Real Time Foam Performance Testing" 34th Annual Technical/Marketing Conference, October 21-24, pp260265.

Crocco, G.L., Kinkelaar, M.R., Neal, B.L., 1997, "Using S.E.A.T. values to improve Atomotive Seating Foam Dynamics and Seat Comfort", Polyurethanes World Congress'97, Amsterdam, The Netherlands pp. 436-446.

Griffin, M.J., "Seating Dynamics and the effect of Ride Comfort", Presented at Comfort Workshop, ARCO Chemical Co., March 1997.

Hilyard, N.C., 1994 "Hysteresis and energy loss in Flexible polyurethane Foams" in *Low Density Cellular*

Plastics: Physical Basis of Behaviour, Hilyard, N.C. and A. Cunnigham, eds., Chapman and Hall, Chapter 8.

Hilyard, N.C., Lee W.L., Cunnigham, A., 1991 "Energy Dissipation in Polyurethane Cushion Foams and Its Role in Dynamic Ride Comfort", Cellular Polymers, Forum Hotel, London, U.K. RAPRA Technology Ltd., pp.187-191.

Kinkelaar, M.R., Cavender, K.D. and Crocco, G.L., 1996 "Vibrational Characterization of Various Polyurethane Foams Employed in Automotive Seating Applications", 37th Annual Polyurethane Technical/Marketing Conference, Las Vegas, Nevada pp. 496-503.

Leenslag, J.W., Huygens, and Tan, A., 1997 "Recent Advances in the Development and Characterization of Automotive Comfort Seating Foam", Polyurethanes World Congress'97, Amsterdam, The Netherlands pp. 436-446.

Runkle, V.A., 1994 "Benchmarking Seat Comfort", SAE Technical Paper 940217.

Shears, J., Bastin, B., Banner, C., 1997 "Sensory Analysis: A new Method for Determining the Feeling and Comfort Characteristics of Flexible Polyurethane Foams", Polyurethanes World Congress'97, Amsterdam, The Netherlands pp. 75-80.

Utsumi, H., M. Isobe, Hiraide, T., Obata, M., Ohkubo, K., Sakai, S., 1997 "Durability of Flexible Molded Polyurethane Foams", Polyurethanes World Congress'97, Amsterdam, The Netherlands pp. 447-465.

Wilson, A.R. and Blair, G.R., 1994, "Polyurethane Automotive Cushioning; Car Durability vs. Foam Properties", 35th Annual Polyurethane Technical/Marketing Conference, October 9-12, Boston, MA. pp. 478-488.

SCHEME 1

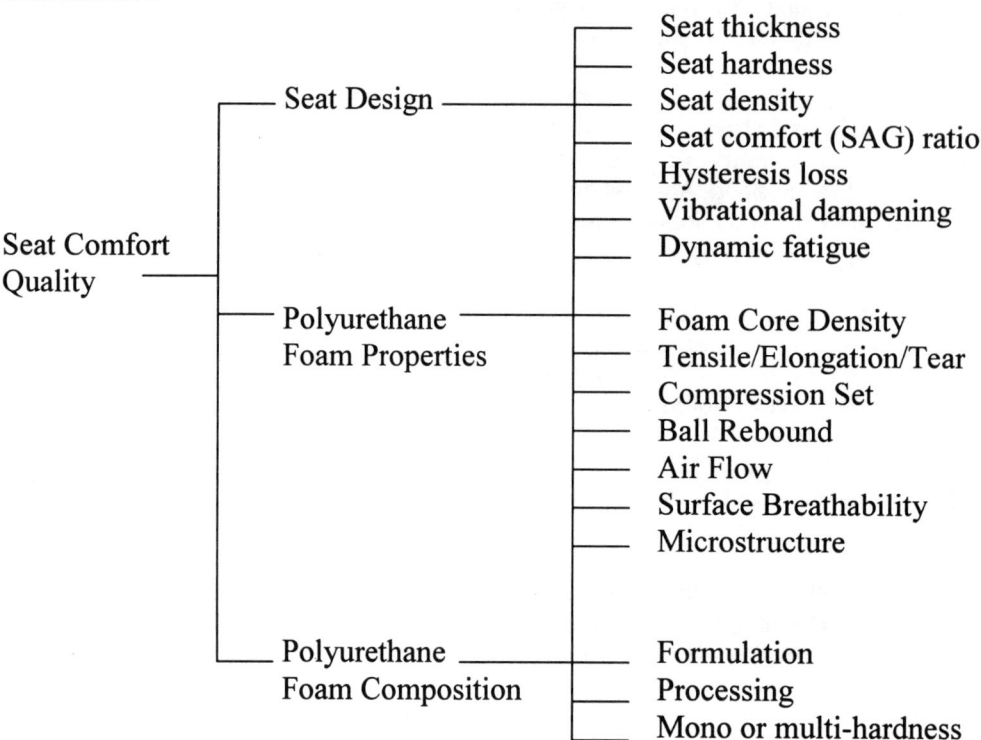

TABLE 1

Vehicle	50% IFD (N)	Core Density (kg/m^3)	Cushion Thickness, mm	Hysteresis Loss %	Ball Rebound %
A1	347	55.5	89.0	20	61
A2	277	40.0	66.4	23	55
B1	353	47.3	48.9	25	60
B2	339	32.9	98.4	30	55
C1	426	38.0	31.7	39	49
C2	261	33.7	34.6	37	50
D1	249	44.4	54.2	28	32
D2	408	36.0	53.7	36	46
E	268	33.6	80.9	28	58
F	327	48.9	34.5	23	54
G1	300	39.2	118.6	25	48
G2	265	31.0	84.2	28	53

TABLE 2

PROPERTY TESTS	UNITS	VEHICLE											
		A1	A2	B1	B2	C1	C2	D1	D2	E	F	G1	G2
Core density	kg/m^3	55.5	40.0	47.3	32.9	38.0	33.7	44.4	36.0	33.6	48.9	39.2	34.0
Tensile Strength	kPa	166	140	130	130	110	100	140	120	130	120	120	120
Elongation	%	130	136	127	134	106	104	139	115	131	116	123	116
Tear Strength	N/m	269	222	220	205	284	285	256	286	212	279	226	205
Compression Set 50% Def. 75% Def.	%	10 9	13 11	12 10	27 36	15 15	15 13	19 20	16 16	13 11	13 12	11 9	14 12
Compression Set 50% H.A.	%	13	18	15	21	18	29	16	18	13	15	14	18
Ball Rebound	%	61	55	60	55	49	50	32	46	58	54	48	53

TABLE 3

VEHICLE	NATURAL FREQUENCY, Hz (ν_r)	A/Ao PEAK HEIGHT(A_p)	ATTENUATION FREQUENCY, Hz (ν_A)	A/Ao HEIGHT @ 6 Hz (A_{6Hz})
A1	4.80	4.86	6.70	1.61
A2	5.65	2.82	8.00	2.67
B1	6.53	3.03	9.34	2.34
B2	5.00	3.99	6.50	1.48
C1	13.6	2.44	>16	0.98
C2	9.63	3.11	13.05	1.11
D1	6.00	3.08	8.70	3.10
D2	7.93	3.34	11.82	1.43
E	5.40	3.42	7.75	2.77
F	6.64	2.67	9.27	2.13
G1	4.63	2.16	6.66	1.38
G2	5.40	3.83	7.67	2.64

TABLE 4

VEHICLE	HEIGHT LOSS %	IFD @ 40% DEF. LOSS, %
A-1	1.4	18.3
A-2	1.9	18.3
B-1	1.7	14.8
B-2	2.2	18.9
C-1	2.2	20.5
C-2	2.2	26.7
D-1	1.6	18.4
D-2	11.6	18.4
E	1.0	17.6
F	1.5	13.7
G-1	2.9	22.5
G-2	1.4	20.0

TABLE 5

VEHICLE	THICKNESS CHANGE, %	CREEP %	IFD CHANGE %	DYNAMIC FATIGUE NUMBER *
A1	3.41	6.68	19.93	73.6
A2	3.67	10.33	25.19	97.5
B1	3.03	8.94	20.73	82.0
B2	5.15	7.99	19.10	86.3
C1	4.51	13.10	36.00	129
C2	15.40	8.00	30.00	154
D1	5.78	9.92	24.90	106
D2	8.44	9.58	24.80	118
E	0.20	9.12	21.00	69.00
F	3.11	6.97	19.4	72.6
G1	5.16	8.92	22.97	96.0
G2	3.67	8.53	22.71	86.6

* Dynamic Fatigue Number =
    1.5x IFD Change + 4x Creep + 5x Thickness Change

TABLE 6

Cushion Type	Formulaton Type	Natural Frequency Hz (v_r)	A/Ao Peak Height (A_p)	Attenuation Frequency Hz (v_A)	A/Ao Height @ 6Hz (A_{6hz})
A2	A2	5.82	3.34	8.53	3.16
A2	A1	5.05	3.81	7.09	1.96
A2	B1	5.33	2.77	7.51	2.29
A2	F	6.80	3.57	9.73	1.86
F	F	6.88	3.39	9.87	1.83
F	A1	5.95	2.88	8.22	2.82
F	A2	6.43	2.88	9.19	2.44
F	B1	5.90	2.77	8.11	2.70

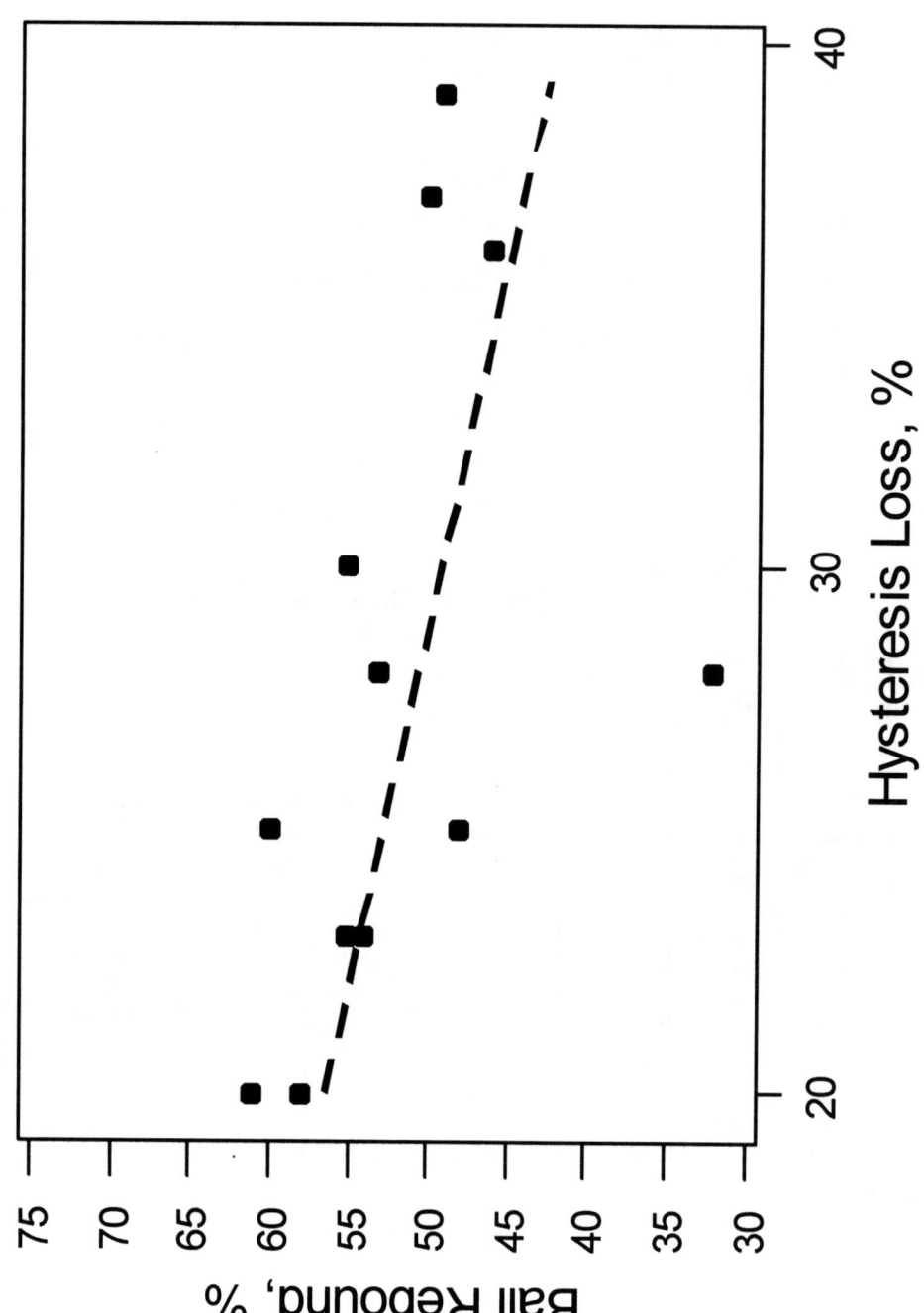

Fig.1 BALL REBOUND VS. HYSTERESIS LOSS %

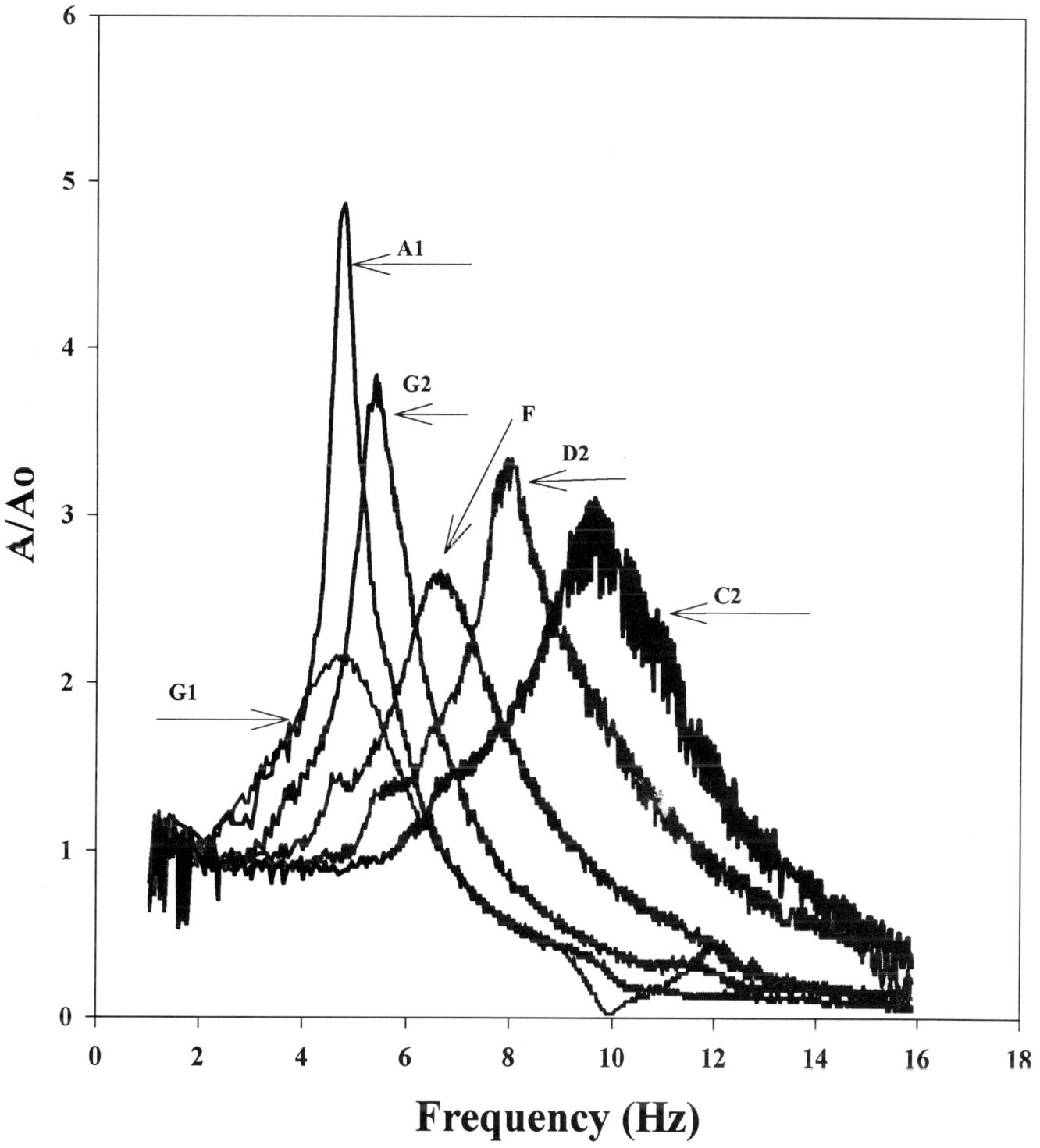

Figure 2. Transmissivity curves for various polyurethane foam cushions

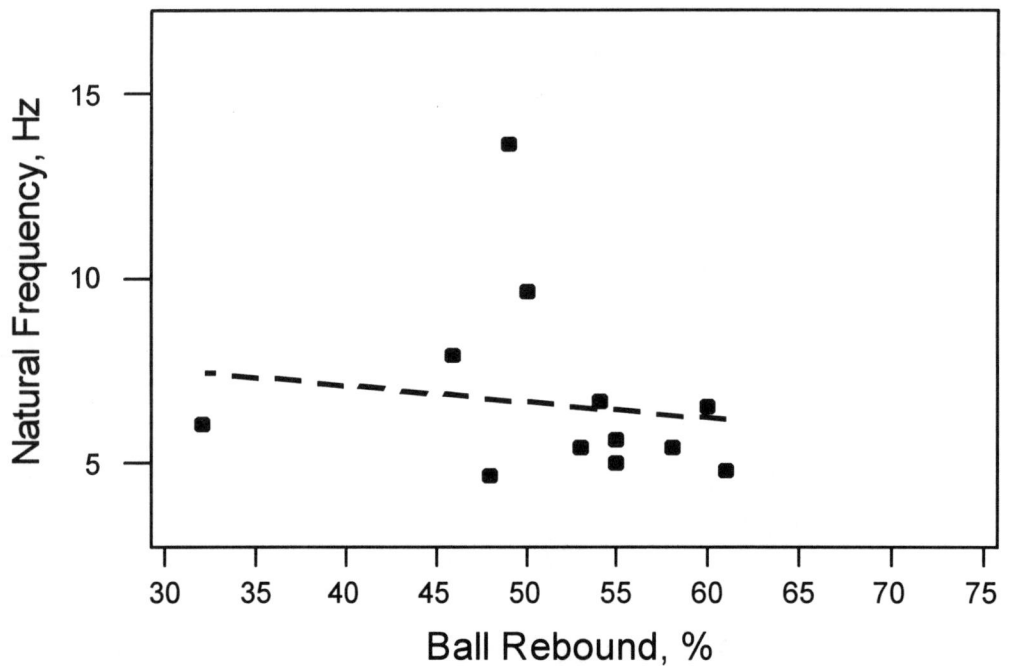

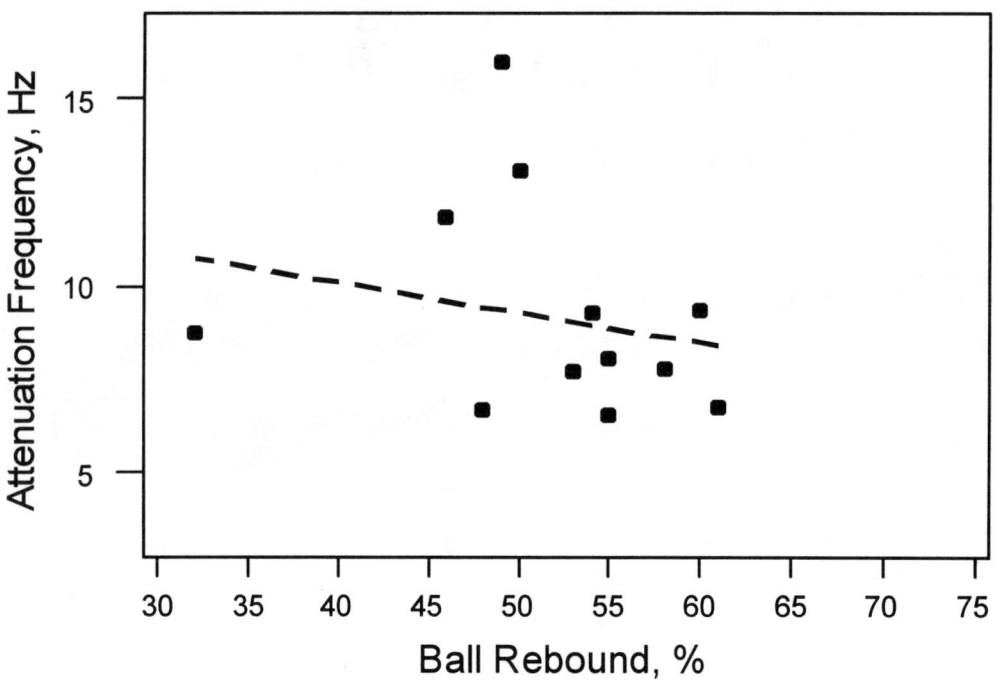

Fig. 3 NATURAL & ATTENUATION FREQUENCY VS. BALL REBOUND

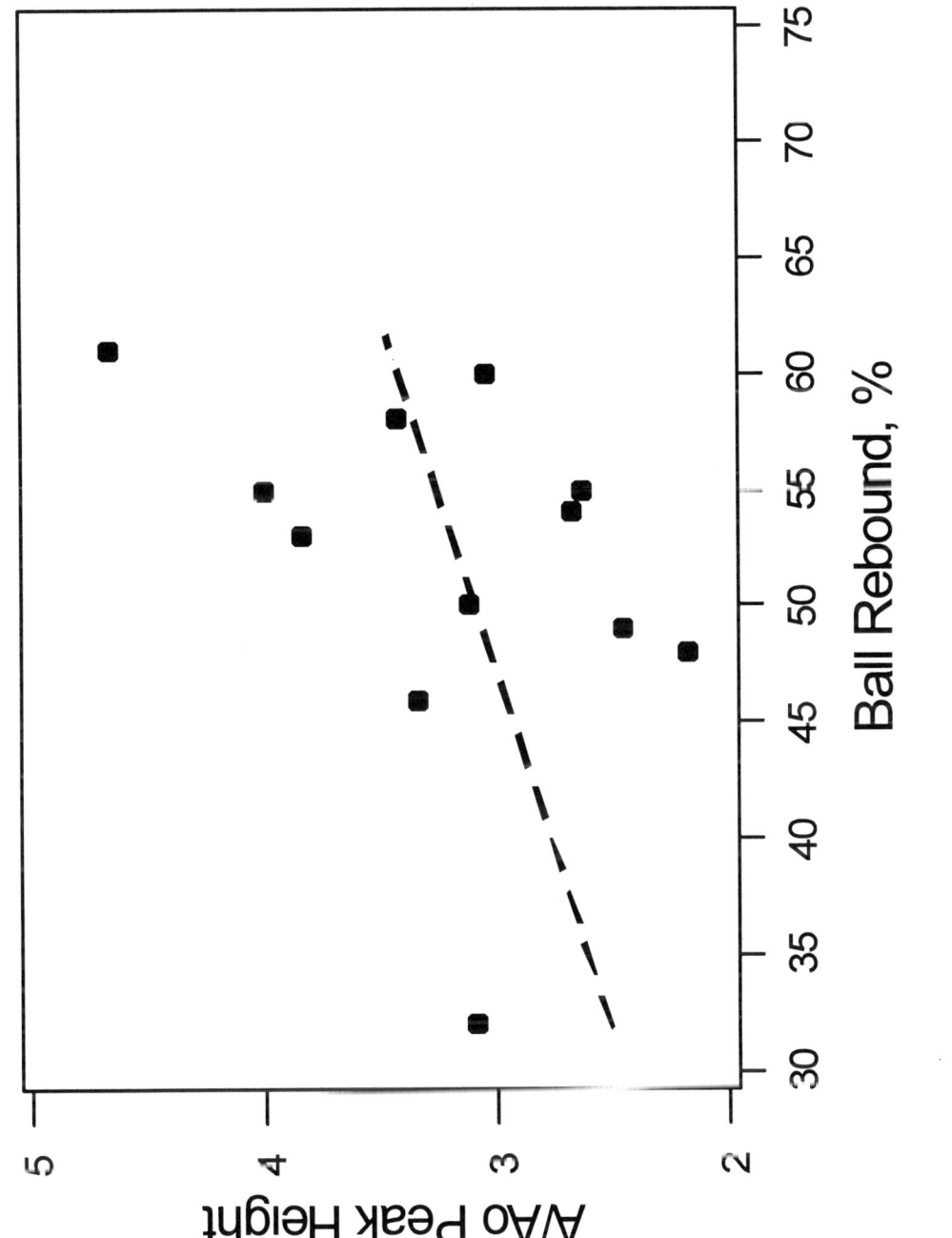

Fig.4 A/Ao PEAK HEIGHT VS. BALL REBOUND

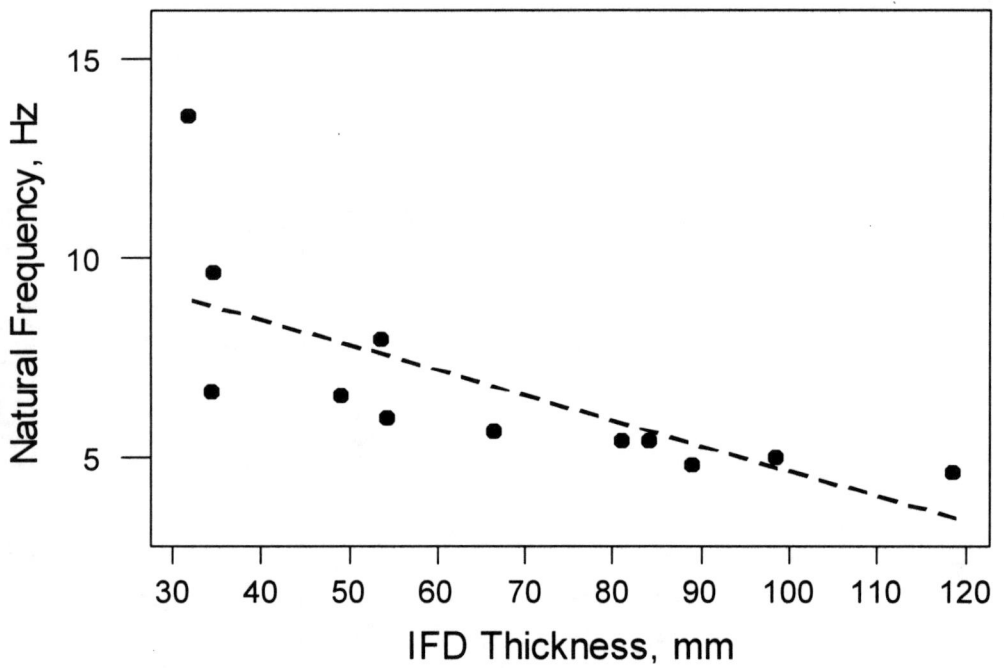

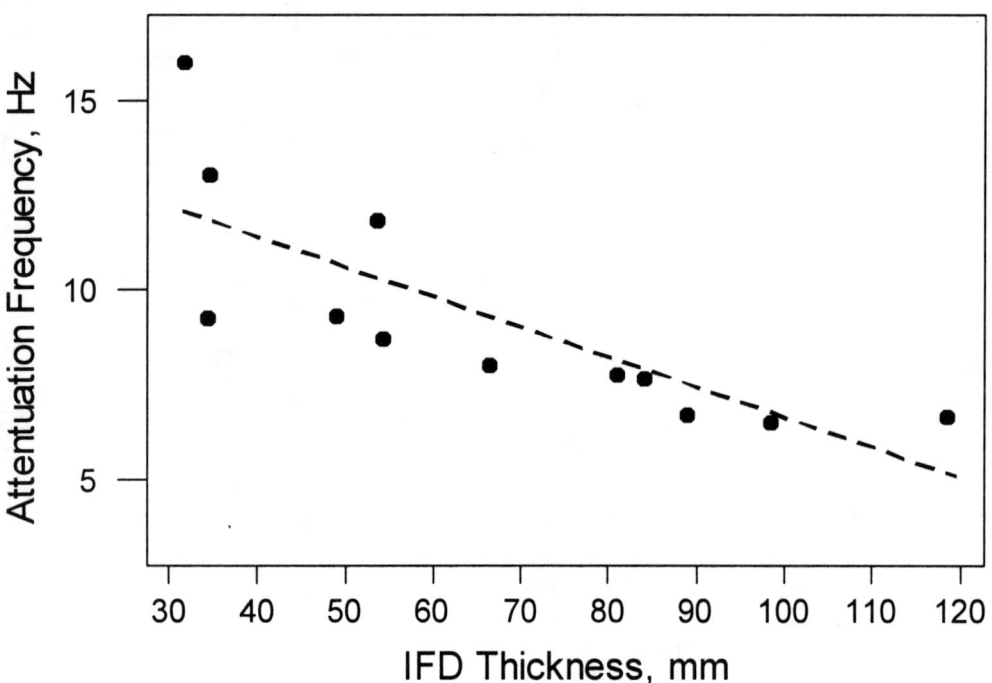

Fig. 5 NATURAL & ATTENUATION FREQUENCY VS. IFD THICKNESS

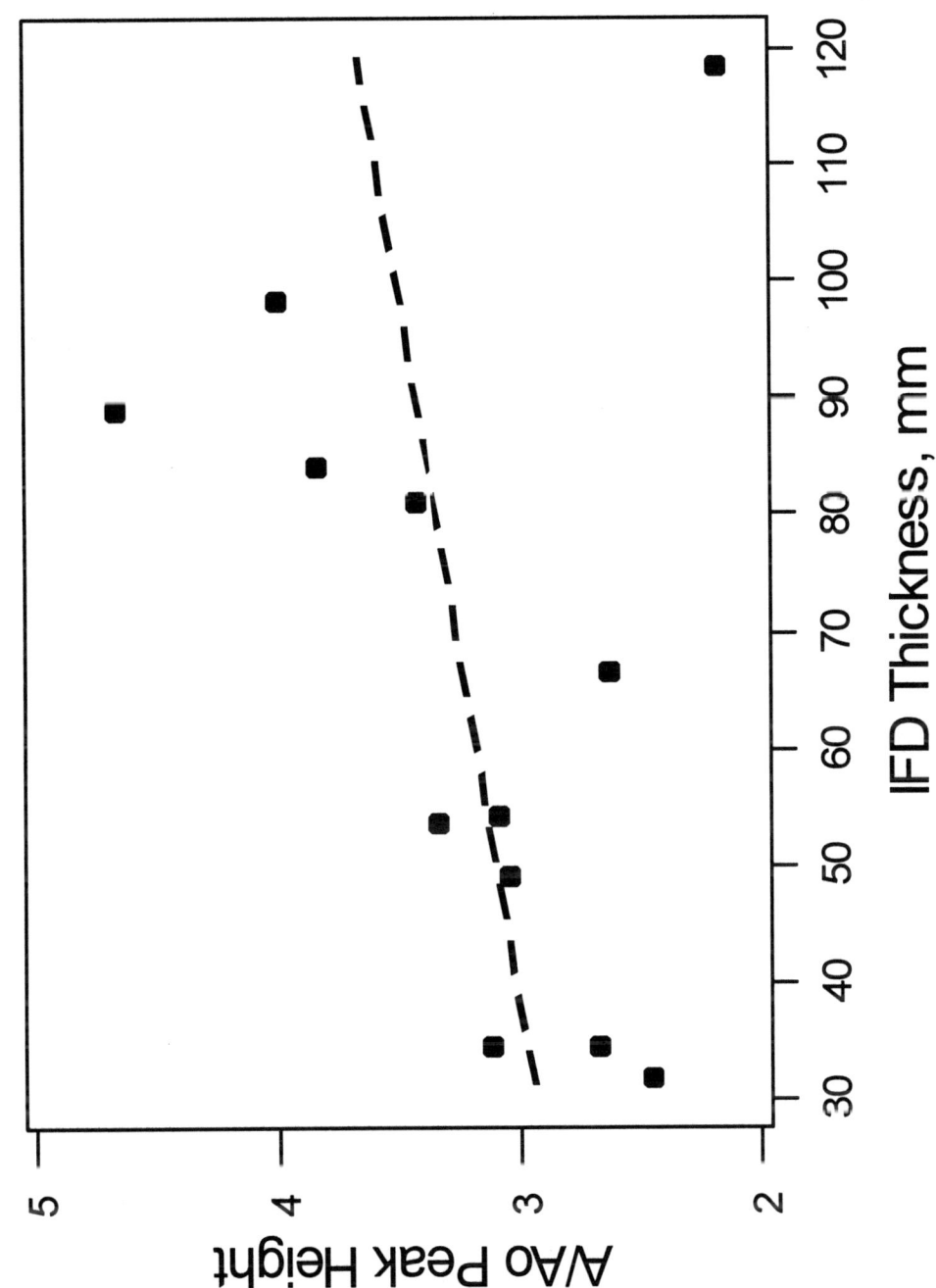

Fig.6 A/Ao PEAK HEIGHT VS. IFD THICKNESS

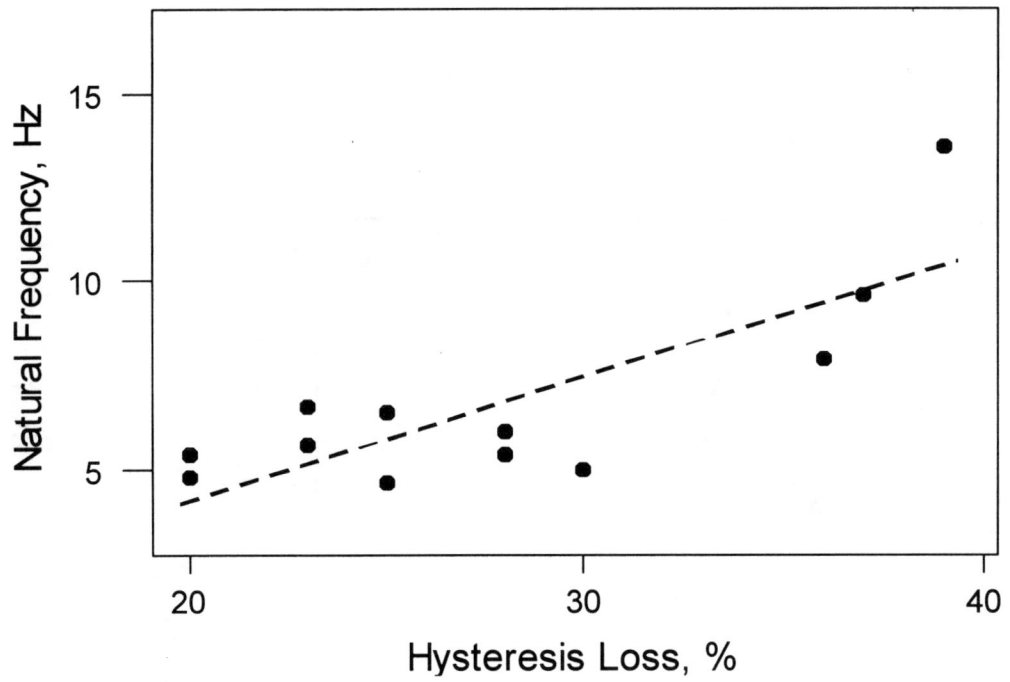
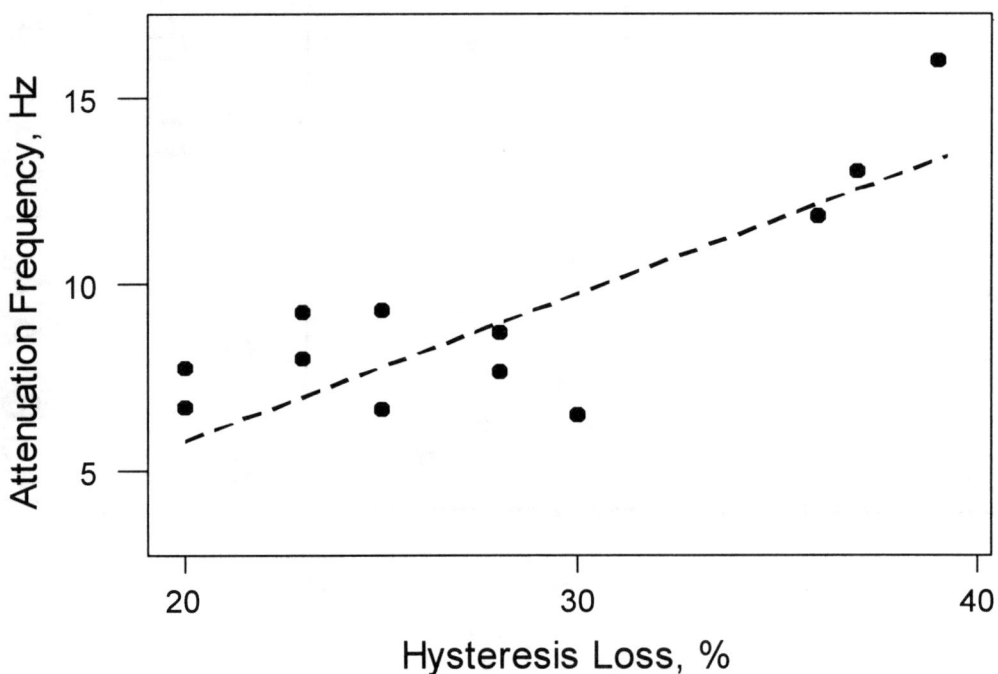

Fig. 7 NATURAL & ATTENUATION FREQUENCY VS. HYSTERESIS LOSS

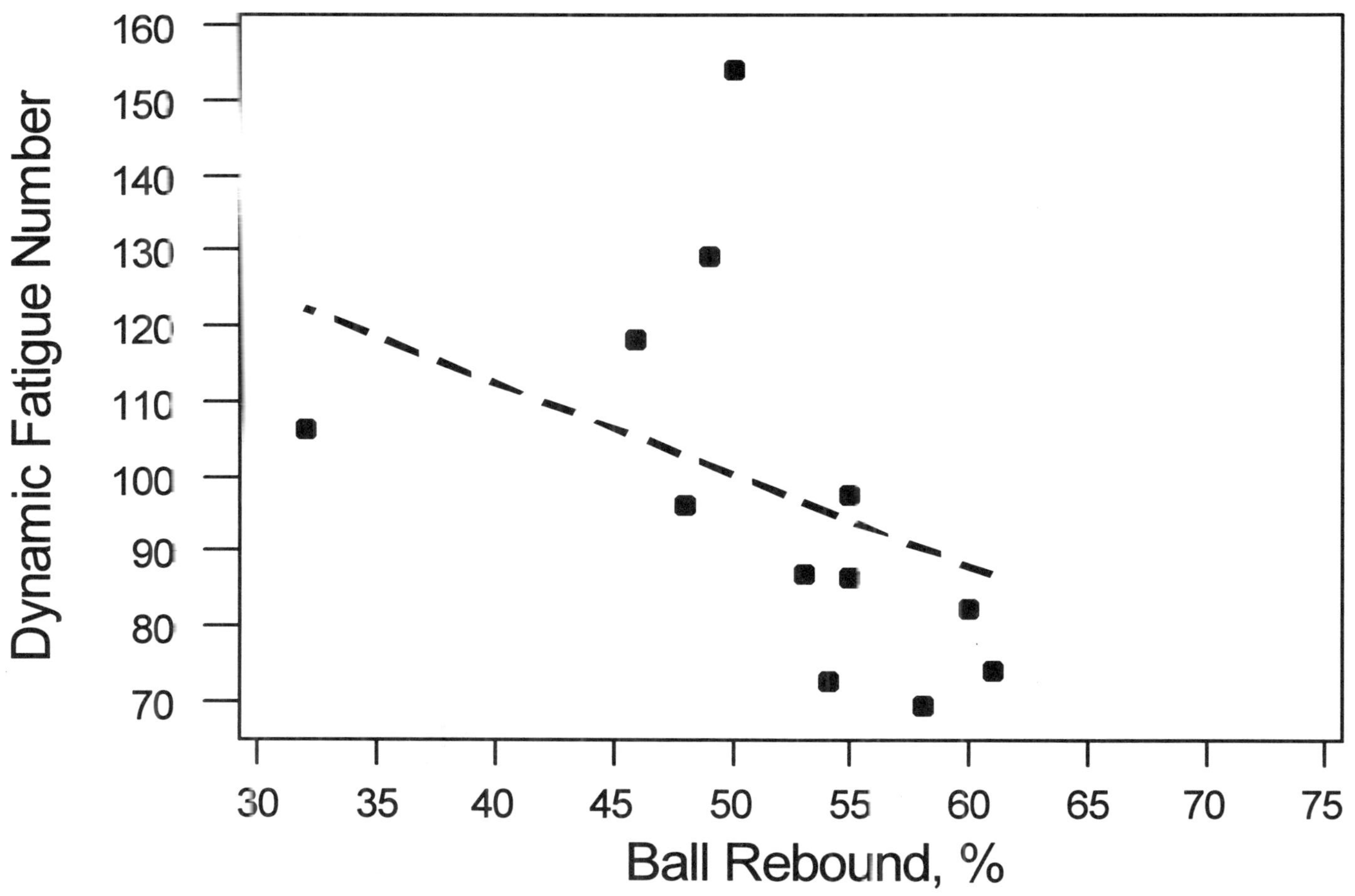

Fig.8 DYNAMIC FATIGUE NUMBER VS. BALL REBOUND

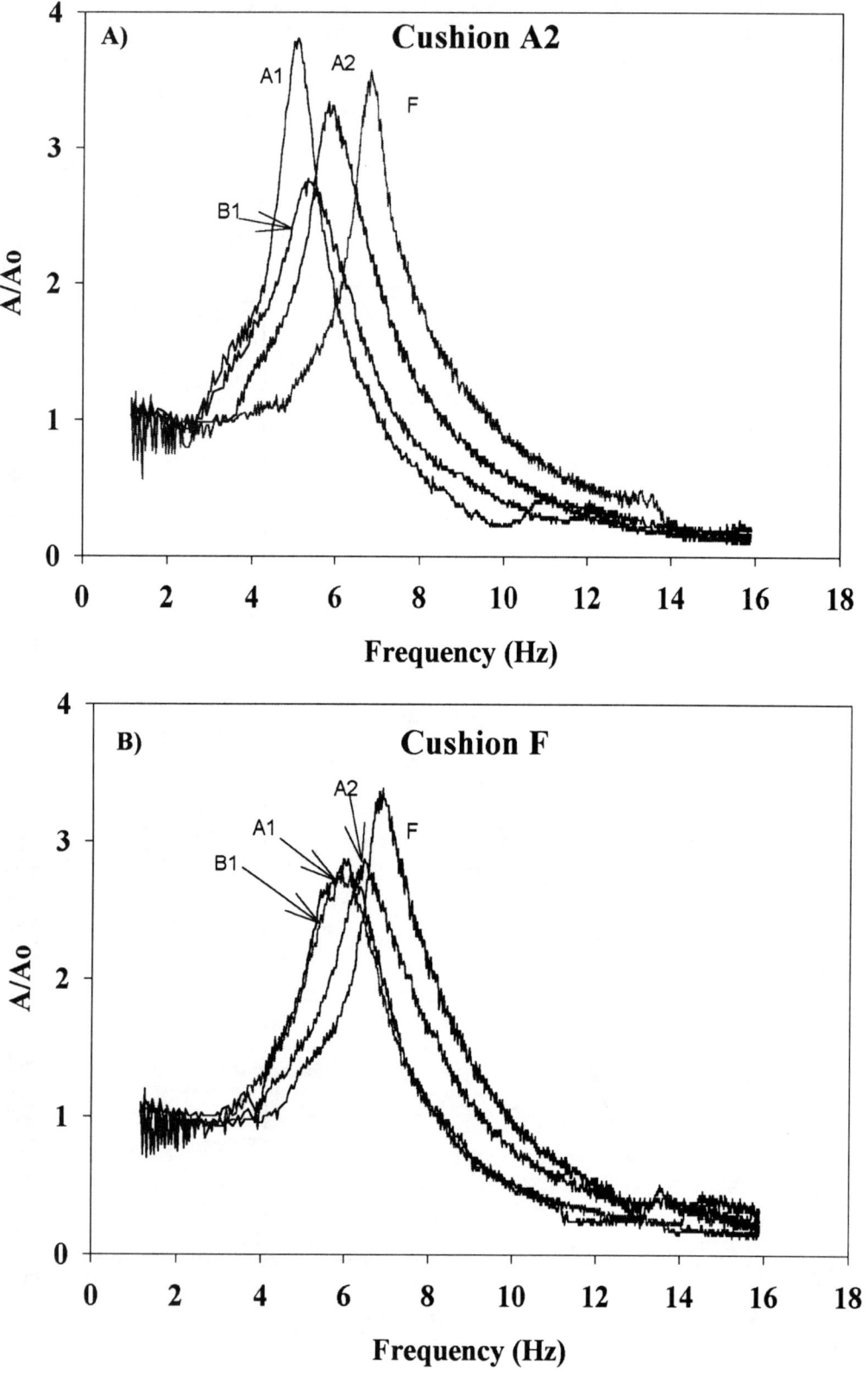

Figure 9. Transmissivity Curves: A) Cushion A2 with various formulations B) Cushion F with various formulations

980657

The Influence of Polyurethane Foam Dynamics on the Vibration Isolation Character of Full Foam Seats

Mark R. Kinkelaar and Brian L. Neal
ARCO Chemical Company

Guy Crocco
ARCO Chemical Products Europe, Inc.

Copyright © 1998 Society of Automotive Engineers, Inc.

ABSTRACT

A recent trend in vehicle seating design is the change in structure from a spring / foam composite to a foam and dead pan structure. For a composite design, the seat vibrational characteristics are tuned for the vehicle by adjusting the spring stiffness. In a dead pan design, however, foam dynamics solely dictate seat vibrational characteristics.

Foam vibrational transmissivities measured by a laboratory test are compared with transmissivities measured with human subjects and road profiles. The in-vehicle vibrational performance of the foam has been quantified using the S.E.A.T. method for several vehicle categories. The role of foam dynamics has been quantified for full foam seats.

INTRODUCTION

One key contribution a vehicle seat provides toward passenger comfort is its ability to isolate the passenger from road and vehicle vibrations. Vibration affects the comfort (or discomfort) experienced by, and also the safety of, the vehicle occupants and is dependent on the frequency, magnitude, and duration of the oscillations. Vibration induced discomfort and fatigue are extensively documented in the literature.[1] Different frequencies affect different parts of the human anatomy[1] to various degrees, and there exists international standards as to how much vibration a human can be exposed to without ill effect.[2]

A vehicle seat has inherent vibrational characteristics based on its materials of construction and design. Seat and automobile manufacturers typically work together to "tune" the vibrational characteristics of the seat to compliment those of the vehicle. A correctly tuned seat has a natural frequency that does not overlap other vehicle natural frequencies and provides attenuation in the frequency ranges that lead to human discomfort. In composite seats, this "tuning" is usually accomplished by manipulation of the spring stiffness; the role of the polyurethane foam in improving vibrational performance has generally been neglected.

A recent trend in the vehicle seating industry is implementation of full foam seating, i.e., a seat design where the foam is placed on a "dead pan" rigidly mounted to the vehicle floor pan. This change in seating design is driven by cost and weight reduction of the assembled seat, and "green" considerations (disassembly for recycle). In a full foam seat there are no springs to adjust and the foam is the sole means of controlling the seat ride dynamics. The polyurethane foam material engineering properties must be developed as has been done for the springs in composite designs. Understanding the vibrational characteristics of the polyurethane foam and how the foam dynamics relate to seat ride dynamics is crucial to "tuning" a full foam seat. For example, the natural frequency of a foam may be specified to protect the occupant from harmful or uncomfortable vibrations. Ultimately, this qualification will lead to changes in the chemical structure or processing of the polyurethane foam.

We have pioneered a laboratory testing procedure to measure the vibrational characteristics of polyurethane foam.[3, 4] Using this test, we have explored some of the fundamental characteristics which contribute to dynamic seating comfort and have performed extensive work to compare the effect of the polyol and isocyanate on the vibrational characteristics of polyurethane foam.[4,5] A brief overview of our test method, typical transmissivity data measured by our test and a brief explanation of how to interpret the resultant data is provided in the Experimental section.

One issue not well covered in the literature is the comparison between laboratory foam vibration characterization data and in-vehicle performance. It would be unrealistic to expect a simple laboratory test to quantitatively predict in-vehicle seat vibrational performance; however, these testing methods should be able to predict relative trends of foam performance in a vehicle. With an understanding of the requisite dynamics for seating applications, one should be able to predict the desired foam vibrational characteristics to improve seat ride dynamic performance of full foam seats using the laboratory method that may generally extend to the in-vehicle performance. To demonstrate that point, this paper describes comparisons made between our laboratory method and a test with seated humans using vibrational input from a vehicle, the Seat Effective Amplitude Transmissibility (S.E.A.T.) test.

EXPERIMENTAL

LABORATORY SCALE METHOD

Figure 1 is a schematic of the laboratory scale transmissivity test used in this study. The apparatus consists of a base plate mounted to an unidirectional, hydraulically actuated load frame (MTS, Minneapolis, Mn). The foam sample being tested sits on the base plate and a 23 kg cylindrical mass (200 mm in diameter) sits on the foam. The movement of the base plate (Ao, or input acceleration) and mass (A, or response acceleration) were monitored via an accelerometer (PCB Piezoelectronics). The test protocol consisted of a sinusoidal frequency sweep from 1 to 16 Hz in 150 seconds with a constant input peak acceleration of 0.2 m/sec^2. Input and response acceleration data were taken and analyzed in real time. The data are reported as transmissivity (or transmissibility), a direct measure of the transmitted vibration relative to the output response divided by input acceleration (A/Ao).

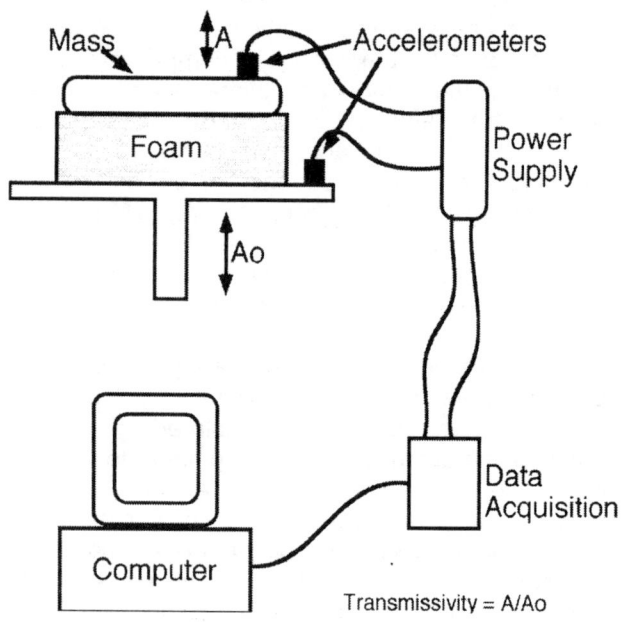

Figure 1. Schematic of laboratory transmissivity equipment.

Figure 2 is an example of typical foam transmissivity data measured by this test method. Shown in Figure 2 is a transmissivity verses frequency plot. There are typically three regions within a foam transmissivity curve. At frequencies significantly lower than the resonance peak, transmissivity is equal to one; that is, the response and input vibration are equal. In the resonance peak region the response vibration is greater than the input vibration, effectively amplifying the input vibration. Finally, at frequencies significantly higher than the resonance peak the transmissivity is less than one, or the vibration is attenuated. Transmissivity results are a function of the mass, geometry, and frequency sweep protocol used in testing and therefore it is difficult to compare transmissivity data from two different test methods. Generally, trends from different test methods agree, but the absolute values of the key measured characteristics such as natural frequency do not agree.

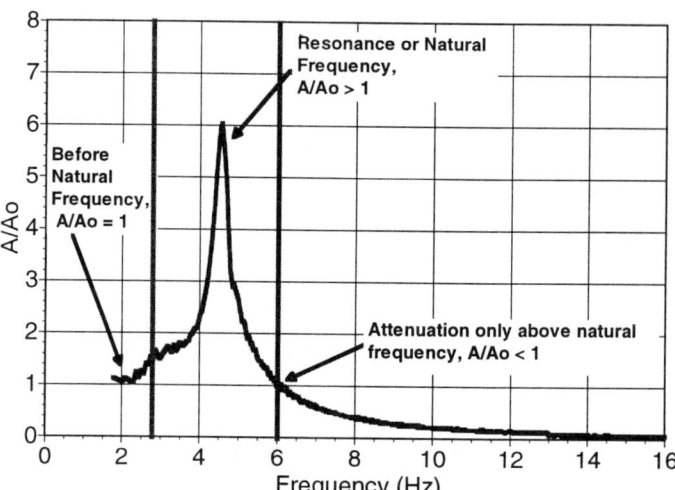

Figure 2. Typical transmissivity data.

The critical vibrational characteristics measured by this test are foam resonance frequency and transmissivity at resonance. With these values, the entire transmissivity curve can be predicted reasonably using the Kelvin spring/dashpot model.[3] The attenuation frequency, or the frequency at which the transmissivity crosses below one after the natural frequency thus marking the beginning of attenuation, is also of interest. By comparing these material responses to vibration, one can compare overall foam performance.

When tuning the final assembled seat, the design engineer must also consider the vehicle vibration spectrum and frequencies that cause human discomfort. With the laboratory test method, foam vibrational performance can be quickly compared and it is useful to consider the desired vibration characteristics based on the laboratory results. Generally, the attenuation properties of the foam are the desired vibrational characteristics for a full foam seat as attenuation is the means to isolate the passenger from the road and vehicle vibration. As attenuation occurs only after the natural frequency, a lower natural frequency is desired to expand the attenuating properties into lower frequencies, i.e., a lower natural frequency correlates with a lower attenuation frequency in the spring/dashpot model. Once a lower natural frequency is obtained, a lower transmissivity at the natural frequency would be beneficial also, as road noise is present to some extent at all frequencies. The optimum balance of precedence between low resonance frequency and low peak transmissivity is currently a matter of debate.

SEAT METHOD

A more in-depth technique to characterize the in-vehicle dynamic comfort of an automotive seat or foam cushion is through the S.E.A.T. method. The S.E.A.T. value, developed by Prof. M. J. Griffin,[1] is simply the ratio of the "ride on the seat" to the "ride on the floor".

$$S.E.A.T.(\%) = \frac{\text{"Ride on the Seat"}}{\text{"Ride on the Floor"}} \cdot 100\% \quad (1)$$

As such, the S.E.A.T. value as represented by Equation 1 provides a single number on which to compare the relative vibration isolation comfort of a particular seat in a vehicle.

The calculation of the S.E.A.T. value, however, is instructive in the type of information that is incorporated into the single value. Equation 1 is more precisely written as -

$$S.E.A.T.(\%) = \left[\frac{\int G_{ff}(f) \cdot |H(f)|^2 \cdot W_i^2(f) df}{\int G_{ff}(f) \cdot W_i^2(f) df} \right]^{1/2} \cdot 100\% \quad (2)$$

The "Ride on the Floor" (denominator in Equation 1) is defined as an integration of the vehicle power spectrum $[G_{ff}(f)]$ measured on the floor multiplied by a frequency weighting factor $[W_i(f)]$ over the frequency (f) range. The weighting factor is derived from British Standard 6841, which was determined from human testing, and is included to represent the relative levels of human discomfort due to vibrations over all frequencies, being higher where humans are most sensitive and lower where humans are less sensitive. The "Ride on the Seat" (numerator in Equation 1) is defined as an integration of the vehicle power spectrum measured on the floor, multiplied by a seat transfer function $[H(f)]$, or seat transmissivity, multiplied by the frequency weighting factor. A visual summary of S.E.A.T. calculation is provided in Figure 3.

Use of the seat transfer function presents the opportunity to consider alternative combinations of vehicles, roads, and seats. The ability to substitute the seat transfer function of a full foam seat into the S.E.A.T. calculation with vibration data from a vehicle equipped with a spring-composite design seat leads to a prediction of the full foam seat performance in that particular vehicle. In our study, this flexibility allows us to compute the performance of various foam chemistries in different vehicles.

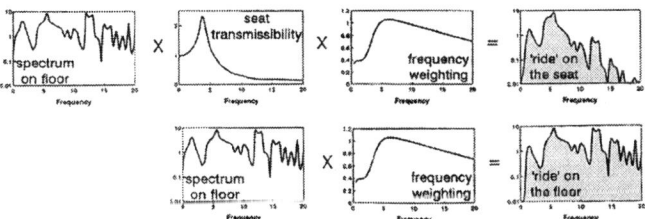

Figure 3. Graphical representation of the S.E.A.T. calculation.

The S.E.A.T. value is an objective quantification of the dynamic comfort of a vehicle seat based on the ability of a seat to protect the occupant from harmful vibrations that are known to produce physical discomfort. If the S.E.A.T. value is 100%, the ride on the seat is comparable to the ride on the floor. This does not mean, however, that the ride is identical; different vibration spectra and seat transfer functions may lead to identical S.E.A.T. values. A S.E.A.T. value greater than 100% implies that the ride on the seat is less comfortable than the ride on the floor. A S.E.A.T. of less than 100% implies that the ride on the seat is more comfortable than the ride on the floor. A 5% difference in S.E.A.T. value should be perceptible by a vehicle occupant.

The vehicle vibration spectra and the seat transfer function are required to calculate S.E.A.T. values. Vehicle vibration measurements were made in several vehicles encompassing a broad range of vehicle classes, as summarized in Table 1.

Table 1. Vehicles used for road testing.

Vehicle	Country	Year	Class
1	Germany	1996	Mid-size Sedan
2	France	1996	Full-size Sedan
3	France	1996	Compact
4	Spain	1996	Subcompact
5	England	1996	Mini-van

Vehicle vibration measurements were made using an EGCS-DO-10 V miniature heavy duty accelerometer attached to the base of the seat. Data acquisition and analysis were done using the HVLab system.[6] Measurements were made for 60 seconds on a cobblestone road at 18 km/h, a country road at 70 km/h, and a motorway at 130 km/h. The dynamic comfort evaluations are dependent on the frequency and intensity of the vibrations. Using Fourier transform techniques, the frequency content of the time history was determined. The power spectral density is the distribution of the mean square value of a time history over frequency.[1]

The seat cushion transmissivity data were measured with seated human subjects at the University of Southampton, Institute of Sound and Vibration Research, Human Factors Research Unit under the auspices of Professor Michael J. Griffin, in safety accordance with British Standard 7085. The input vibrations that were measured in the vehicle were reproduced on an electro-hydraulic moving platform equipped with an Entran EGCSY-240D-10 accelerometer. Human transmissivity was measured only with vehicle 1 road data. The cushions were placed in a pan used in the assembled seat and the ensemble was mounted on the platform. An accelerometer pad, meeting the specifications set out in ISO 10326-1, containing an Entran EFCS-DO-10V accelerometer was placed on the surface of the cushion. The HVLab system was used to control the platform and acquire signals from the accelerometer on the cushion. The various road type vibrations were reproduced. The measurements were made with three male subjects (referred to as Subject 1, 2 and 3) weighing 75, 67, and 83 kg, respectively, seated on the foam. Analysis of the data was done using the HVLab system.

The vibrational time histories were transferred to a moving platform capable of testing cushions with seated humans. There are three advantages to testing in this way. It is a controlled analysis of the foam under realistic conditions. Also, it allows the vibrational characteristics of the foam itself to be determined without interference from the other seat components. For example, the seat covering dampens the resonance peak vibrations.[7] Finally, it allows comparisons to be made between different foam technologies without the necessity of constructing a fully assembled seat.

Vibrational measurements on the cushion with a seated human produce an output time history that is different from the input time history by the influence of the polyurethane foam. The relationship between the two time histories is expressed as the ratio of the amplitude of the output vibrations to the amplitude of the input vibrations as a function of the vibration frequency to give the seat transfer function. The seat transfer function at any frequency includes the magnitude and phase. The magnitude alone is called the transmissibility[1] or transmissivity. The transmissivity of the simple laboratory method is analogous to that of the S.E.A.T. method. In the S.E.A.T. method, the rigid mass is replaced by a human subject and the vehicle power spectra is used instead of a constant acceleration input. The average of the transmissivity of the three subjects was used for calculations in this study.

FOAM SAMPLE PREPARATION

Three cushions of various foam chemistries were utilized in this study and they are described in Table 2.

Compared in this study were foams of identical static firmness, density and molded geometry with significantly different vibrational characteristics. Foam A was made with methyl diphenyl diisocyanate (MDI) chemistry, foam B was made with toluene diisocyanate (TDI) chemistry and foam C was made using TDI/MDI chemistry and low-monol polyol. All foams included in this study were prepared with high pressure metering equipment. The critical characteristics being compared are foam transmissivity properties. Table 3 summarizes the foam natural frequency and transmissivity at natural frequency as measured by our laboratory scale test. Foam A had a higher natural frequency and lower transmissivity at the natural frequency (low ball rebound, high frequency foam). Foam B had a lower natural frequency with a high transmissivity at the natural frequency (low frequency high ball rebound) and foam C had a lower natural frequency with a lower transmissivity at the natural frequency (low frequency, low ball rebound).

Table 2. Polyurethane foam cushions' properties.

Foam ID	Density (kg/m^3)	IFD @ 40mm DIN 53579 (N)	Isocyanate	Polyol
A	60	302	MDI	Conv.
B	60	291	TDI	Conv.
C	60	303	TDI/MDI	Low Monol

RESULTS AND DISCUSSION

LABORATORY SCALE TRANSMISSIVITY

Figure 4 is a plot comparing the foam vibrational performance of the various foam technologies as measured by the laboratory scale transmissivity test. Table 2 summarized the foam natural frequency and transmissivity at natural frequency measured for these foams. Foam A had the highest natural frequency, whereas foams B and C had about the same natural frequency. Comparing the transmissivity at natural frequency data, foam B had the highest transmissivity at the natural frequency. Foams A and C both had lower transmissivities at resonance. Based on the understanding that lower natural frequency and lower transmissivity at the natural frequency are both desired, foam C would be predicted as having the best overall vibration properties balance for ride comfort.

Comparing the transmissivities in Figure 4, it is clear that the attenuation frequency (the frequency where transmissivity crosses 1) for foam A was the highest. The foam B cushion followed by the foam C cushion exhibited lower attenuation frequencies.

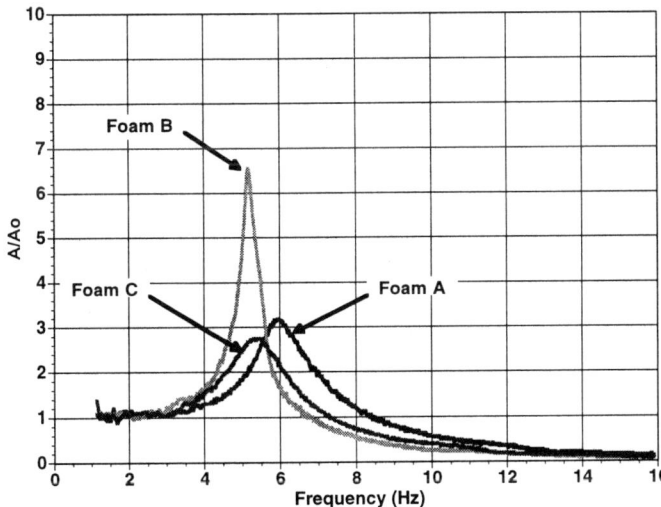

Figure 4. Laboratory transmissivities of the three foams.

Table 3. Summary of transmissivity data from human and laboratory testing.

Foam ID	Natural Frequency		Transmissivity at Natural Frequency	
	Laboratory	Human	Laboratory	Human
A	5.95	3.91	3.20	1.43
B	5.15	3.13	6.54	1.69
C	5.30	3.13	2.75	1.49

HUMAN TRANSMISSIVITY

The human transmissivity data is provided in Figure 5 and the natural frequency and transmissivity at natural frequency data is provided in Table 3. From Table 3, the human transmissivity test provided lower natural frequencies and lower transmissivity at the natural frequency than the laboratory scale method. This difference is due to many factors, the most significant of which are the use of a rigid mass instead of a viscoelastic (human) mass and the input vibration profile (frequency sweep verses road noise). Even though the measured responses are different, the general trends in vibrational characteristics between the laboratory and human methods seem to be in agreement for these cushions.

Human discomfort due to vibration is a function of the measured human transmissivity curve and the frequency weighting factor used in the S.E.A.T. calculation. One means of further investigating the basis of the S.E.A.T. values is to consider the multiplication of the human transmissivity with the frequency weighting factor for human discomfort. Figure 6 is a plot of this frequency weighting factor and Figure 7 is a plot of the product of this factor and the human transmissivity data in Figure 5. Over the entire vibration spectrum studied, foam A exhibited the highest discomfort weighted transmissivity. The frequency weighting factor accentuated foam A's higher natural frequency, leading to a higher peak near the natural frequency relative to the other two foams. These factors all contribute to a higher S.E.A.T. value.

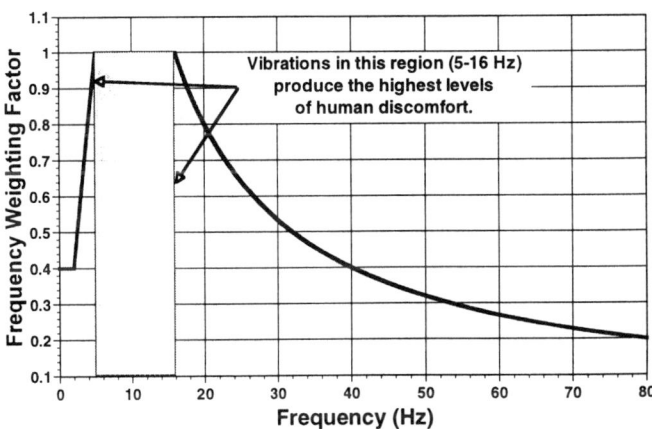

Figure 6. Frequency weighting factor for human discomfort.

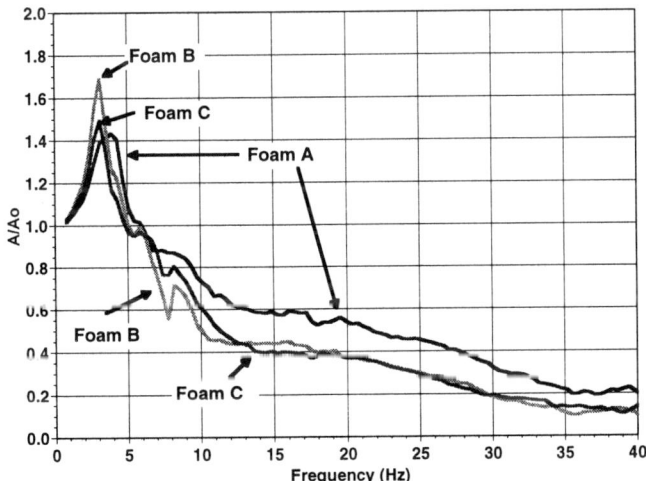

Figure 5. Human transmissivities of the three foams.

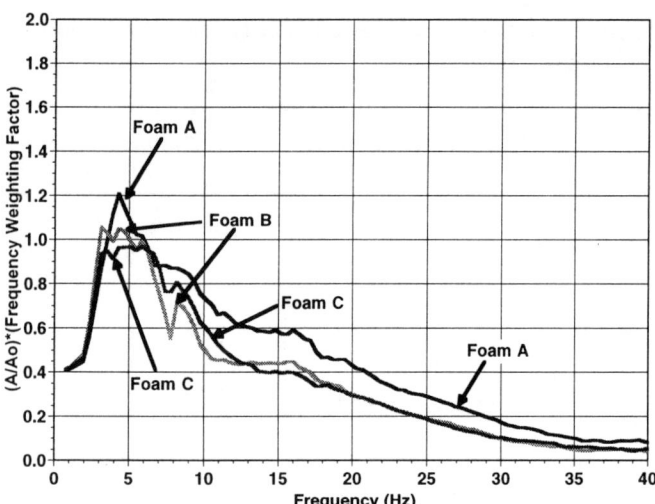

Figure 7. Human transmissivities multiplied by the frequency weighting factor.

Near the natural frequency, foam B had a higher weighted transmissivity than foam C. This result was due to the higher transmissivity of foam B relative to foam C in this frequency region. In the attenuation frequency range, foams B and C had similar overall weighted transmissivity values, whereas foam A clearly had a higher weighted transmissivity.

SEAT VALUE COMPARISON

The value of the S.E.A.T. method calculation is that the mathematics permit the calculation and consideration of the performance of a particular seat in a vehicle that the seat has never physically occupied. Therefore, permutations of vehicle and seat combinations can be explored via a computer (i.e., without a wrench). We have taken the human transmissivity curves of foams A, B, and C and "tested" them in vehicles 1 through 5 on the three road surfaces previously described. Summaries of this exercise are reported in Table 4. This is the mathematical equivalent of placing the foam cushions on the floor pan of each of the vehicles and riding on the various road surfaces.

Table 4. S.E.A.T. values.

Road	Cobblestone			Country			Motorway		
Vehicle	A	B	C	A	B	C	A	B	C
1	84.6	79.9	75.4	78.3	70.6	69.1	79.3	77.4	72.3
2	72.1	67.1	62.9	89.9	80.2	78.5	38.9	37.1	34.6
3	63.6	59.3	55.6	71.2	64.4	62.5	65.6	62.5	58.4
4	113.3	105.6	99.0	104.8	92.1	90.2	72.6	69.3	64.7
5	91.3	85.1	79.8	85.6	76.3	74.8	101.0	96.2	89.8

For all vehicles on all road types, foam C provided the lowest S.E.A.T. value, followed by foam B, and finally foam A. These results are not unwarranted, given the frequency weighted transmissivity results discussed above. However, there existed a possibility that a particular vehicle and roadway combination would have favored different foam cushions. This exercise reveals that the differences in the cushions are significant enough to produce the same trend in S.E.A.T. values across the range of vehicles and road surfaces studied.

The human-derived S.E.A.T. values and the laboratory scale transmissivity test results are compared in Figure 8. Plotted in Figure 8 are change in S.E.A.T. value as a function of changes in natural frequency and transmissivity at natural frequency as measured by the laboratory test. Comparing foams with similar transmissivities at the natural frequency, S.E.A.T. values decrease with decreasing natural frequency. Comparing foams with similar natural frequencies, S.E.A.T. values decrease with decreasing transmissivity at the natural frequency. Therefore, the region of desired vibrational characteristics are lower resilience, low natural frequency foam. A 0.65 Hz change in the natural frequency led to, on average, a 9.6 unit change in the S.E.A.T. value. Likewise, a 3.8 unit change in the laboratory transmissivity at natural frequency led to a 3.7 change in the S.E.A.T. value. This implies that the S.E.A.T. value is more sensitive to changes in the natural frequency than the transmissivity at natural frequency of the polyurethane foam.

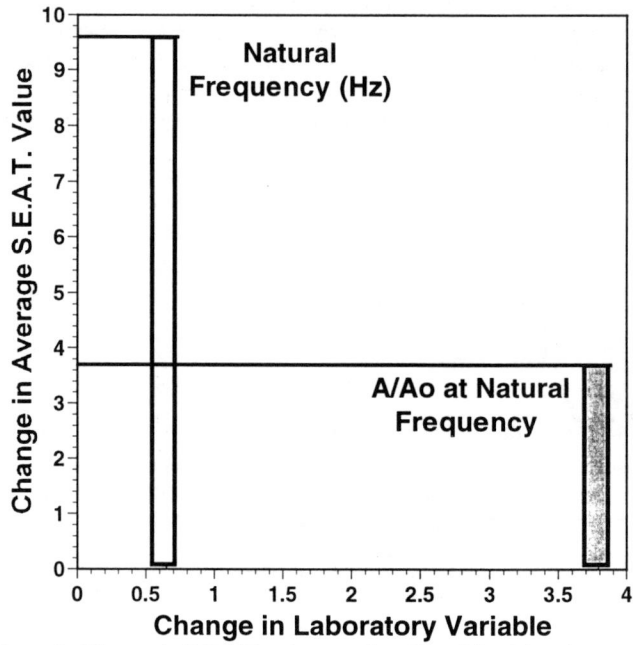

Figure 8. Change in S.E.A.T. value as a function of the laboratory measured natural frequency and transmissivity.

CONCLUSION

The laboratory test method, using a rigid mass, was unable to reproduce the depth and quality of information obtained from the S.E.A.T. method. However, this simple laboratory test provided information that was appropriate to qualify the relative vibration performance of different foam chemistries. The process of benchmarking the effects of chemistry and processing on the vibration performance is reduced to an experiment that is easy to implement, reproducible, and does not involve the exposure of a human subject to vibration.

The ability to utilize separate or different transfer functions in the S.E.A.T. method calculations provides a unique opportunity to predict the performance of different seats on different roadways or, better yet, different vehicles. Once the seat transfer function is determined, it can be used to calculate how that particular seat will perform with a variety of vehicles. In this study, we have considered the implications of using a full foam seat in a variety of different vehicles fitted with spring-composite design seats. The calculated performance of the various foam chemistries used in full foam seats in various vehicles demonstrates that polyurethane foam provides those vibrational characteristics necessary to protect the occupant from harmful or uncomfortable levels of vibration.

ACKNOWLEDGMENTS

The authors wish to thank B. Britt, R. Carter, J. Beckner, D. Hair and P. Williams of ARCO Chemical Company's S. Charleston, WV Technical Center for their assistance in preparing and testing foams for this work.

REFERENCES

1. Griffin, M. J., **Handbook of Human Vibration**, Academic Press: London, pp. 404-408, 1990.
2. ISO standard 2631.
3. Kinkelaar, M. R., and Cavender, K. D. "Real Time Dynamic Comfort and Performance Factors of Polyurethane Foam in Automotive Seating", SAE paper 960509, 1996.
4. Kinkelaar, M. R., Cavender, K. D., and Crocco, G. L., "Vibrational Characterization of Various Polyurethane Foams Employed in Automotive Seating Applications", Polyurethanes Division SPI Conference, pp. 496-503, 1996.
5. Crocco, G. L., and Kinkelaar, M. R. "Improving the Driving Comfort Quality of Automotive Seating", ISATA Conference, 1997.
6. The HVLab Data Acquisition and Analysis System was developed by the Human Factors Research Unit, Institute of Sound and Vibration Research, University of Southampton, Highfield, Southampton, SO9 IBH.
7. Ebe, K., Griffin, M. J., "Effect of Sample Shape and Seat Cover on the Vibration Transmissibility of Automotive Seats", IPC-8 Conference, 1996.

980658

The Effects of Regional Compliance and Instantaneous Stiffness on Seat Back Comfort

Eric C. Hughes, Wenqi Shen, and Alicia Vértiz
Delphi Interior & Lighting Systems

Copyright © 1998 Society of Automotive Engineers, Inc.

ABSTRACT

To facilitate the design and validation of comfortable automotive seats, it is necessary to thoroughly understand how the occupant-seat interface affects the perception of comfort. Previous studies indicated that the local supporting properties of the seat back, namely the regional compliance and regional stiffness, play a significant role in the perception of seat comfort. The purpose of this study was to investigate the effects of these properties on perceived comfort. The study used five production seats as test samples. The regional force-deflection properties of the seats were tested, and eighteen participates subjectively evaluated the seats for the perception of seat back compliance and stiffness. The quantitative and subjective test data was used to examine the relationships between the local supporting properties of the seat back and the perception of seat comfort. Based on the testing procedures, recommendations for regional stiffness and lumbar prominence were made.

INTRODUCTION

A comfortable seat back must support the occupant's back in a way that matches his or her natural seated posture. A mismatch between the occupant's back and the seat may produce uncomfortable pressure concentrations (Reed and Schneider, 1996). These uncomfortable pressure concentrations can be minimized by appropriately designing the regional compliance and lumbar prominence to better match the occupant's natural seated posture. Of course, the seat must accommodate a wide range of occupants with differing body structures, which makes designing a comfortable seat back system a challenging task. The seat's role in comfortably supporting the occupant's posture is typically referred to as static seat comfort.

The seat back must also allow the occupant to adjust his or her position comfortably, which was referred to as transient comfort by Shen and Vértiz (1997) and described as follows:

"Transient comfort is perceived as the subject adjusts posture or fidgets around on the seat. Riding tasks often require frequent change of posture, so do the muscle groups when some relaxation is in need. A comfortable seat should not constitute any local impact or interfere with such changes, but provide gentle damping effect on the transient loading."

Transient comfort can be optimized by appropriately designing the instantaneous stiffness of both the lumbar and thoracic regions. Essentially, the goal of designing a comfortable seat back system requires a delicate balance between lumbar prominence and regional compliance while maintaining an acceptable instantaneous stiffness throughout the entire seat back.

There has been a significant amount of published research focusing on how a comfortable seat back should be designed. Most of the research examined the amount of lumbar prominence, but the effects of regional compliance and instantaneous stiffness have received little attention. As previously mentioned, regional compliance and instantaneous stiffness are important characteristics of a comfortable seat back system, which typically consists of a trim cover, trim pad, principal foam pad, lumbar adjustment system and/or suspension system. A thorough understanding of these characteristics is useful in designing a comfortable seat back. Particularly, this knowledge would assist in selecting an effective lumbar adjustment system, as stated by Shen and Vértiz (1997):

"A number of lumbar adjustment systems are commercially available. Unfortunately, when prominence is increased, compliance tends to reduce. A close examination of seat-back load-deflection property should be carried out. This property may change dramatically by using different lumbar adjustment mechanisms. To date, most effort in lumbar support has been on the amount of prominence, and the regional force-deflection property has hardly been investigated."

The aim of this study is to develop a better understanding of how compliance, prominence, and instantaneous stiffness influence static and transient seat comfort with the hope that the acquired information may aid in the design of comfortable seat backs and in the selection of effective lumbar adjustment systems. Specifically, this paper examines: 1) the effects of instantaneous stiffness on transient comfort; and 2) the trend with which lumbar prominence and regional compliance influence static seat comfort.

SEAT BACK CHARACTERISTICS & THEIR OPERATIONAL DEFINITIONS

This section contains a brief overview and description of the seat back characteristics that are examined within this paper. It explains what each characteristic physically represents and the methods used to measure them. This section is separated into two parts. The first part identifies the regions of the seat back referred to as the lumbar and thoracic regions. The second part describes and defines lumbar prominence, regional compliance, and regional instantaneous stiffness.

LUMBAR AND THORACIC REGIONS OF THE SEAT BACK

The terms lumbar and thoracic refer to the regions of the seat back that support the occupant's lumbar and thoracic spinal segments. Figure 1 shows the human spine with the spinal segments and seat back regions labeled. The lumbar and thoracic regions were identified by locating the center of the respective region using the seat back's centerline profile. The centerline profile of each seat back was digitized with the vertical axis parallel to the seat back's frame and the horizontal axis perpendicular to the seat back's frame. The center of the lumbar region is generally located between 150 mm to 200 mm, vertically, from the bite line and is the most pronounced point along the seat back profile. The center of the thoracic region is approximately 200 mm, vertically, from the center of the lumbar region. The centers of these regions are illustrated on the digitized seat back in Figure 2.

SEAT BACK CHARACTERISTICS

To date, there are no standard procedures for measuring lumbar prominence, regional compliance, and regional instantaneous stiffness. In order to reduce confusion, the procedures used to measure these characteristics will be explained here.

Lumbar Prominence

Lumbar prominence will be defined geometrically using the seat back's digitized profile, an approach similar to that used by Reed, Schneider, and Eby (1995). Figure 2 illustrates how the lumbar prominence is measured. Note that, this approach uses only the seat back's digitized profile, the center of the lumbar region, and the center of the thoracic region to measure lumbar prominence.

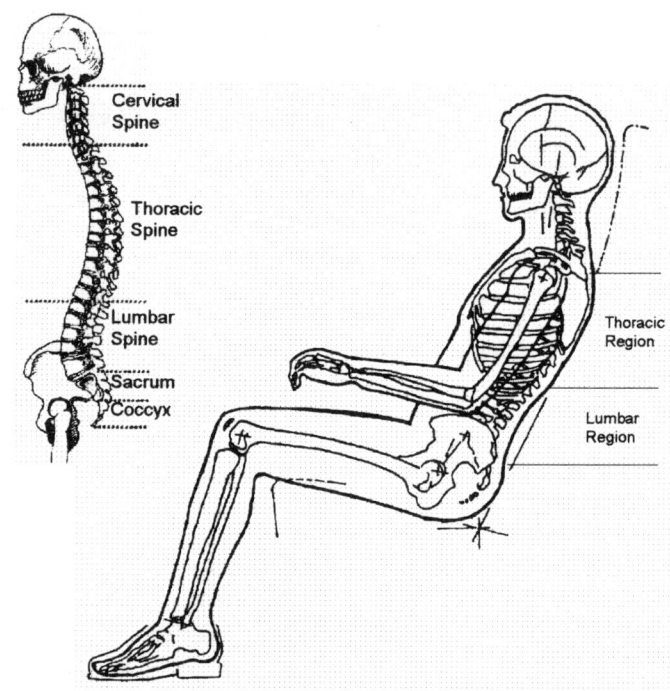

Figure 1. An illustration of the lumbar and thoracic regions of the seat back

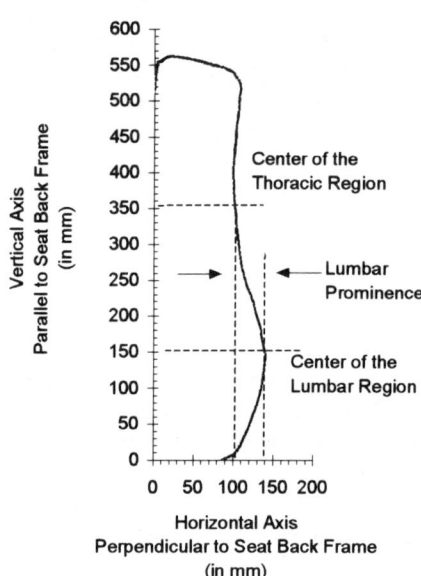

Figure 2. A digitized seat back profile illustrating how the lumbar prominence was measured

Definition of the Lumbar Prominence:
- Lumbar prominence is defined as the distance, perpendicular to the frame, between the center of the lumbar region and the center of the thoracic region.

Regional Compliance

Regional compliance is the characteristic that dictates how the seat back regions deflect in response to an applied load. The applied load is the weight distribution of the subject as he or she reclines onto the seat back. However, directly measuring the seat back deflection due to the weight distribution of a subject is a difficult task, because any method of measuring deflection that changes the subject-seat interface may inadvertently affect the subject's posture and the seat back deflection. As a result, the deflection was approximated using the seat back's regional force-deflection properties and the subject's body pressure distribution (BPD). The regional force-deflection properties were determined experimentally by force-deflection tests of the lumbar and thoracic regions. The subject's BPD was used to determine the weight or force applied by the subject to specific areas of the lumbar and thoracic regions. These forces, referred to as the regional operating loads, were then used to locate the associated regional compliance along the force-deflection curves, as shown in Figure 3 (b).

Definition of Regional Compliance:
- Regional compliance is defined as the amount of deflection due to the subject's regional operating load. It is easily inferred that regional compliance is associated with: 1) the operating load which may differ among small female, medium male, large male, etc.; and 2) the area of the region based on which the compliance is quantified.

Regional Instantaneous Stiffness

Stiffness and compliance are both related to the seat back's force-deflection properties. Compliance is the amount of seat back deflection due to an applied load, which affects the loaded seat back contour and the static support of the occupant. However, stiffness is the rate at which the seat back deflects, which influences how rigid/flexible the seat back feels. Stiffness is defined as the gradient of the seat back's regional force-deflection curves and is expressed in units of N/mm. Typically, the force-deflection curves are somewhat non-linear, meaning that the regional stiffness depends upon the applied load. As a result, the operating load and force-deflection curves were used to determine the regional instantaneous stiffness, which is an estimate of the regional stiffness perceived by the subject. The tangent line in Figure 3 (b) illustrates the instantaneous stiffness for a subject applying a regional load of 60 N to a circular area with a 15cm diameter.

Definition of Regional Instantaneous Stiffness:
- Regional instantaneous stiffness is defined as the rate of change or gradient of the seat back's force-deflection curves at the subject's regional operating load.

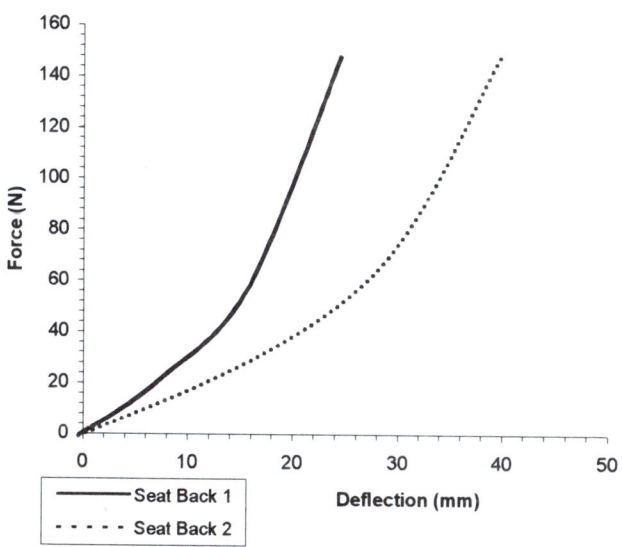

(a) Regional force-deflection tests

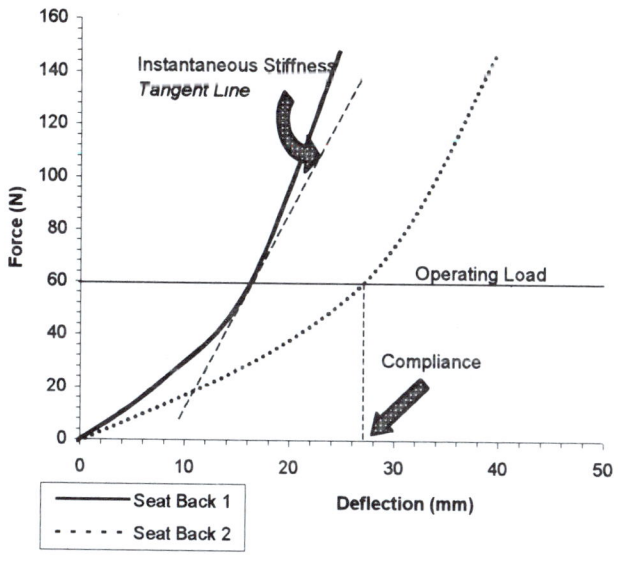

(b) Compliance and instantaneous stiffness (depicted by the tangent line) for an operating load of 60 N

Figure 3. An illustration of how compliance and instantaneous stiffness were determined.

TEST METHODS

This section outlines the force-deflection, BPD, and subjective test procedures. It is separated into three parts. The first part contains general information about the test samples and includes each of the test sample's digitized profile. The last two parts describe the quantitative and subjective testing methods. The quantitative tests include the force-deflection and BPD test procedures, and the subjective test explains how perceived comfort was assessed for each test sample. All of these tests were performed with the test samples fully intact (i.e. without removing the trim cover or any other component of the seat back).

TEST SAMPLE SELECTION AND BENCHMARKING

Five test samples were selected from production seats. These seats were grouped according to seat back firmness and are listed below.

Level 1 (Hard):
  Sample No. 1
  Sample No. 2

Level 2 (Medium):
  Sample No. 3

Level 3 (Soft):
  Sample No. 4
  Sample No. 5

Samples 2 and 5 are equipped with lumbar adjustment systems and were tested with the lumbar in the following positions: full-in, 20 mm from full-in, and full-out. The centerline profile of each seat back was digitized and a number of other seat back characteristics were documented. Figure 5 shows each of the test sample's digitized profile. These digitized profiles provide general information about the differences in contours among the test samples. Table 2 contains a summary of the seat back characteristics documented during this study.

QUANTITATIVE TESTS

Force-deflection tests were conducted on both the lumbar and thoracic regions of each test sample. As previously mentioned, samples 2 and 5 were tested with the lumbar adjustment systems in the full-in, 20 mm from full-in, and full-out positions. In addition, the body pressure distributions of six subjects, one subject from each of the categories in Table 1, were also recorded for each test sample. This information was used to estimate the load applied by different size subjects (i.e. small female, medium male, large male, etc.) to specific areas of the lumbar and thoracic regions.

Category	Height (mm)
1. Small Male	1690 - Less
2. Medium Male	1691 - 1790
3. Large Male	1791 - Over
4. Small Female	1610 - Less
5. Medium Female	1611 - 1710
6. Large Female	1711 - Over

Table 1. Categories of subjects and their respective heights

Force-Deflection

The regional force-deflection tests were performed with the sample in the "in car position" and the load applied vertically, as shown in Figures 4 (b) and (c). The indentor was free to rotate such that the indentor plane matched the contour plane of the test sample. The angle of rotation was recorded during each test to verify the consistency of this parameter among the different test samples. The measured angle varied less than 5 degrees among the test samples and was therefore neglected in the analysis. The indentor was positioned at either the center of the lumbar region or the center of the thoracic region for the lumbar regional tests and the thoracic regional tests, respectfully, as shown in Figure 4 (a). Test specifications for the regional force-deflection tests are as follows:

- Preload: 11 N
- Indentor Size: 15 cm diameter circular indentor
- Load Rate: 2 mm/s
- Maximum Load: 156 N

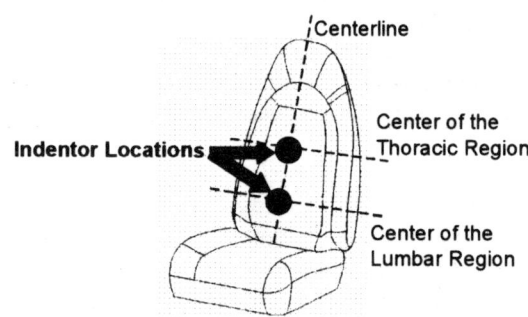

(a) The indentor locations for the lumbar and thoracic regional tests

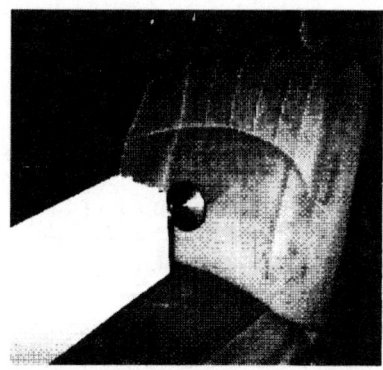

(b) A photo of the experimental setup

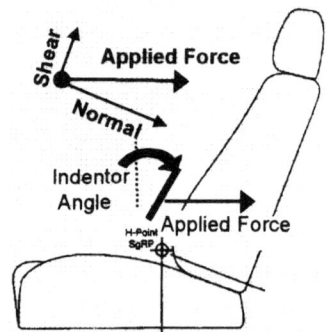

(c) A diagram of the experimental setup

Figure 4. The experimental setup for the regional force-deflection tests

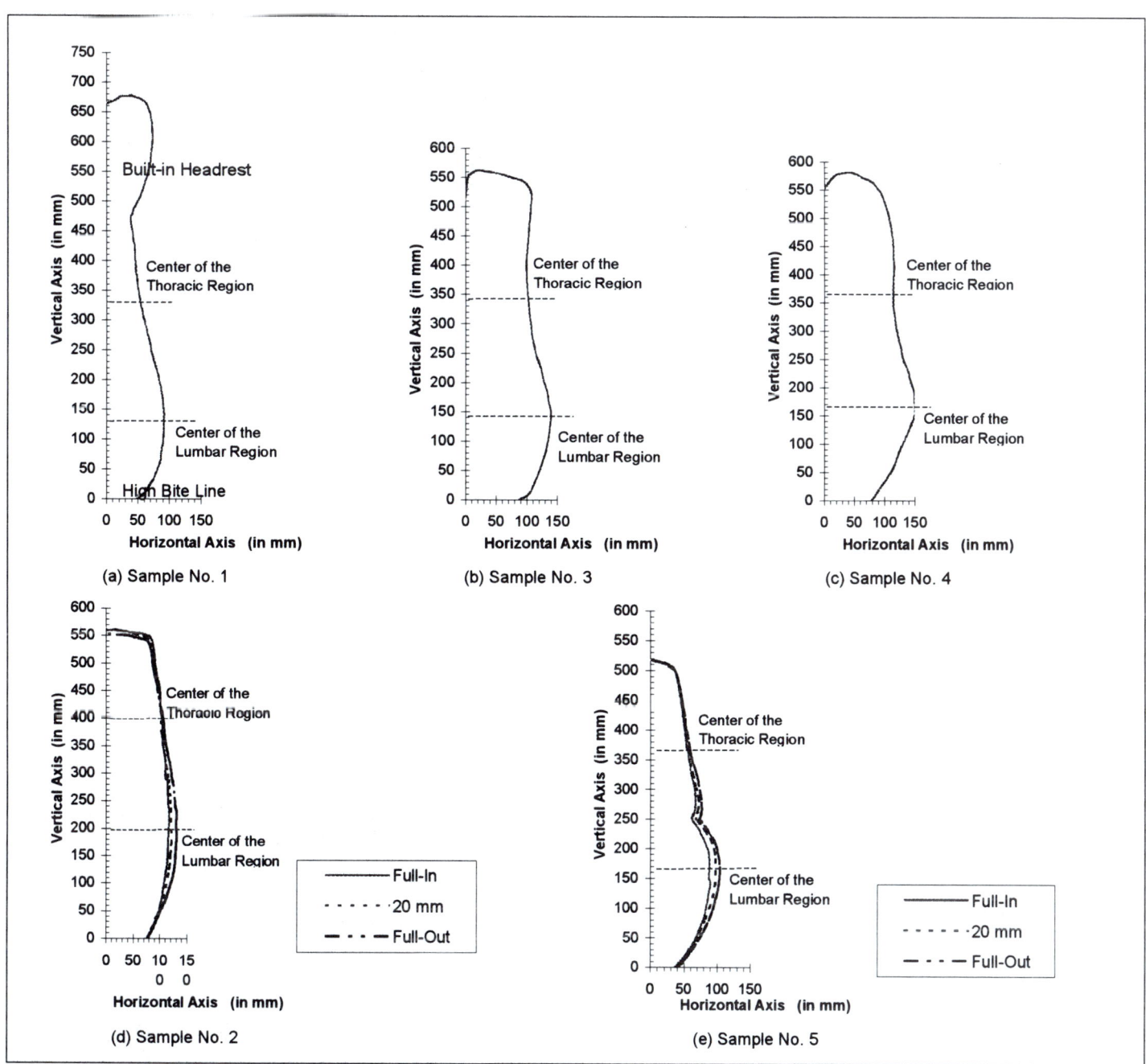

Figure 5. The digitized seat back profile of each test sample

Test Sample	Trim Cover	Principal Foam Pad	Suspension System	Adjustable Lumbar System	Adjustable Lumbar System	Lumbar Prominence	Lumbar Prominence	Lumbar Prominence	Lumbar Prominence
	Type	ILD (N)	Type	Type	Travel Length (mm)	Fixed Lumbar (mm)	Full-in (mm)	20 mm (mm)	Full-out (mm)
Seat 1	Cloth	300	Flex-o-lator	None	--	30	--	--	--
Seat 2	Leather	200	None	Schukra	35	--	18	22	30
Seat 3	Cloth	280	Flex-o-lator	None	--	34	--	--	--
Seat 4	Cloth	203	Flex-o-lator	None	--	39	--	--	--
Seat 5	Leather	195	None	Plastic Band	30	--	35	39	44

Table 2. General information about each of the test samples

Body Pressure Distribution

BPD tests were performed using the Tekscan Pressure System, manufactured by Tekscan, Inc. This system uses pressure sensor mats, which include resistance and capacitance type sensors. These sensors are arranged in a matrix with the centers of any two adjacent sensors 10mm apart. For these tests, the seats were placed in a test buck in the "in car position" with a back angle of twenty-six degrees. The subject was asked to sit in the test sample with his or her hands on a steering wheel to simulate the driving condition. After approximately five minutes of sitting in the test sample, the subject's BPD was recorded.

SUBJECTIVE TEST

Eighteen subjects, three for each of the categories in Table 1, were recruited for subjective testing. These tests were performed using the same setup as the body pressure distribution tests. In order to focus the subject's attention on different aspects of seat comfort, these tests were divide into four parts. The first part was designed to obtain general information about the subject, his or her current vehicle, and his or her likes/dislikes regarding automotive seat backs. This information was used to compare the subject's current vehicle and seating preferences with his or her evaluation of the test samples. The last three parts were designed to evaluate the overall, static, and transient comfort of each test sample. While sitting in the test sample, the subject was asked to rate the seat back in the following categories:

Overall Comfort:
Q1: overall seat back comfort (uncomfortable...comfortable)
Q2: lumbar region (uncomfortable...comfortable)
Q3: thoracic region (uncomfortable...comfortable)

Static Comfort:
Q4: surface hardness (Too Soft...Too Hard)
Q5: the amount of lumbar support (Too Little...Too Much)

Transient Comfort - The subject was asked to fidget or move around in the seat; while doing so rate the:
Q6: the feel of the lumbar region (Too Flexible...Too Rigid)

Questions 1, 2, and 3 were rated on a continuous 9 point scale from uncomfortable to comfortable (one being uncomfortable and nine being comfortable). Questions 4, 5, and 6 were rated on a 9 point bipolar scale with a five representing a comfortable or "just right" rating.

RESULTS

This section examines the data collected during the quantitative and subjective tests. The first part of this section explains how regional operating loads were determined using BPD data. The second part examines the relationships between the quantitative and subjective test results for all the test samples. In the third part, a case study of sample 5 was performed using the same procedures as in part two of this section.

REGIONAL OPERATING LOAD

Body pressure distribution data was used to estimate the regional operating loads for each of the categories in Table 1. The maximum loads in the lumbar and thoracic regions were determined for each of the recorded BPD's. Typically, the areas of the seat back under the subject's lower back and shoulder blades experience the greatest pressure concentrations. These areas are located in the lumbar and thoracic regions, respectfully, and were chosen as the locations for the loads resolved from the subject's BPD. The loads were calculated with the same surface area as the indentor used in the regional force-deflection tests. Regression was then used to estimate the regional operating loads as shown in Figures 6 (a) and (b).

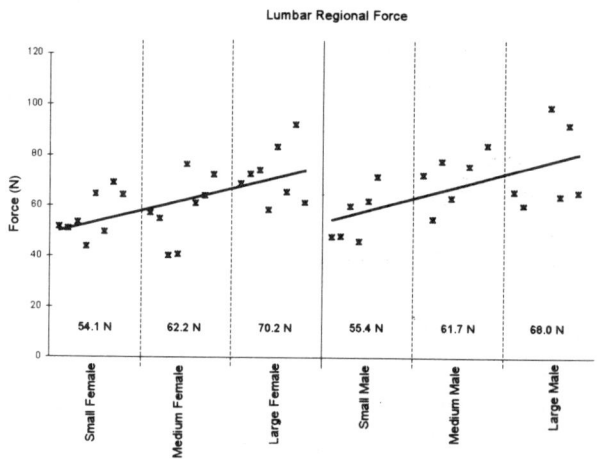

(a) The lumbar regional operating loads for small females, medium males, large males, etc.

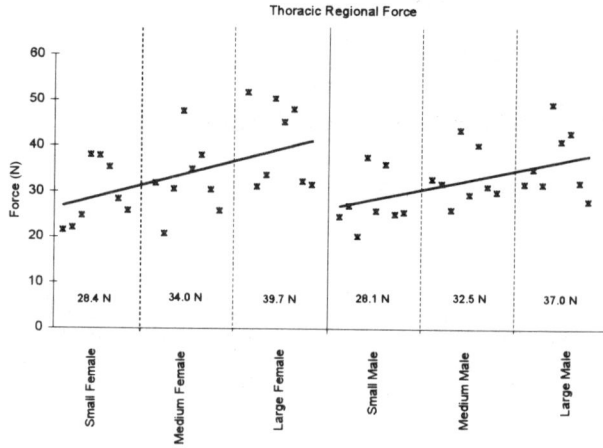

(b) The thoracic regional operating loads for small females, medium males, large males, etc.

Figure 6. Analysis of the BPD data

ALL SAMPLES

The first step in this analysis was to examine the correlation matrix shown in Figure 7 to determine if the correlation between the subjective and quantitative data physically made sense. The main concern in this analysis was subjective questions 5 and 6, because the subjects' response to Q4 indicated that there was some difficulty in rating the seat back surface hardness. Question 4 was therefore disregarded in the analysis. The matrix shows that Q5 was correlated to prominence, thoracic compliance, and lumbar stiffness. Thoracic compliance affects the degree to which the subject's shoulder blades sink into the upper seat back. By increasing the thoracic compliance, the subject may perceive an increase in lumbar support because of the more rearward position of his or her shoulders. Therefore, a correlation between Q5 and thoracic compliance would be excepted. Prominence and lumbar stiffness are directly related to the lumbar support and would also be excepted to correlate with Q5. We anticipated that Q5 would show significant correlation to lumbar compliance. After examining the data, however, it was apparent that the correlation between Q5 and lumbar compliance was low, possibly resulting from inconsistent test samples with differing characteristics. The matrix shows that Q6 was correlated to lumbar compliance and lumbar stiffness. Again, this was excepted because of the direct influence of these characteristics on the lumbar support.

Figure 7 shows a negative correlation between prominence and lumbar stiffness. This seems to indicate that lumbar stiffness decreases as the prominence increases. The negative correlation between these two characteristics was also a result of using test samples with different characteristics. It does not indicate that lumbar stiffness would decrease as the lumbar adjustment system moves from full-in to full-out, increasing the prominence. In fact, the opposite occurs in most commercially available lumbar adjustment systems.

To determine if there was a relationship between subjective questions, a different correlation matrix was calculated, as shown in Figure 8. Q4, Q5 and Q6 were converted from a bipolar scale to a continuous scale to match the rating system used in questions 1, 2 and 3. The matrix shows that Q1 was significantly correlated with Q5 and Q6. This indicates that both the amount and the stiffness of lumbar support influences the perception of overall seat back comfort. Naturally, both the lumbar and thoracic regions contribute to the perception of overall seat back comfort, as shown by the correlation between Q2, Q3 and Q1.

The median ratings of the subjective questions were compared to the quantitative measurements of stiffness, prominence, and compliance. This analysis showed that there was not a direct relationship between most of the subjective questions and the local supporting properties of the seat back. However, there was a distinct relationship between the subjects' response to Q6 and lumbar stiffness as shown in Figure 9. Based on the quantitative and subjective test results, it appears that for each seat the local supporting properties of the seat back should be engineered to an appropriate value in order to achieve a "just right" perception.

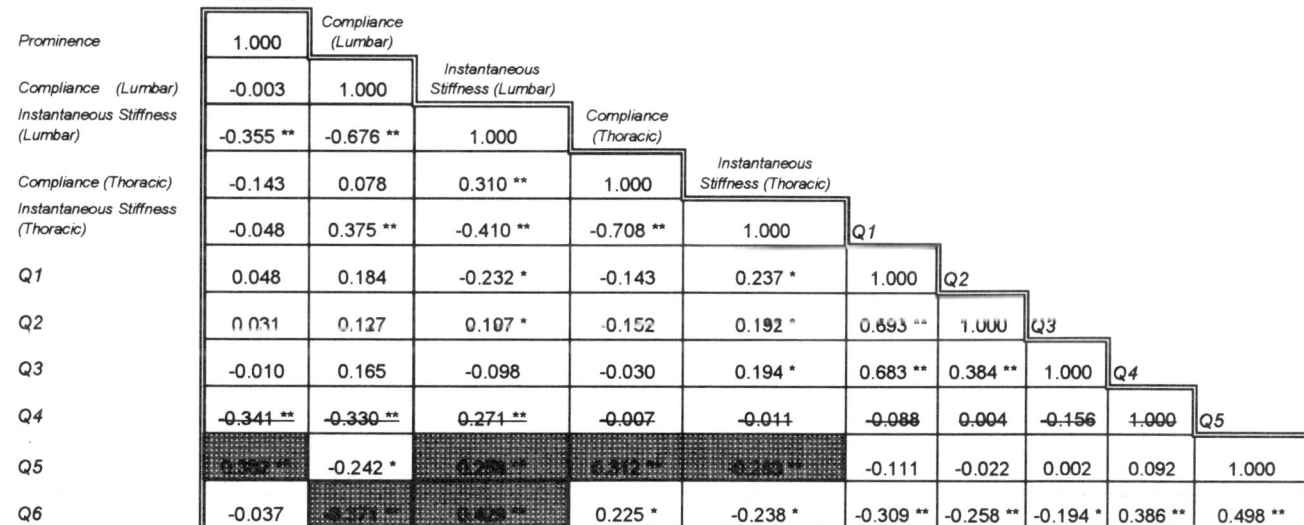

** Correlation is significant at the 0.01 level (2-tailed).
* Correlation is significant at the 0.05 level (2-tailed).

Figure 7. Correlation matrix for all test samples

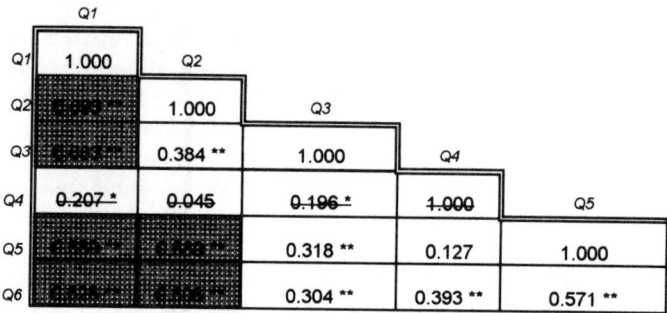

** Correlation is significant at the 0.01 level (2-tailed).
* Correlation is significant at the 0.05 level (2-tailed).

Figure 8. Correlation matrix for the subjective questions (Questions 4, 5, and 6 were converted from a bipolar scale to a continuous scale, matching the rating system used in questions 1, 2, and 3)

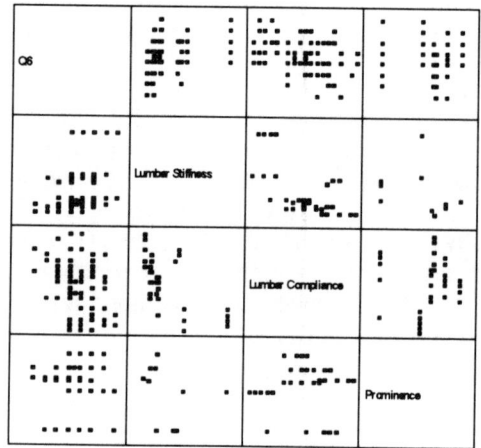

(a) A Scatter plot of Q6 verses lumbar stiffness, lumbar compliance, and prominence

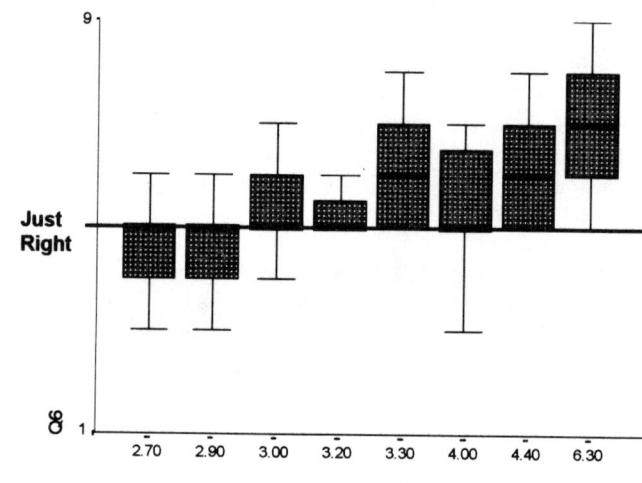

(b) A boxplot of Q6 verses lumbar stiffness

Figure 9. An examination of question 6 for all test samples

CASE STUDY

A case study was performed on sample 5 using the same analysis methods. This seat was selected because it has a lumbar adjustment system and is considered a best-in-class seat. Figure 10 shows the correlation matrix for sample 5. Notice that the correlation between prominence and lumbar stiffness is positive. This shows that as the prominence increases, by moving the lumbar adjustment system from full-in to full-out, the lumbar stiffness also increases, which is typical of most lumbar adjustment systems. In this matrix, Q5 and Q6 are more significantly correlated to compliance, prominence and stiffness than in the previous correlation matrix. This is because other seat back characteristics, like the seat back contour, are more controlled in the case study. Figure 11 describes some of the relationships between the median ratings of Q1, Q6 and the quantitative measurements of stiffness and prominence. Notice that there were no drastic changes between Figures 9 (b) and 11 (d).

Pearson Correlation

	Prominence	Compliance (Lumbar)	Instantaneous Stiffness (Lumbar)	Compliance (Thoracic)	Instantaneous Stiffness (Thoracic)	Q1	Q2	Q3	Q4	Q5
Prominence	1.000									
Compliance (Lumbar)	-0.790 **	1.000								
Instantaneous Stiffness (Lumbar)	0.908 **	-0.888 **	1.000							
Compliance (Thoracic)	0.756 **	-0.322 *	0.683 **	1.000						
Instantaneous Stiffness (Thoracic)	-0.831 **	0.969 **	-0.849 **	-0.356 *	1.000					
Q1	-0.198	0.134	-0.091	-0.104	0.235	1.000				
Q2	-0.317 *	0.125	-0.089	-0.229	0.252	0.581 **	1.000			
Q3	-0.133	0.309 *	-0.126	0.188	0.360 *	0.676 **	0.268	1.000		
Q4	0.155	-0.129	0.233	0.257	-0.121	0.089	0.156	-0.007	1.000	
Q5						-0.181	-0.079	-0.040	0.282	1.000
Q6						-0.146	-0.175	-0.105	0.400 **	0.642 **

** Correlation is significant at the 0.01 level (2-tailed).
* Correlation is significant at the 0.05 level (2-tailed).

Figure 10. Correlation matrix for sample 5

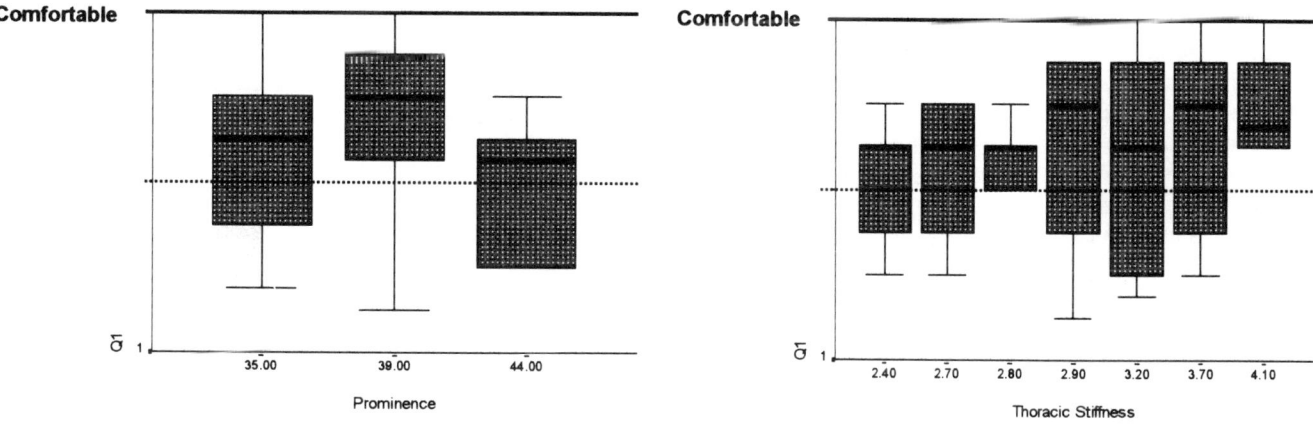

(a) A boxplot of Q1 verses prominence

(b) A boxplot of Q1 verses thoracic instantaneous stiffness

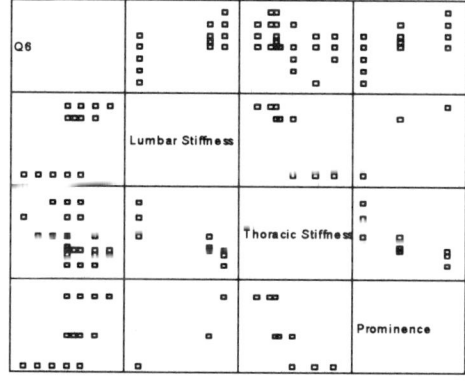

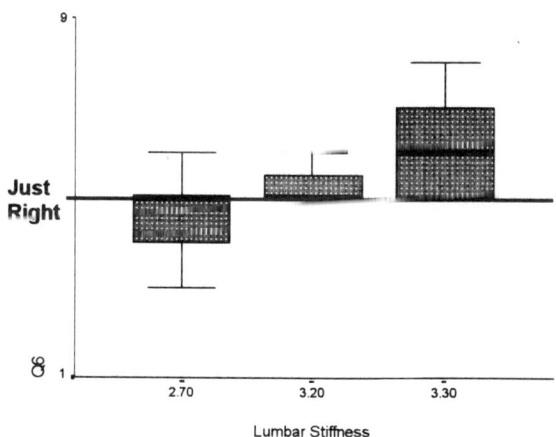

(c) A Scatter plot of Q6 verses stiffness and prominence

(d) A boxplot of Q6 verses lumbar stiffness

Figure 11. An examination of Q1 and Q6 for sample 5

DISCUSSION

The purpose of this study was to investigate the seat back characteristics that contribute to static and transient comfort. For static comfort, the relationship between compliance, prominence, and the subjective ratings of Q5 was examined. This examination led to the defining of a new term, the effective lumbar prominence. The effective lumbar prominence takes into account the regional compliance of the seat back and how it affects the loaded seat back profile (see Figure 12). It was calculated, as shown in Equation 1, using the seat back's nominal compliance and the lumbar prominence.

Effective Lumbar Prominence = LP – DLP

Where, DLP = NLC – NTC

LP ⇒ Lumbar Prominence
DLP ⇒ The change in lumbar prominence due to loading
NLC ⇒ Nominal lumbar compliance
NTC ⇒ Nominal thoracic compliance

Equation 1. The calculation of effective lumbar prominence

The nominal seat back characteristics were defined at the medium males operating loads (61.7N for the lumbar region and 32.5N for the thoracic region). The relationship between the effective lumbar prominence and the median ratings of Q5 was examined (see Figure 13). The results of this examination indicated that the subjects preferred an effective lumbar prominence between 23mm and 27mm. These results correlate well with research conducted by Porter and Norris (1987); Dowell (1995); and many others. Porter and Norris used a wooden test seat to determine the preferred lumbar prominence. They found that 20mm prominence was preferred to 40mm and 50mm for both reclined and vertical back angles. Dowell measured the lumbar curvature of 773 seated persons and found that the mean lumbar depth was 25mm for males and 22mm for females.

For transient comfort, the relationship between stiffness and the subjective ratings of Q6 was examined. The results from the study showed that lumbar stiffness was related to the subjects' response to Q6, as shown in Figures 9 (b) and 11 (d). From this data, it is recommended that the nominal lumbar stiffness not exceed 3.2N/mm.

SUMMARY OF RECOMMENDATIONS

The following recommendations are based upon the test procedures used in this study and may or may not be valid for other testing methods.

1. The effective lumbar prominence (calculated using Equation 1) should be between 23mm and 27mm

2. The nominal lumbar stiffness should not exceed 3.2N/mm

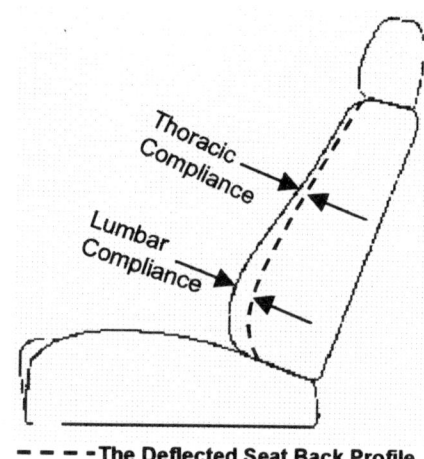

(a) An illustration of lumbar and thoracic compliance

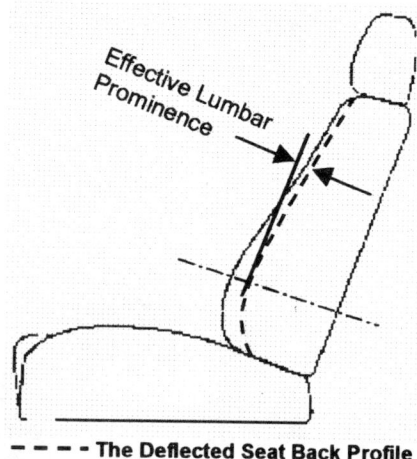

(b) An illustration of the effective lumbar prominence

Figure 12. The effective lumbar prominence

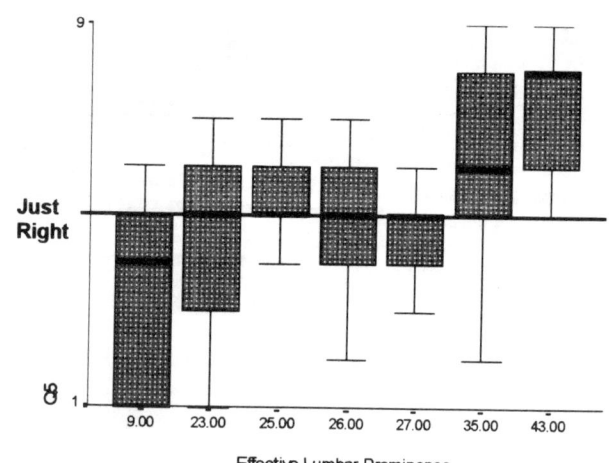

Figure 13. The relationship between effective lumbar prominence and the subjective ratings of Q5

CONCLUSIONS

1. The selected measures of seat back characteristics are appropriately related to the perception of comfort.

2. Results suggest that for each seat, the characteristics of prominence, compliance, and stiffness should be engineered to a "just right" level.

3. The effective lumbar prominence, which takes into account the compliance and lumbar prominence, can be used to estimate the amount of lumbar support perceived by an occupant.

4. The examination of two lumbar adjustments systems showed an increase in lumbar stiffness as the lumbar moved from full-in to full-out. This is an unwanted feature of a lumbar adjustment system, because the lumbar stiffness should never exceed the tolerance threshold of the occupant. The lumbar stiffness should be controlled by engineering the padding thickness in the lumbar region, the padding ILD, and type of lumbar adjustment system.

ACKNOWLEDGMENTS

The authors would like to thank Raghu Gurram, John Bruce, and Laurie Newell for their assistance in the BPD, force-deflection, and subjective tests.

REFERENCES

Dowell, W. R. (1995) An estimation of lumbar height and depth for the design of seating. *Proceedings of the Human Factors and Ergonomics Society 39th Annual Meeting* (pp. 409-411). October 9-13, San Diego, California

Porter, J. M. and Norris, B. J. (1987) The effects of posture and seat design on lumbar lordosis. In *Contemporary Ergonomics* (Edited by E. D. Megaw), pp. 191-196, London: Taylor & Francis.

Reed, M. P., and Schneider, L. W. (1996) Lumbar support in auto seats: conclusions from a study of preferred driving posture. SAE Paper No. 960478

Reed, M. P., Schneider, L. W., and Eby, B. A. H. (1995) Some effects of lumbar support contour on driver seated posture. SAE Paper No. 950141

Shen, W., and Vértiz, A. M. (1997) Redefining seat comfort. SAE Paper No. 970597

980659

Noise Absorption of Automotive Seats

Pusheng Chen and Gordon Ebbitt
Lear Corporation

Copyright © 1988 Society of Automotive Engineers, Inc.

ABSTRACT

Seat covers made from textiles, leather and vinyl were evaluated for noise absorption. The textiles included woven velours, pile knits and flat wovens. The noise absorption of the covers and the corresponding seat assemblies was tested by the reverberation room method per ASTM C423. The effect of different foams was also tested. For the leather and vinyl covers, the effect of perforation was evaluated. Test results showed distinctive differences between textiles and leather/vinyl with cloth seats having superior noise absorption. Even among the textiles, there are significant differences. Core foam densities affect the characteristics as well. For pile fabrics (woven velours and pile knits), the size of the pile fiber does not affect the acoustic characteristics of the seat. Also, no significant difference was observed between a bonded seat and a conventional (cut and sew) seat.

INTRODUCTION

The prevalent design goal for many automotive programs is to provide as much absorption in the interior of the vehicle as possible. This decreases the interior sound pressure level for a given amount of energy entering the interior. If the absorption is increased too much, there may be problems with communication between the front and the rear of the vehicle. However, this is generally not considered as important as reducing the sound pressure level.

It is desirable for all soft trim components in the interior of vehicles to absorb noise, but only a few components have the capability. For a porous material to absorb sound, the thickness of the material must be large relative to the wavelength of the sound. Typically, the thickness needs to be greater than 1/10 of the wavelength. This requirement makes the seats the major noise absorbing components, especially at lower frequencies.

In one estimate, the seats account for nearly half the absorption in a typical vehicle [1]. This is more than twice as much as the next most important absorber, the headliner. Optimizing the acoustic absorption of the seats is critical for quieter interiors. To accomplish the optimization, we need to know the characteristics of seat components. For a given seat design, the noise absorbing characteristics depend on the seat cover materials and the type of foam used in the cushion and back.

A seat cover is typically a laminate (i.e., face material bonded to slab foam). Face materials range from textiles to leather and vinyl. Within the textile group, face materials can be further classified as woven velours, pile knits and flat wovens. There is little variation in leather and vinyl except that permeability can be imparted to the materials through perforation.

The noise absorbing capability of the seats and covers was tested by the reverberation room method per ASTM C423 [2]. This technique is described later.

Test results show that textiles are drastically different from leather and vinyl in noise absorption. Even within the textile group, there are significant differences. Perforation of leather/vinyl, as well as foam density, affects the acoustic characteristics of seats. Also, the difference between a bonded seat and a conventional seat is negligible.

NOMENCLATURE

Pile - tufts of fibers in the perpendicular direction with respect to the plane of fabric;
dpf (denier per filament) - an indication of fiber size (higher dpf indicates larger fiber);
Woven velour - pile fabric made through a weaving process;
Pile knit - pile fabric made through a knitting process;
Slab foam - foam used to make a laminate with fabric;
Core foam - foam used to make a cushion or back pad;
Laminate - bonded product of fabric and slab foam.

EXPERIMENTAL

SAMPLES

Several textile laminates, a leather and a vinyl material were included in the study. Seat cover materials are listed in Table 1.

Cushion and back pads were made from three core foams: Foam A, Foam B-I and Foam B-II. They all were polyurethane foam. Chemicals for Foam A may differ slightly from that for Foam B-I and Foam B-II, which were made from the same chemicals. The densities for Foam A, Foam B-I and Foam B-II are 56.8, 46.1 and 38.9 kg/m^3, respectively.

All seat cushions and seat backs had identical shapes and dimensions.

Table 1. Seat covers included in the test

Seat Cover ID	Cover Material Description	Note
SC1	Woven velour laminate with 2.2 dpf pile fiber	Same slab foam as in SC2
SC2	Woven velour laminate with 4.4 dpf pile fiber	Same slab foam as in SC1
SC3	Knit pile laminate	
SC4	Laminate with flat woven A and low density slab foam	
SC5	Laminate with flat woven B and high density slab foam	
SC6	Leather with slab foam	SC6-9 have same slab foam
SC7	Perforated* leather with slab foam	
SC8	Vinyl with slab foam	
SC9	Perforated* vinyl with slab foam	
SC10	Flat woven laminate (Two sets)	Used to compare bonded seat and conventional seat

* Perforation was achieved by running the leather and vinyl through a non-threaded sewing needle with 6 stitches per inch. Hole lines were one half inch apart.

REVERBERATION ROOM TEST

Random incidence absorption test by reverberation room method is perhaps the most relevant acoustic test for seats. The test is simple to perform and it includes the entire seat assembly. Besides, random incidence better simulates actual conditions in the vehicle where noise is likely to arrive from many directions. Measurement of in-vehicle decay times is another way to test absorptive properties of seats, but these measurements are not standardized and are more complex.

Absorption by reverberation room is calculated from reverberation time using an empirical formula. First, the amount of absorption in the empty room is calculated from the reverberation time of the empty room. Then a seat is placed in the room and the reverberation time is measured and the amount of absorption calculated. This absorption is due to both the seat and the room. Subtracting the amount of absorption in the empty room from the absorption yields the absorption due to the seat alone. The absorption is expressed in metric sabins.

One metric sabin is the amount of absorption provided by one square meter of a perfect absorber (100% absorption coefficient). The measured absorption of the sample in metric sabins can be divided by the surface area of the sample to yield its absorption coefficient. Since it is difficult to determine the effective acoustic surface area of a seat, all random incidence data in this paper is reported in metric sabins.

For these measurements a single seat was measured in a 200 cubic meter reverberation room. The size of the room is dictated by both ASTM and ISO standards to provide good diffusion (required for the measurements) down to approximately 100 Hz. While it is desirable to measure an entire set of seats in a room of this size, it is difficult to measure a single seat in such a large room because a single seat does not contribute much absorption relative to such a large room.

Measuring a single seat introduces some random variation in the measurement results. To facilitate comparisons of test results, each group of data in this study has been fitted with a third order polynomial curve.

Figure 1 shows the set-up of the reverberation room. Three loudspeakers, two on the floor and one near the ceiling (not shown in the figure), were positioned in three trihedral corners. The microphone mounted on a tri-pod measures sound pressure decay at five positions. In this study, the microphone takes 40 measurements at each position.

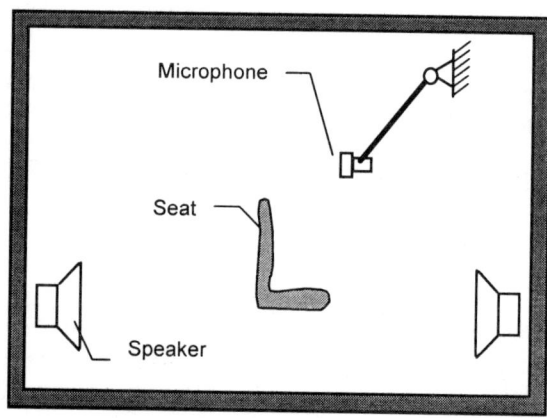

Figure 1. Reverberation room.

TESTS

Tests were conducted in three set ups to evaluate the trim components and the seats (i.e., the assemblies).

Set-up 1: Cover only

Seat covers were tested without the core foams. The covers were slipped over the seat frame. The sample looked like a seat except there were no core foams. This set-up reveals how much noise the seat cover alone can absorb.

Set-up 2: Foam only

In this set-up, the core foams were attached to the seat frame, but were not covered with any face material. Test results indicate how much noise the core foam pads alone can absorb.

Set-up 3: Seat assembly

Covers and core foams were assembled into seats. Seat covers (except ones in the bonded seat) were pinned down to the core foams to simulate a conventional seat build. Different combinations were tested to reveal the effect of fabric construction, the effect of pile fiber size, the effect of bonding, etc.

Throughout the testing, the temperature and relative humidity were controlled in ranges of 18-21°C and 62-66%, respectively.

RESULTS

The results are grouped to show the effects of various parameters on noise absorption.

EFFECT OF FABRIC CONSTRUCTION

Figure 2 compares covers SC2-SC5 (no core foams, refer to Table 1 for more information about the covers). Significant differences among the covers were exhibited. SC2 had the best noise absorption. This may be due to the tight construction of the fabric. On the other hand, the loose construction of SC3 imparted the cover with less noise absorption.

Figure 3 compares four seats made with core foam A and covers SC2-SC5. Interestingly, the seat with cover SC3 (pile knit) performed well, although the cover alone did not perform well. Conversely, the seat with SC5 (flat woven B w/ high density slab foam) did not perform well although the cover itself did well. This may be due to the high density slab foam in the cover which may have reduced noise absorption at frequencies over 315 Hz.

Seat with cover SC2 (4.4 dpf pile fiber) performed best at lower frequencies (below 630 Hz). Again, this is attributed to the tight construction of the fabric.

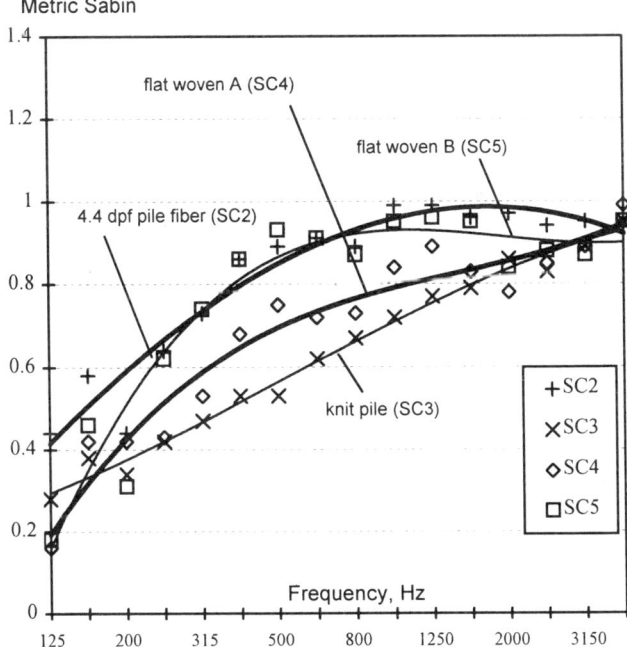

Figure 2. Noise absorption of textile covers only.

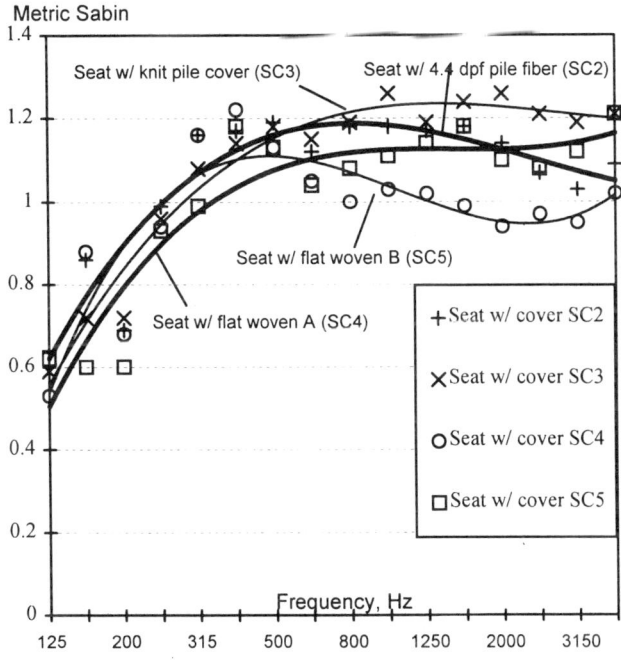

Figure 3. Noise absorption of seats with various textile covers.

EFFECT OF PILE FIBER SIZE

Pile fiber size governs the appearance retention of seat covers [3]. Larger fibers impart the covers with better crush resistance and less marking. One of the concerns of using larger pile fibers is that it may sacrifice the acoustic performance of the seats.

Figure 4 compares covers SC1 and SC2 along with their corresponding seats. The only difference between the two covers is the size of the pile fibers which, as shown in the figure, does not affect the noise absorption of the seats.

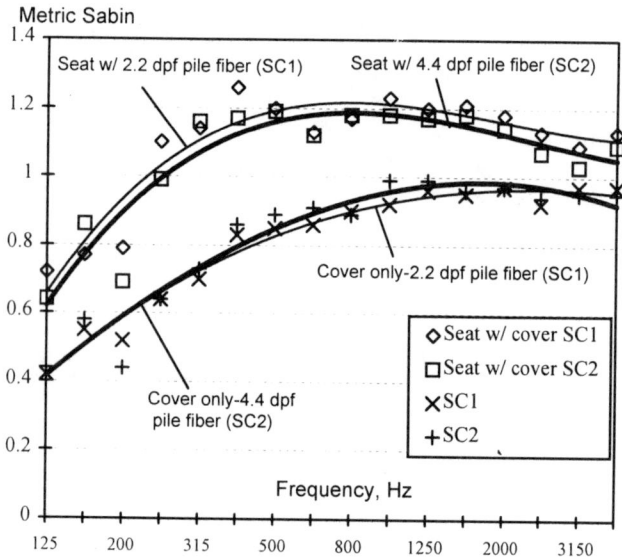

Figure 4. Noise absorption of seat with different pile fiber sizes.

COMPARISON BETWEEN FABRICS AND LEATHER/VINYL

Tests revealed that leather and vinyl have very similar noise absorption performance. They are, therefore, treated as equivalent to each other. Figure 5 shows that the noise absorption of leather or vinyl seats is significantly lower than that of the cloth seats.

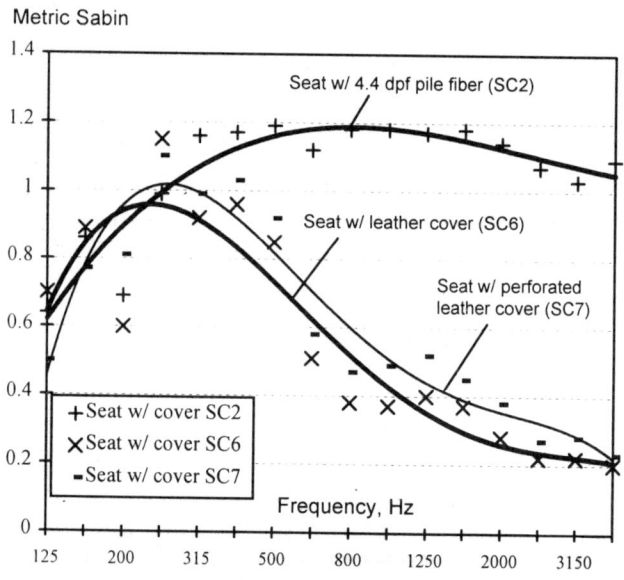

Figure 5. Comparisons of cloth vs. leather seats and leather vs. perforated leather seats.

EFFECT OF PERFORATION ON LEATHER/VINYL

Also shown in Figure 5, the seat with perforated leather exhibited noticeable improvement.

EFFECT OF FOAM TYPES

Figure 6 compares the three core foams. Below 200 Hz, Foam A performed best followed by Foam B-II. Above 200 Hz, the order of performance was reversed.

In general, foam B-II performed better than Foam A. It is interesting to note that the density of Foam B-II was 31% lower than that of Foam A.

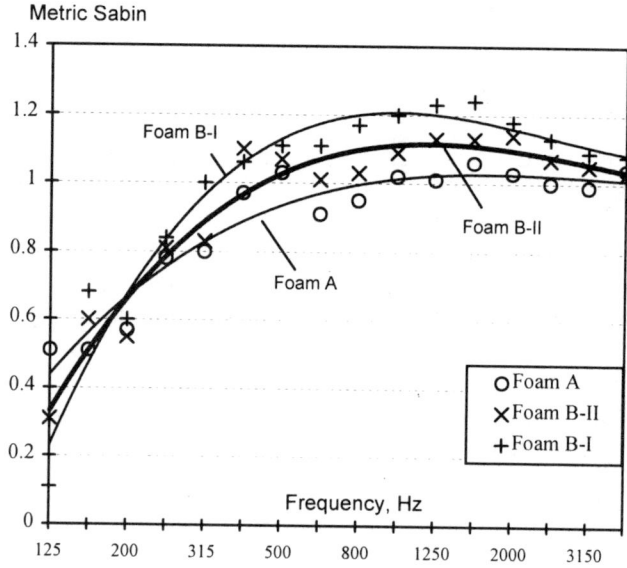

Figure 6. Noise absorption of core foams only.

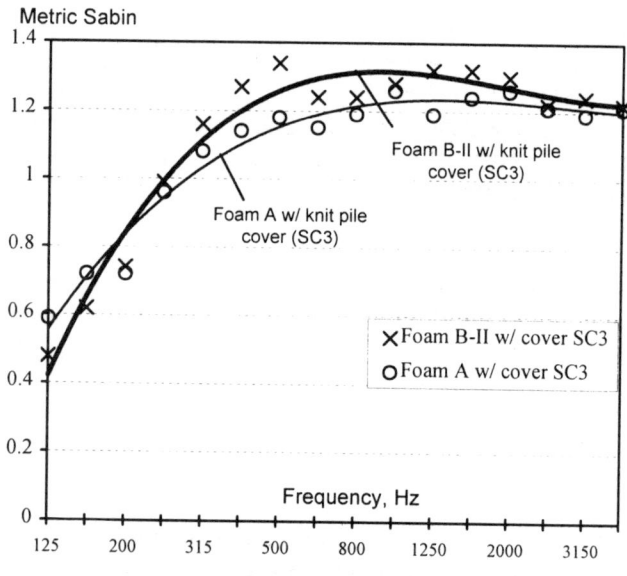

Figure 7. Noise absorption of seats with different core foams.

The difference between Foam A and Foam B-II was reflected when the foams were assembled into seats with a knit pile cover (SC3), as shown in Figure 7.

COMPARISON BETWEEN BONDED SEAT AND CONVENTIONAL SEAT

Two seats consisting of identical core foam and covers were included for comparison. One set of covers and one set of core foams were bonded together using the SureBond® process. In this process a film placed between the cover and core foam is melted by steam, thus serving as an adhesive and bonding the cover to the pad. When the adhesive film melts, it leaves a permeable bond. A conventional seat was also made from identical covers and core foams. The difference, as indicated in Figure 8, between the bonded and the conventional seats is negligible.

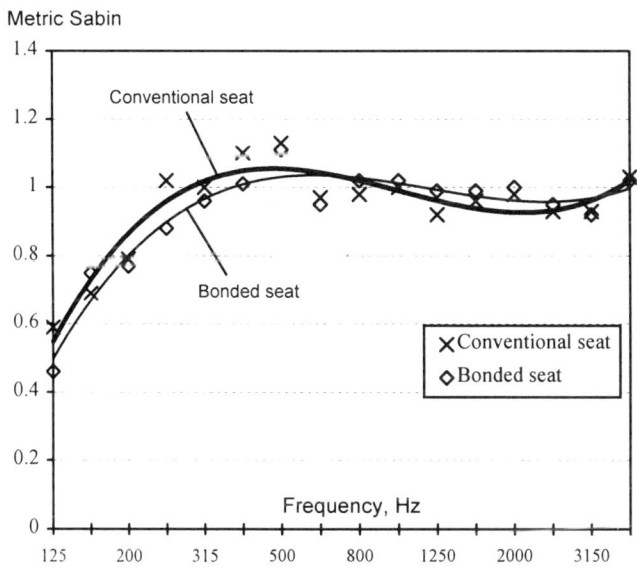

Figure 8. Noise absorption of seats: bonded vs. conventional.

CONCLUSIONS

- Noise absorption of different textile seat covers varies significantly;
- The size of the pile fiber does not affect noise absorption;
- Cloth seats have superior noise absorbing capability compared to leather/vinyl seats;
- Perforation on leather/vinyl can improve the noise absorption of leather/vinyl seats;
- Foam density affects the acoustic performance of cloth seats;
- The difference between bonded and conventional seats is negligible.

ACKNOWLEDGMENTS

The authors sincerely thank their colleagues for their supports in this study.

REFERENCES

1. R. Wentzel and E. Green, "Assessing Headliner and Roof Assembly Acoustics," 971926, SAE International, Proceedings of the 1997 Noise and Vibration Conference, Traverse City, MI.
2. ASTM C423, "Standard Test Method for Sound Absorption and Sound Absorption Coefficients by the Reverberation Room Method," 1990.
3. M. Cordonnier and P. Chen, "Newness Retention of Textile Automotive Seat Covers," 960510, SAE International Congress & Exposition, Detroit, Michigan, February 26-29, 1996.

980660

Dynamic Ride Quality Investigation for Passenger Car

Se-Jin Park, Wan-Sup Cheung
Korea Research Institute of Standards and Science

Young-Gun Cho, Yong-San Yoon
Korea Advanced Institute of Science and Technology

Copyright © 1998 Society of Automotive Engineers, Inc

ABSTRACT

The ride values of passenger cars are investigated for Korean subjects based on the vibration of the human bodies. When three subjects are excited by driving a vehicle on road, their responses of acceleration are measured at 12 points on their bodies according to Griffin's 12 axis system (3 translational axes on a seat surface, 3 rotational axes on a seat surface, 3 translational axes at the seat back and the 3 translational axes at the feet). Since one of the most important parameters for ride comfort is the level and duration of the root mean square acceleration experienced, the ride values, such as the seat effective amplitude transmissibility, the component ride value, and the overall ride value based on acceleration root mean square are evaluated for different four vehicles using frequency weighing functions and axis multiplying factors. The ride indices are also studied considering to the seat dynamic characteristics with subjects.

INTRODUCTION

Ride comfort is a very important problem in vehicle design. Seating comfort is associated with the various factors such as dynamic, postural, visual, sonic, thermal comfort. Among these types of comfort, one of the principal components of a vehicle environment which can affect passenger's comfort is vibration. To improve ride quality from the standpoint of vibration, lots of studies have been attempted to identify factors contributing to the ride quality and to match the ride quality with the subjective rating[2,9,12,13], even though the ride quality is inherently subjective measure, and the perceived comfort level of different subjects exposed to the same stimulus varies. Because the level of perceived vibration is different according to the frequency and the axis of vibration, the equivalent comfort contours based on the subjective rating are formed and adapted in ISO 2631 and BS 6841. Griffin suggested some ride values regarding that the level and duration of the root mean square acceleration of the whole-body is closely related with the subjective ride quality by frequency weighing functions which are the inverse of equivalent comfort curve[1,2].

The major components contributing to the ride quality of vehicle are tire, suspension, and seat. The optimization schemes have been developed to increase the ride quality with such design variables as tire, suspension, and seat property[7,15]. The seat dynamics may become significant in the ride quality since the seat directly transmits the vibration to passenger.

In this study we evaluated the ride value of four different vehicles from the exposed vibration based on the root mean square of acceleration driving on the two different roads, and seat dynamics is investigated for the transmissibility in the seats. The experimental procedure, the ride values, and the comparison of the ride values for vehicles are presented in this paper.

EXPERIMENTAL PROCEDURE

The exposed vibration on human is transmitted to the whole-body through the contact area between human and seat. Figure 1 shows the experimental setup for measuring the whole-body vibration at the feet, hip, and back. The vibration is measured at 12 axis - three translations (X_f, Y_f, Z_f) at feet, 3 translations (X_s, Y_s, Z_s) and rotations (R_x, R_y, R_z) at hip, 3 translations at back (X_b, Y_b, Z_b). Such measuring scheme proposed by Griffin[1] is well accepted in the vehicle industry, and adopted by BS 6841.

The measuring sensors are the 3-axis servo acceleration sensor (Columbia Research Model : SA-307TX) for measuring the translational accelerations at feet and hip, rotational acceleration sensor(Columbia Research Model : SR-207HP) for hip, and 3-axis translational acceleration sensor (Entran EGCS-A-10) for back. The output signal is recorded for 3 minutes with 6kHz sampling rate by 16 digital recorder(Sony DAT 216A).

The subject was seated at the front passenger seat, and the experimenter and the equipment were placed in the rear seat. The measuring devices were installed in a passenger car as shown in Figure 2. The subject sits on a sit-bar in which 3 translational and 3 rotational sensors are attached whiling laying his feet on the foot-pad having 3 translational sensors, and a plate having 3 translational sensors is fastened on the driver's back.

This experiment was conducted for four vehicles on unevenness road with three Korean expert drivers. The velocity was maintained at a constant speed of 40km/h for keeping the same vibration condition of road. The age, height, and weight of subjects are described in Table 1. Since the posture of the subject is important for measuring the human vibration, the change in posture may alter the human response, so the subjects was asked to keep up their posture during collecting the signal and not to fasten seat belt for the pure seat dynamics.

Table 1. The list of the subjects

	Age (year)	Height (cm)	Weight (kg)
Person 1	39	181	85
Person 2	33	169	75
Person 3	35	165	70

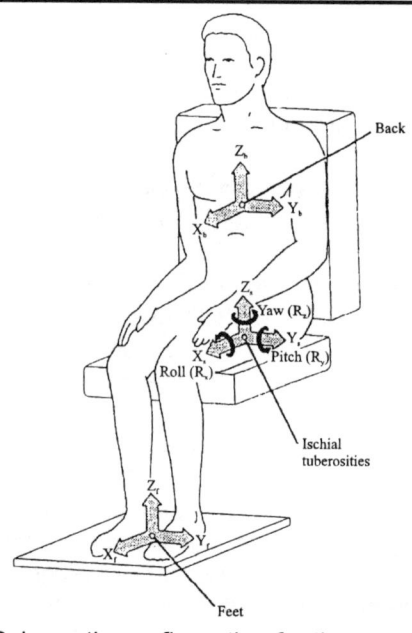

Figure 1. Schematic configuration for the measurement of whole-body vibration[1]

Figure 2. Experimental setup for the field test.

RIDE VALUE

The exposure of the human to vibration may invoke discomfort, motion sickness, or disease. As the ride quality is essentially a subjective measure, a given human's comfort level may be different for the same vibration. It is, therefore, not easy to represent the ride value as a quantity. However, many researches have tried to quantify the ride value, and so developed many ride values. In this paper, we evaluate the ride values through the component ride value, the overall ride value, and the seat effective amplitude transmissibility[1,2].

FREQUENCY WEIGHTING FUNCTIONS AND AXIS MULTIPLYING FACTORS

As ISO 2631,8041 and BS 6841 indicate, human response to vibration differs for the axes and frequency of vibration. The weightings have been derived from Griffin[1]. The subjects' responses are used to determine the equivalent comfort contours. The inverse of such curves form the 'frequency weighting functions'. In order to minimize the number of frequency weightings for axes, some are used for more than one axis with different 'axis multiplying factors'. Therefore, frequency weightings are composed of 4 functions : W_b, W_c, W_d, W_e. The weighting function W_b is defined as Eq. (2) and W_c, W_d, W_e as Eq. (1). The detailed parameters are described in Table 2. The frequency weighting functions and axis multiplying factors are described in Figure 3 and Table 3.

$$H_w = \frac{(s+2\pi f_3)}{(s^2 + 2\pi f_4 s/Q_2 + 4\pi^2 f_4^2)} \times \frac{2\pi K f_4^2}{f_3} \quad (1)$$

$$H_b = \frac{(s+2\pi f_3)(s^2 + 2\pi f_5 s/Q_3 + 4\pi^2 f_5^2)}{(s^2 + 2\pi f_4 s/Q_2 + 4\pi^2 f_4^2)(s^2 + 2\pi f_6 s/Q_4 + 4\pi^2 f_6^2)} \times \frac{2\pi K f_4^2 f_6^2}{f_3 f_5^2} \quad (2)$$

Table 2 Filter coefficients[2]

Coefficients / Weighting	f_1 / f_6	f_2 / Q_2	Q_1 / Q_3	f_3 / Q_4	f_4 / K	f_5
W_b	0.4 / 4	100 / 0.55	0.71 / 0.90	16 / 0.95	16 / 0.4	2.5
W_c	0.4 / -	100 / 0.63	0.71 / -	8 / -	8 / 1.0	-
W_d	0.4 / -	100 / 0.63	0.71 / -	2 / -	2 / 1.0	-
W_e	0.4 / -	100 / 0.63	0.71 / -	1 / -	1 / 1.0	-

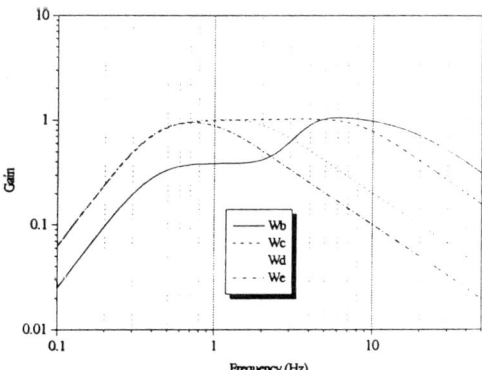

Figure 3. Response of weighting functions

Table 3. Weighting functions and axis multiplying factors in BS 6841

Measurement Position	Weighting function (w_i)	Axis multiplying factor (m_i)
X_f	W_b	0.25
Y_f	W_b	0.25
Z_f	W_b	0.40
X_s	W_d	1.00
Y_s	W_d	1.00
Z_s	W_b	1.00
R_x	W_e	0.63
R_y	W_e	0.40
R_z	W_e	0.20
X_b	W_c	0.80
Y_b	W_d	0.50
Z_b	W_d	0.40

COMPONENT RIDE VALUE

The contribution of each axis can be quantified as the ride value according to contact points between the human and the seat, and the 'component ride value' is defined as the acceleration r.m.s. value of each axis. To obtain the component ride value, power spectral density($P_{ii}(f)$) is multiplied by weighting function($w_i(f)$) for each axis(i), then its square root is multiplied by axis multiplying factor(m_i) as Eq. (3). These values enable to evaluate the relative contributions of each axis.

$$\text{Component Ride Value}_i = m_i \times \left[\int \{P_{ii}(f) w_i(f)^2\} df \right]^{1/2} \quad (3)$$

OVERALL RIDE VALUE

The overall ride value is evaluated as the 2-norm of the component ride value as Eq. (4). Since this value gives total vibration level, it is adequate for making simple comparison between vehicles with different suspension and seat, etc. We can say the vehicle having the highest overall ride value would be expected to be the most uncomfortable one from the viewpoint of vibration.

$$\text{Overall Ride Value} = \left[\sum_{i=1}^{12} (\text{Component Ride Value}_i)^2 \right]^{1/2} \quad (4)$$

SEAT EFFECTIVE AMPLITUDE TRANSMISSIBILITY (SEAT)

The SEAT value is defined as the weighted vibration ratio between Z_f of the floor and Z_s of the hip as Eq. (5), where $P_{ss}(f)$ and $P_{ff}(f)$ are the floor and the seat acceleration spectra, and w_b is frequency weighting function. The SEAT value greater than 1 means the vibration at the seat is greater than the vibration at the floor. The SEAT lower than 1 indicates that the level of vibration is decreased by the seat.

$$\text{SEAT} = \left[\frac{\int \{P_{ss}(f) w_b(f)^2\} df}{\int \{P_{ff}(f) w_b(f)^2\} df} \right]^{1/2} \quad (5)$$

ANALYSIS

ROAD CONDITION

This experiment was conducted in a passenger car at a speed of 40km/h, running on two different road conditions : road1(endurance road), road2(unevenness road). The road conditions are shown in Table 4 to compare these two roads in view of acceleration at the foot z-axis. The acceleration of vehicle 2 is increased by 13% on road 2 and that of vehicle 3 by 49%. This shows that the vehicle is vibrated more severely on unevenness road. Vehicle 2 and 3 have similar acceleration r.m.s on road 1, but the acceleration of vehicle 3 becomes larger on road2 than that of vehicle 2.

Table 4 Z_f - axis acceleration averaged for 3 persons

Vehicle	Acceleration RMS (m/s^2)		(Road2-Road1) / Road1 × 100 (%)
	Road 1	Road 2	
V 1	2.79	3.72	33
V 2	2.12	2.39	13
V 3	2.14	3.19	49
V 4	2.38	3.34	40

ACCELERATION R.M.S AT MEASUREMENT AXES

The results of twelve acceleration r.m.s including 3 subjects' data acquired from vehicle 1 are presented in Figure 4. These figures include accelerations of fore-and-aft (x-axis), lateral (y-axis) and vertical (z-axis) vibration inputs to the subjects' feet, ischial tuberosities and back and the roll(r_x), pitch(r_y) and yaw(r_z) vibration at the subjects ischial tuberosities. The higher acceleration level will significantly contribute to passenger's discomfort.

In these figures of translational axis the range of frequency having the major power is below 20Hz, on the other hand the frequency range in rotational axis is more higher. The r.m.s level around 10Hz is the most dominant in z-axis of feet, hip, and back, and other axes also have peaks around 10Hz. Though the three subjects' mass, height, and posture are different, their response shows the similar pattern for the same vehicle and road, so it can be said that the effect of the subjects on vibration is small.

These r.m.s functions showing the exposed vibration level are associated with tire, chassis, seat, engine, road type, wheel base filtering effect, human dynamics, and so on. These modes are also correlated with each other, so it is not easy to identify whose dynamics is directly connected with each mode. However the dynamics of vertical acceleration (Z_f axis of foot, Z_s of hip, and Z_b of back) is relatively well known from the numerous studies[7,15]. The r.m.s. level at Z_f of feet doesn't include seat and human dynamics, so we can observe only chassis dynamics which represents the suspension and tire mode at 1.2 and 13Hz. On the other hand the r.m.s. level at Z_s of hip includes seat dynamics. Figure 5 demonstrates the gain and phase of transfer function between Z_f of the feet and Z_s of the hip. This acceleration transfer function has the similar characteristics of second order system composed of mass, damper, and spring, and the well-known fundamental seat mode is observed at 3~4Hz. As the gain is above 1 at the range of 3~8Hz, the acceleration at hip is more amplified than at the floor in that frequency range.

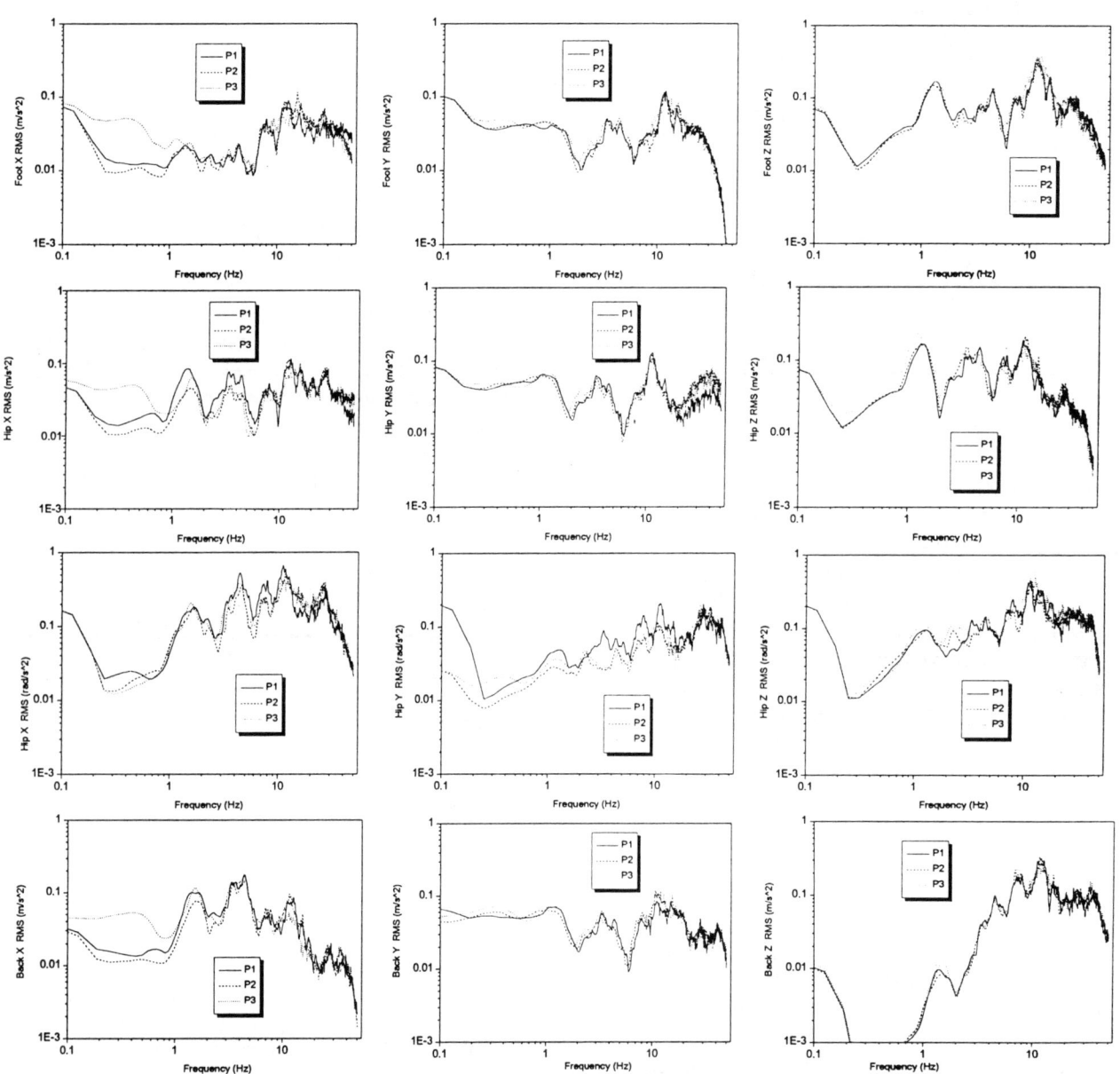

Figure 4. 12-axis vibration level about vehicle 1

Table 5 Acceleration R.M.S. of 12 axis for person 1 and road 1
(UW : unweighted r.m.s, W : weighted r.m.s, MW : weighted r.m.s multiplied by axis multiplying factor)

Vehicle	Vehicle 1			Vehicle 2			Vehicle 3			Vehicle 4		
R.M.S Axis	UW	W	MW	UW	W	MW	UW	W	MW	UW	W	MW
X_f	1.04	0.73	0.18	1.37	0.91	0.23	1.33	1.06	0.26	1.93	1.49	0.35
Y_f	0.95	0.78	0.19	0.75	0.57	0.14	0.69	0.53	0.13	1.13	0.96	0.24
Z_f	2.66	2.23	0.89	2.07	1.76	0.70	2.08	1.83	0.73	2.37	1.86	0.75
X_s	1.33	0.35	0.35	1.59	0.33	0.33	1.34	0.30	0.30	2.34	0.46	0.46
Y_s	1.10	0.33	0.33	1.33	0.32	0.32	0.73	0.25	0.25	1.49	0.31	0.31
Z_s	1.59	1.39	1.39	1.80	1.55	1.54	1.94	1.76	1.76	2.55	2.20	2.20
R_x	5.81	0.77	0.49	6.76	0.78	0.49	7.09	0.73	0.46	10.6	1.07	0.68
R_y	2.67	0.24	0.10	5.38	0.29	0.11	1.86	0.19	0.08	3.04	0.26	0.10
R_z	4.43	0.44	0.09	8.74	0.63	0.13	4.29	0.44	0.09	7.14	0.56	0.11
X_b	1.07	0.97	0.78	1.04	0.93	0.74	0.99	0.90	0.72	1.45	1.29	1.04
Y_b	1.10	0.33	0.17	1.14	0.33	0.16	0.76	0.30	0.15	1.38	0.37	0.18
Z_b	2.82	0.49	0.19	3.03	0.50	0.20	2.38	0.48	0.19	3.10	0.55	0.23

Table 6 Acceleration R.M.S. of 12 axis for person 1 and road 2
(UW : unweighted r.m.s, W : weighted r.m.s, MW : weighted r.m.s multiplied by axis multiplying factor)

Vehicle	Vehicle 1			Vehicle 2			Vehicle 3			Vehicle 4		
R.M.S Axis	UW	W	MW	UW	W	MW	UW	W	MW	UW	W	MW
X_f	1.42	0.94	0.24	1.86	1.23	0.31	1.98	1.63	0.41	2.45	1.74	0.44
Y_f	1.39	1.142	0.29	1.00	0.72	0.18	1.03	0.76	0.19	1.54	1.26	0.32
Z_f	3.70	3.29	1.29	2.46	2.06	0.82	3.29	2.87	1.15	3.57	2.93	1.17
X_s	1.74	0.51	0.51	1.93	0.54	0.55	1.85	0.54	0.54	2.24	0.73	0.73
Y_s	1.60	0.46	0.46	1.72	0.46	0.46	1.12	0.37	0.37	1.94	0.38	0.38
Z_s	2.14	1.94	1.94	1.97	1.70	1.70	2.70	2.51	2.51	3.30	3.00	3.00
R_x	7.88	1.06	0.67	2.35	0.30	0.19	10.04	1.04	0.66	13.03	1.16	0.73
R_y	3.76	0.36	0.14	6.61	0.37	0.15	2.60	0.27	0.11	4.32	0.28	0.11
R_z	6.91	0.67	0.13	4.13	0.29	0.06	5.75	0.58	0.12	11.24	0.89	0.18
X_b	1.37	1.23	0.98	1.30	1.16	0.93	1.30	1.17	0.93	1.53	1.30	1.04
Y_b	1.54	0.51	0.26	1.51	0.48	0.24	1.24	0.49	0.24	0.65	0.28	0.14
Z_b	3.72	0.69	0.27	3.70	0.62	0.25	3.31	0.68	0.27	4.12	0.71	0.28

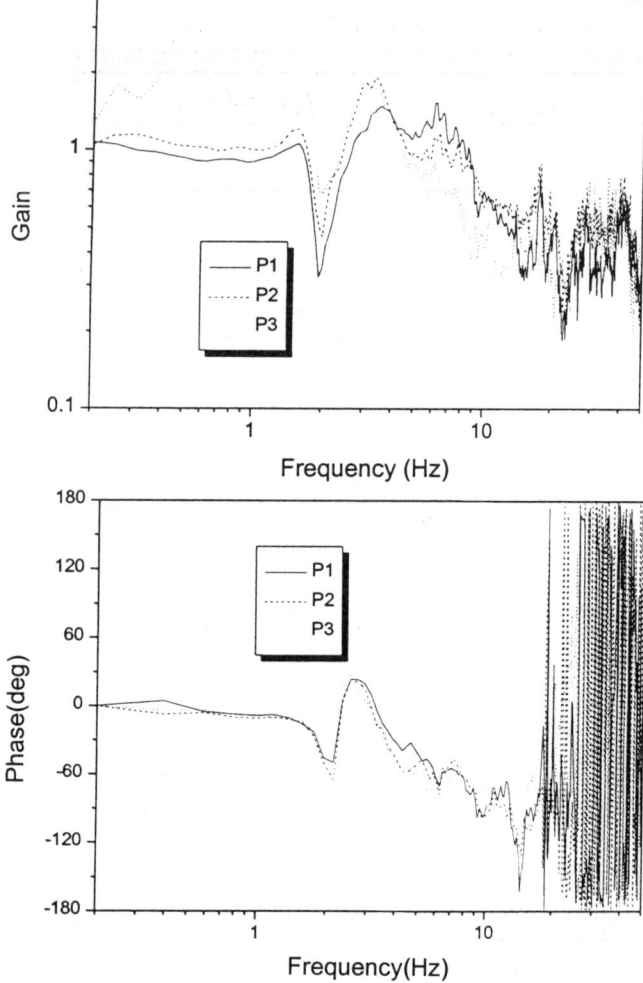

Figure 5. Transfer Functions between Z_s axis and Z_f axis (Vehicle 1)

RIDE VALUE ANALYSIS

Component ride value

The component ride value is calculated as Eq. (3) with the frequency weighting functions and the axis multiplying factors. As Figure 3 represents, the gain of function W_b and W_c has the value of 1 at the range between 5 ~ 16Hz and 0.5 ~ 8Hz respectively, while the gain of W_d and W_e have the value of 1 between 0.5 ~ 2Hz and 0.5 ~ 1Hz. In driving a passenger car, the P.S.D. is concentrated on the range of 1~15Hz as depicted in Figure 4, so the axes(X_f, Y_f, Z_f, Z_s, X_b) using W_b and W_c have more significance on the ride quality. But the axes (X_s, Y_s, R_x, R_y, R_z) using W_d and W_e have less contributions on the ride quality. We can see these results in Table 5 and Table 6. In these tables the value UW means r.m.s without frequency weighting functions, the value of W means the r.m.s with frequency weighting functions, and the value of MW makes use of weighting function and axis multiplying factor as depicted in Table 3. If UW and W are compared, those axes using W_b and W_c are decreased about 10~30% due to weighting functions, but axes using W_d and W_e are decreased about 70~90%. The comparison of the component ride value (WM in Table 5~6) reveals that Z_s of hip is the most important axis for the overall ride value and it contributes 50~70% to the overall ride value. The secondary axes are Z_f and X_s which contribute 10~23% and 11~14% to the overall ride value respectively. These three major axes' portion is 80~90% of ride value.

Overall ride value

The overall ride value was evaluated for the different 4 vehicles. Figure 6 is the averaged value for 3 persons. Because the road condition of road 2 is more severe than the condition of road 1, the overall ride value is greater than that of road 1 by 20 ~ 83%. The ranking of the overall ride value doesn't vary by different road conditions, so we can say that the overall ride value mainly depends on the vehicle, not on the road condition. The vehicle 4's overall ride value is the highest, from this fact ride quality of vehicle 4 is the worst in the point of human vibration. In addition, vehicles 1 and 2 have a similar ride quality on road 1, but vehicle 2 has better ride quality in severe road condition like road 2 due to chassis nonlinearity.

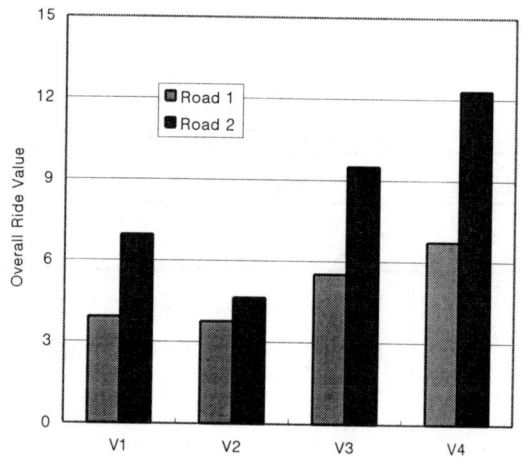

Figure 6. Overall ride value averaged for 3 persons

SEAT value

SEAT value depicted in Table 7 is defined as the ratio of weighted acceleration between the hip and feet like Eq. (5). During acquisition of acceleration signal, we acknowledge that each subject's posture should remain constant in comfortable posture, so that one's seating posture is maintained but each subject's posture may be different. From Table 7 we can't find the effect of the mass of driver because so many other factors except mass can influence the SEAT value. But even though 3 persons' seat posture and mass are different, the SEAT value show only a small variation within 10%. Therefore, we can conclude the SEAT value is mainly influenced by the vehicle and seat properties.

Table 7. Listing of SEAT results of road 1

Vehicle	SEAT value			
	Person 1	Person 2	Person 3	Mean
Vehicle 1	0.62	0.67	0.46	0.58
Vehicle 2	0.88	0.78	0.79	0.82
Vehicle 3	0.96	1.10	1.10	1.05
Vehicle 4	1.20	1.10	1.00	1.10

Figure 7 shows the SEAT value for two different roads. In this figure we can see that the SEAT value stays constant irrespective of different road with small deviation of 1 ~ 7% due to linear characteristics of seat cushion. Since the SEAT value of vehicle 1 is 0.6 and that of vehicle 2 is 0.8, seat can transmit reduced acceleration to hip by 60 ~ 80%. But as the SEAT value is close to 1 for vehicle 3 and 4, vibration isn't decreased by seat. As especially SEAT value in road 1 is greater than 1, vibration may be rather increased by seat. Vehicle 1 has the lowest SEAT value, the seat of vehicle 1 reduced the vibration greater than any other seat of vehicle.

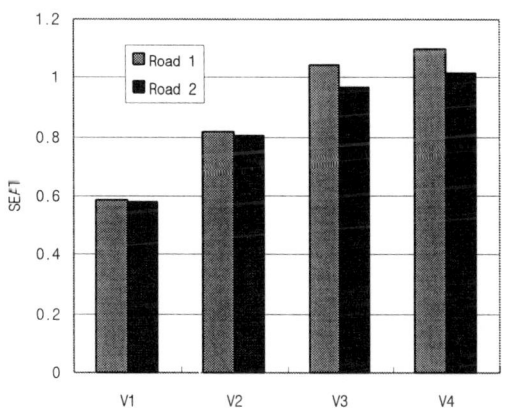

Figure 7. SEAT Value averaged for 3 persons

CONCLUSION

To evaluate ride quality of vehicle, this experiment was conducted with four vehicles and three Korean drivers on unevenness and endurance road for three Korean expert drivers. We evaluated the ride values (component ride value, overall ride value, SEAT value) for the purpose of comparison of ride quality in view of vibration exposed on human, and we could rank the vehicle by those values. It was also identified that the Z_s axis of the hip, Z_f axis of the foot, and X_s axis of back are a great important axis for ride quality, and SEAT value shows a low deviation for road condition.

REFERENCES

1. M. Griffin, 1990, "Handbook of Human Vibration", Academic Press, London
2. K.C. Parsons and M.J. Griffin, "Methods for Predicting Passenger Vibration Discomfort", SAE 831029
3. Griffin, M.J.,"Evaluation of Vibration with respect to human response", SAE 860047
4. Griffin, M.J.,"Duration of Whole-Body Vibration Exposure : Its Effects on Comfort", Journal of Sound and Vibration, Vol. 48, pp.333-339, 1976
5. Griffin, M. J., "The Evaluation of Vehicle Vibration and Seats", Applied Ergonomics, Vol. 129, pp. 143-154, 1989
6. Measurement and evaluation of human exposure to whole-body mechanical vibration and repeated shock, British standard BS 6841:1987
7. E. Berger and B.J. Gilmore, "Seat Dynamic Parameters for Ride Quality", SAE 930115
8. Yoshihiko Kozawa and Gunji Sugimoto et al, "A New Ride Comfort Meter", SAE 860430
9. Prako, F., Lee, R., "Vibration Comfort Criteria", SAE 660139
10. Jean-Marc Judic and John A. Cooper, "More Objectives Tools for the Integration of Postural Comfort in Automotive Seat Design", SAE 930113
11. John H. Varterasian and Richard R. Thompson, "The Dynamic Characteristics of Automobile Seats with Human Occupants", SAE 770249
12. Reed, M.P., Kakishima, Y., Lee, N.S., Satio, M., Schneider, L.Y., "An Investigation of Driver Discomfort and Related Seat Design Factors in Extended Driving", SAE 910117
13. Wambold, J. C., "Vehicle Ride Quality Measurement and Analysis", SAE 861113
14. Brian Peacock and Waldemar Karwowski, "Automotive Ergonomics", Taylor & Francis, 1993
15. F.M.L., Amirouche, "Optimization of the Contact Damping and Stiffness Coefficients to Minimize Human Body Vibration", Journal of Biomedical Engineering, November 1994, Vol. 116, pp. 413-420
16. S.J. Park, Chae-Bogk Kim, "The Evaluation of Seating Comfort by the Objective Measures", SAE 970595

980661

Stirling Air Conditioned Variable Temperature Seat (SVTS) & Comparison with Thermoelectric Air Conditioned Variable Temperature Seat (VTS)

Steve Feher
Feher Research Co.

Copyright © 1998 Society of Automotive Engineers, Inc.

ABSTRACT

The thermoelectric Variable Temperature Seat, (VTS), offers automotive OEMs a new approach to occupant comfort enhancement in both hot and cold weather. The VTS is capable of both cooling and warming the vehicle occupant relatively efficiently and relatively quietly through the use of conditioned air flow through the seat pad structure.

The subject of this paper, the Stirling Variable Temperature Seat, (SVTS), is the latest advancement in seat cooling and heating technology, and significantly improves upon the thermoelectric, (Peltier), VTS in several significant ways.

INTRODUCTION

This paper will explain the differences between the SVTS and the VTS, and how the SVTS successfully addresses the following concerns associated with the VTS:

1- Moderate air ΔT in cooling mode.

2- Low efficiency in cooling mode, particularly as air ΔT increases.

3- Less than normal orthopedic seat support and comfort.

Drawings and tables, based upon prototype test results and use experiences with both technologies, are included to illustrate the conclusions drawn in this paper.

BACKGROUND

CONCERNS NO. 1 AND 2: Moderate air ΔT increases perceived cool-down time. Moderate air ΔT is particularly significant for leather covered seats since leather is a fairly good thermal insulator, so perceived cool-down time is increased even further over body cloth covered seats. The greater the available ΔT in cooling mode, the faster the seat will in fact cool down and will be perceived to cool down, especially under conditions of relatively very high vehicle interior air and surface temperatures, for example above ~120.0° F/49.0° C.

The basic reason for the difference in thermal perception is that vehicle occupant body skin surface temperature averages around 96.0° F/35.6° C, and perceived seat temperature is relative to occupant skin surface temperature, not ambient air temperature. If the available air ΔT is too small, then as vehicle interior air and surface temperatures rise, the temperature of the air flowing into the seat will be closer and closer to occupant skin temperature, resulting in a reducing perception of cooling power because of the reduced difference in temperature, or ΔT, between seat air and occupant skin surface temperature.

The definition of adequate, or sufficient, ΔT in cooling mode would then appear to be that ΔT which would provide an effective thermal transfer, and hence a perception of effective cooling, even under the worst case of high vehicle interior air and surface temperatures. The SVTS is capable of meeting this requirement, even at vehicle interior air and surface temperatures of 150.0° F/65.6° C, because it is capable of an air ΔT of ~76.0° F/42.0° C, (see Table 1). This ΔT enables SVTS seat air at ~74.0° F/23.3° C, at and ambient vehicle interior air temperature of 150.0° F/65.6° C.

Low cooling mode efficiency is a concern because the electrical load placed upon the average car is increasing, and ever-increasingly more expensive alternators are becoming necessary to supply the demand for electrical power. At maximum cool-down power in cooling mode, the VTS operates at approximately 40-50% electrical efficiency. The VTS operates at much higher efficiency in heating mode, because the input power I2R is additive in heating mode, however heating mode efficiency is not nearly as important as cooling mode efficiency, because in cold weather, prodigious amounts of waste heat are available from the internal combustion engine to heat the vehicle interior. It's cooling power that is most needed, at the highest efficiency possible, to offer the opportunity for rapid cool-down and fuel savings by not having to always use central AC in hot weather.

CONCERN NO. 3: Less than normal orthopedic seat support and comfort. This has been a concern in the past because of an effect referred to as the "Snowshoe Effect" in SAE paper 931111, entitled <u>Thermoelectric Air Conditioned Variable Temperature Seat (VTS) and Effect Upon Vehicle Occupant Comfort, Vehicle Energy Efficiency, and Vehicle Environmental Compatibility</u>, by the same author. The Snowshoe Effect results from the use of an array of ~12.7 mm diameter helically wound steel coils forming an air flow layer or spacer assembly, placed between the body cloth covering the seat and the support foam underneath.

The steel coil assembly is made by coiling steel wires and interlocking adjacent rows of coiled wire, similarly to the way chain link fencing rows are intertwined with adjacent rows.

Because the steel coil assembly spreads the occupant's weight over the load bearing support foam, the foam doesn't deform as it ordinarily would, particularly in the areas of the ischial tuberosities, often resulting in a sensation or sense of not quite settling into the seat, or of sitting on top of the seat instead of sitting in the seat. Those occupants who feel that they're sitting on top of the seat instead of in it will often also experience a lack of orthopedic support.

STIRLING VARIABLE TEMPERATURE SEAT (SVTS)

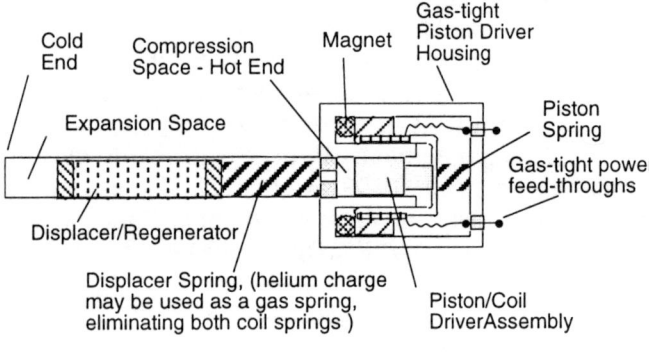

Fig. 1

CONCERNS NO. 1 AND 2: Figure 1 shows the basic components of a free-piston Stirling cycle heat pump. Although the Stirling cycle has been known for many years, as both a prime mover and a heat pump, until Sunpower, Inc., Athens, Ohio produced the first free-piston type machines, Stirling cycle heat pumps and engines were generally considered to be of limited practicality. The free-piston concept eliminates sliding or rotating seals and the friction, wear, and leakage associated with them, resulting in a much more reliable, efficient, and cost effective mechanical interpretation of the Stirling cycle. Other developments, including materials and assembly techniques have resulted in a new machine that has unique advantages over other technologies, in certain applications. One of the more significant applications is the SVTS. The Stirling SVTS is up to 600% more efficient than the Peltier thermoelectric VTS, and the potential efficiency of Stirling in the future is even higher.

Figure 2 shows the general configuration of the actual SVTS Stirling machine, set up to deliver cooled or heated air to the seat. A PTC heater is shown in contact with the cold end or "cold head" of the machine, and is only used for heating mode. The Stirling machine is not energized when the system is in heating mode. It is not feasible to reverse the direction of heat flow in this Stirling machine, so a resistive heater is used for heating mode. By attaching the heater to the cold end, it is possible to make use of the main heat exchanger in both modes, saving space, weight, complexity, and expense over having a totally separate heating sub-system.

In practice, the PTC heater is clamped between the cold end and another small heat exchanger, Figure 6, in order to allow the PTC heater to dissipate heat from both sides into the seat air stream. If this is not done, the PTC heater will reach it's Curie, or anomaly point, too easily, in effect, overheating at a lower than desired output air temperature.

Figure 2 is based upon the first SVTS prototype, Type M223a/SVTS, which is a free-piston Stirling machine with a maximum rating of 100.0 watts of cooling power, which was originally developed for a completely different application, and which was modified for the SVTS application. The next SVTS machines will be designed and built specifically for the SVTS application, and will have a maximum cooling capacity of 40.0 watts, and will be much smaller and lighter.

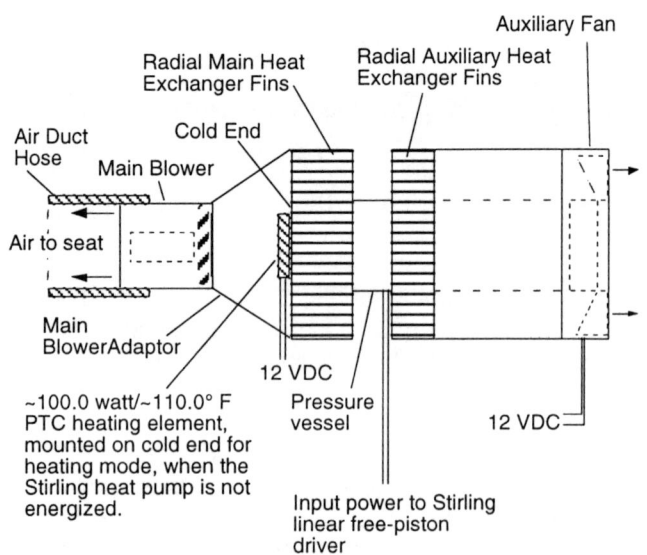

Fig. 2

Figure 3 illustrates how fast the SVTS cools down in cooling mode compared to the VTS in a vehicle interior cooled by conventional A/C under hot, sunny weather conditions. Before the VTS output air drops to a comfortable level, and long before the average interior air temperature drops to a comfortable level, the SVTS has reached a more effective cooling ΔT.

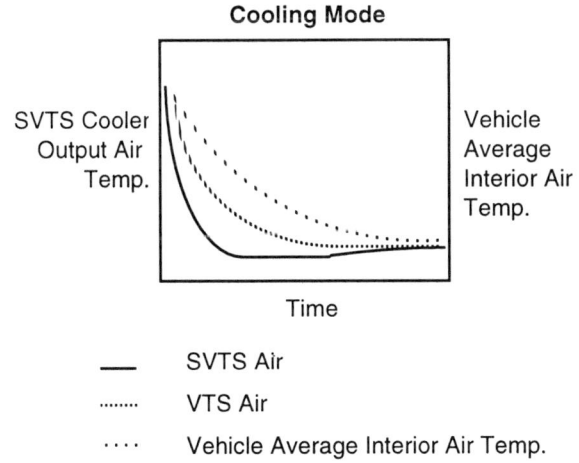

Fig. 3

Since the SVTS gets cooler faster in cooling mode, the occupant perceives the cooling effect sooner, especially with a leather covered seat, hence, the occupant is satisfied faster with the SVTS than with the VTS. The SVTS provides more immediate gratification, and relieves the occupant from a hot seat sooner than the VTS. Since the SVTS is capable of greater air ΔTs than the VTS, it provides much more effective occupant cooling at higher vehicle interior air temperatures as well.

SVTS power consumption is easier and less expensive to control. The reason for this is primarily that the Stirling machine is so much more efficient. Because the SVTS is so much more efficient, it takes less input power for a given unit of cooling power output, therefore for a given cooling power level, the input power is much less. Since SVTS input power is approximately 16-17.0% of VTS input power for the same net cooling power, (and the SVTS can produce a much greater ΔT simultaneously), the SVTS control system can be much smaller, lighter, and less expensive.

SVTS power is controlled by varying the amplitude of a sine wave, or by varying the duty cycle of a pulse width modulator operating at a harmonic of the piston oscillation frequency, driving the linear electric motor that oscillates the piston back and forth. Stirling machine piston frequency is constant, only the amplitude is varied to control cooling power. At low power levels, the amplitude of the input power pulse is small, and at modest ΔTs, the COP, (Coefficient of Performance), will rise above the already high rated COP of 2.5-3.0.

There are a number of ways of controlling SVTS cooling power:

1- Variable air flow to the seat.

2- Variable seat air temperature, as a function of the variable amplitude piston driver setting.

3- Both variable air flow and temperature.

In cooling mode, a thermistor or thermocouple is used to sense the temperature of the cold end of the Stirling machine, or the temperature of the air downstream of the cold end main heat exchanger. This may be used in a closed loop to control piston driver amplitude as a function of a preselected temperature. An open loop may also be used, allowing the occupant to adjust the amplitude of the piston directly. In most instances, practically speaking, a closed loop with temperature select is more appropriate for general use, however, it is also possible to design the controller for both capabilities.

With a closed loop control, as the air in the interior of the vehicle cools down from the use of central AC and/or as a result of opening the windows and using the IP blower to bring outside ambient air into the interior of the vehicle, SVTS heat pump output air temperature will not drop below the selected temperature, because, as the thermal sensor reads a drop in cold end temperature or output air temperature, the amplitude of the piston driver is automatically adjusted to maintain the preselected temperature.

With an open loop control, under the above conditions, output air temperature will drop somewhat proportionally to any drop in ambient air temperature, so the occupant must adjust the piston driver amplitude manually to maintain the same seat air temperature/cooling power.

In heating mode, the PTC resistance heating element automatically maintains a maximum preselected design point temperature, given a constant input voltage to the PTC heater.

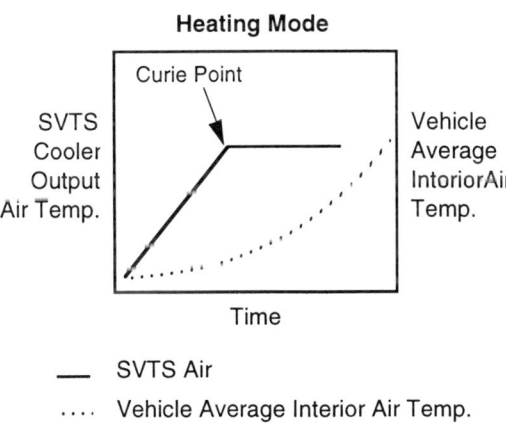

Fig. 4

Figure 4 shows how relatively fast the SVTS heats up in heating mode versus the interior of a vehicle being heated with engine heat. A comparison with VTS heating mode is not shown because both are very similar in heating mode, except for control simplicity provided by the PTC heater in the SVTS. The SVTS is not quite as efficient in heating mode as the VTS, however, as mentioned at the beginning of this section on Concerns No. 1 & 2, heating mode efficiency is far less significant than cooling mode efficiency because of the availability of engine heat for interior space heating purposes, and the small loss in heating efficiency, (~15-25%), is a very small price to pay for such a significant gain in cooling efficiency, (up to 600%).

The SVTS does offer the enhanced simplicity of PTC heater control in heating mode. This means that it is virtually impossible for the heating element to overheat as long as input voltage is within specification. The design maximum air temperature will never be exceeded over the entire range of seat air flow, which is a function of main blower speed, and which is used to regulate net heating power in heating mode, and/or as a result of interior air temperature rise from the use of engine heat, or changes in ambient conditions.

In heating mode, VTS output air temperature increases as main blower air flow is reduced, and/or as interior air temperature rises from the use of engine heat, necessitating control of electrical power, manually or automatically, to the thermoelectric VTS in order to limit maximum air temperature over the entire range of seat air flow.

Figure 5 is an end view of Figures 2 and 6, showing the PTC heater attached to the cold end, which is surrounded by the main heat exchanger radial fins.

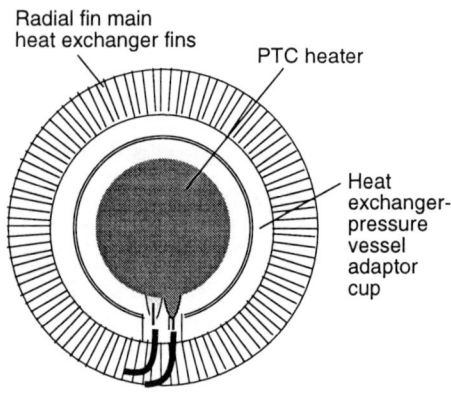

Fig. 5

Figs. 5 & 6 also show views of the main heat exchanger-to-pressure vessel adaptor cup. This cup fits over the cold end of the Stirling machine and serves as the baseplate for the main heat exchanger radial fins, the PTC heater, and the PTC auxiliary heat exchanger. The adaptor cup is threaded in order to enable relatively quick and easy assembly and disassembly of the PTC heater assembly with thermal grease, preferably of the silver bearing type. Very importantly, the adaptor cup allows the cold end to absorb heat from process air from the PTC auxiliary heat exchanger, via the PTC heater and PTC auxiliary threads and circumferential contact face, in cooling mode, thereby increasing the efficiency of the Stirling machine in cooling mode. This can be seen more clearly in Figure 6.

The thermal conductivity of the PTC heater is only about 2.0 w/cm/°C, however, in addition to providing a balanced area on both sides of the PTC heater for better heating mode operation, the PTC auxiliary heat exchanger increases the heat transfer area of the cold end in cooling mode. In addition to this, the thermal path from the center of the cold end to the center of the PTC auxiliary heat exchanger is shorter and more efficient in both cooling mode and heating mode.

The PTC auxiliary heat exchanger addresses an important thermal efficiency problem which has been an issue with respect to the practical use of Stirling coolers for air conditioning purposes for many years.

Because Stirling coolers are dry machines, and do not use a circulating refrigerant, such as R-134a or R12 for example, but use helium or other gases or combinations of gases as working fluids, the heat exchanger area is relatively small compared to Rankine cycle refrigerators, which can circulate refrigerant through serpentine paths in relatively large heat exchangers that allow for efficient thermal transfer from relatively large amounts of air with little pressure drop and with little ΔT across the heat exchanger as a whole.

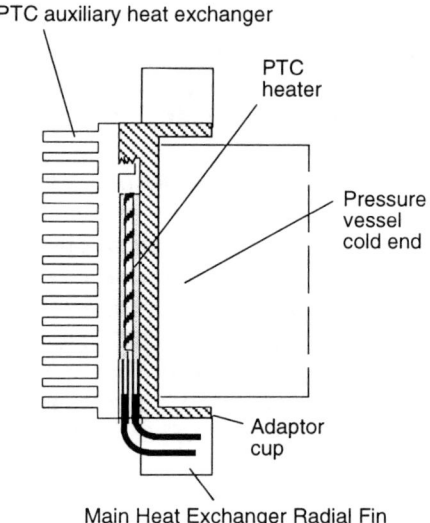

Fig. 6

The Stirling main heat exchanger has to be relatively small because all of the heat absorbed by the machine is being absorbed by a relatively small area at the cold end of the machine. If the main heat exchanger is made too big, it will become inefficient in terms of r-theta, or degrees rise or fall/watt absorbed or rejected.

By extending the area of the main heat

exchanger, and reducing the total thermal path length between the process air and the cold head, Stirling machine efficiency is improved.

It's important to note that the Stirling machine is still much more efficient than Peltier devices in cooling mode, even without the PTC auxiliary heat exchanger, because cooling mode air flow and thermal load, of even a high performance air conditioned seat, lie within the capacity range of a small Stirling cooler with a relatively small radial finned main heat exchanger. The charts at the end of this paper are the result of early prototype tests without the PTC auxiliary heat exchanger. When new SVTS machines are built with the PTC auxiliary heat exchanger, performance and efficiency in both cooling and heating mode will be enhanced.

The PTC auxiliary heat exchanger is preferably of the radial fin or "pin-fin" type, which allow air to flow in circumferentially around the perimeter of the heat exchanger through the fins, or pins, to the main blower. Since seat air first travels through the main heat exchanger, which is annular, it's more efficient to draw the air into the PTC auxiliary exchanger from around it's larger area perimeter because that also requires the smallest change in the direction of air flow that is established through the radial main heat exchanger fins.

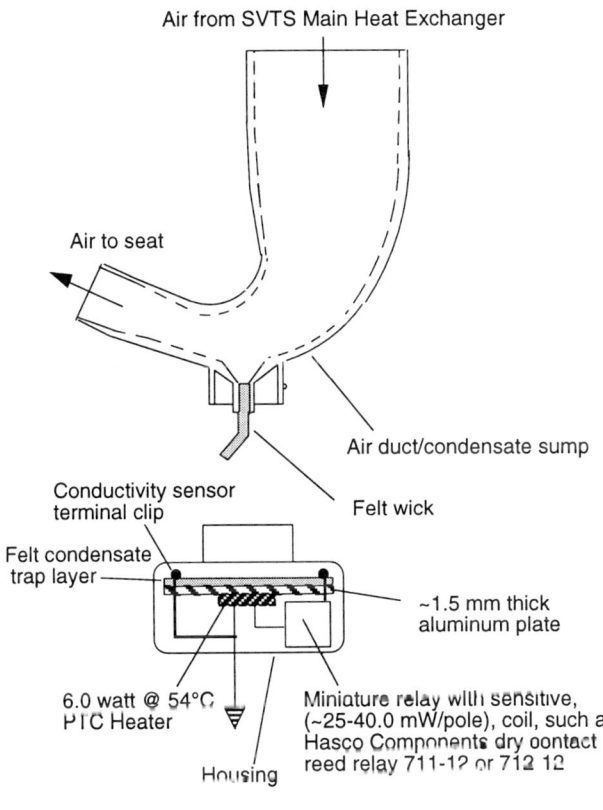

Fig. 7

Figure 7 illustrates the SVTS condensate control, which is crucial to the operation of the SVTS in hot and humid environments. Because the Stirling cooler is so much more efficient, it pumps more heat for a given power input. The Stirling machine is also capable of pumping heat at temperatures considerably further below ambient temperature than the Peltier thermoelectric VTS. Because the SVTS operates at lower temperatures, or greater ΔTs in cooling mode, it is capable of condensing more moisture out of unconditioned ambient air that is being cooled and then blown through the seat air flow pads. Seat air at lower humidity is also more comfortable for the occupant in a humid environment than seat air that has been cooled, but not cooled enough to effect a more desirable level of dehumidification. Drier air is also able to absorb more moisture from the occupant's posterior, resulting in a more comfortable seat initially, especially if the occupant is already perspiring upon entering the vehicle.

Because the Stirling machine is capable of condensing more water out of the seat cooling air in cooling mode, the SVTS requires a condensate management system that has the capacity to handle the worst of circumstances, within the smallest possible space, and with minimal weight, complexity, and cost.

The VTS uses a system of condensate control that involves the use of a wick to move modest levels of condensate under humid conditions to the warm side of the thermoelectric heat pump assembly in cooling mode.

The SVTS condensate control system is different from the VTS control in that it is electrically powered, specifically to re-vaporize condensate, and will only energize, automatically, at a condensation rate requiring active condensate control.

Condensate accumulates in the condensate sump, Figure 7, which also forms the main air duct downstream from the Stirling machine main blower outlet. In this configuration, the SVTS machine is assumed to be mounted a few degrees from vertical within the seat backrest. If the machine is mounted horizontally under the seat, the sump/air duct has to be shaped accordingly.

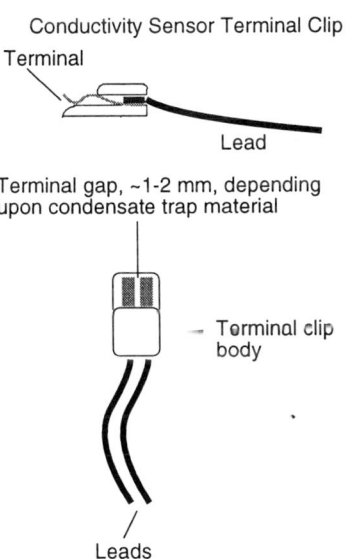

Fig. 8

If enough condensate accumulates to soak the short wick that is embedded within the lower wall of the sump, condensate drips down onto a felt pad that is bonded to a thin aluminum plate. If the felt pad is wetted with enough condensate to make a circuit between the two metal terminals of the conductivity sensor clip, shown in Figure 8, the sensitive relay shown in Figure 7 closes, which energizes a small PTC heating element that heats the aluminum plate, vaporizing the condensate back into the ambient air stream flowing through the vehicle.

If the condensation rate never reaches a level sufficient to fully wet the felt condensate trap, the PTC condensate vaporizer will not energize.

The sensor clips, Figure 8, are simple and easy to assemble by clipping over the edge of the felt condensate trap and aluminum heat spreader plate.

Figure 9 is a plan view of the air duct/condensate sump of Figure 7, looking down into the inlet. The rectangular box located asymmetrically on the left is the snap-on housing for the felt trap and heater section of the condensate control system, with a slotted vent to allow re-vaporized water to escape to ambient.

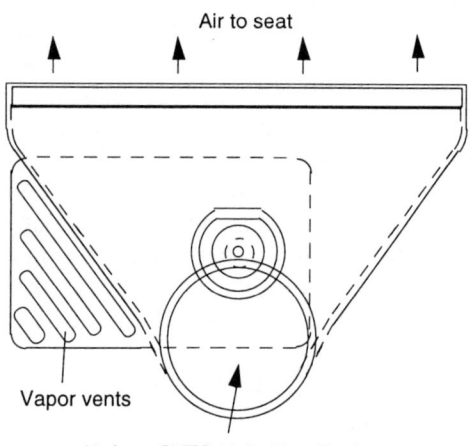

Fig. 9

CONCERN NO. 3: Figure 10 is an end elevation view of TriLock Spacer Fabric, woven by Pittsfield Weaving Co., in Pittsfield, NH. This material may be woven using different fibers and combinations of fibers, including polypropylene, saran, nylon, etc. The structure of TriLock allows air to flow longitudinally through the woven tubes as well as laterally through the woven tube walls, without collapsing when sat upon. The tubes deflect a small amount when bearing a load, giving the SVTS contact surface added resilience.

TriLock is available in different thicknesses, with correspondingly different tube diameters. For Stirling Variable Temperature Seat use, 6-10 mm is best, because it is very flexible as a sheet, conforming readily to the shape of the user. It also allows for a thinner seatrest and backrest assembly, which is important for optimum vehicle interior space efficiency. It is possible to use a thin layer of Trilock for seating because the seatrest and backrest lengths are usually between 330-406 mm. Mattress pads, for example, require a 12-13 mm diameter TriLock pad because the air flow path is considerably longer, between 1.83-2.03 m in length.

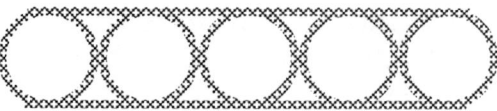

Fig. 10

An excellent 6.35 mm diameter TriLock for the SVTS air flow layer is 9006-007-1, and is made of saran and polyethylene fibers, roughly 70-30. Fiber diameters are .533 mm and .305 mm. Mesh is 8.27/cm with the warp and 8.66/cm across the warp throughout the structure of the material, except for where adjacent tubes interconnect, where the mesh doubles both with and across the warp. TriLock comes off of the weaving machine with the tubes going across the machine, so the warp is perpendicular to the normal axis of the tubes.

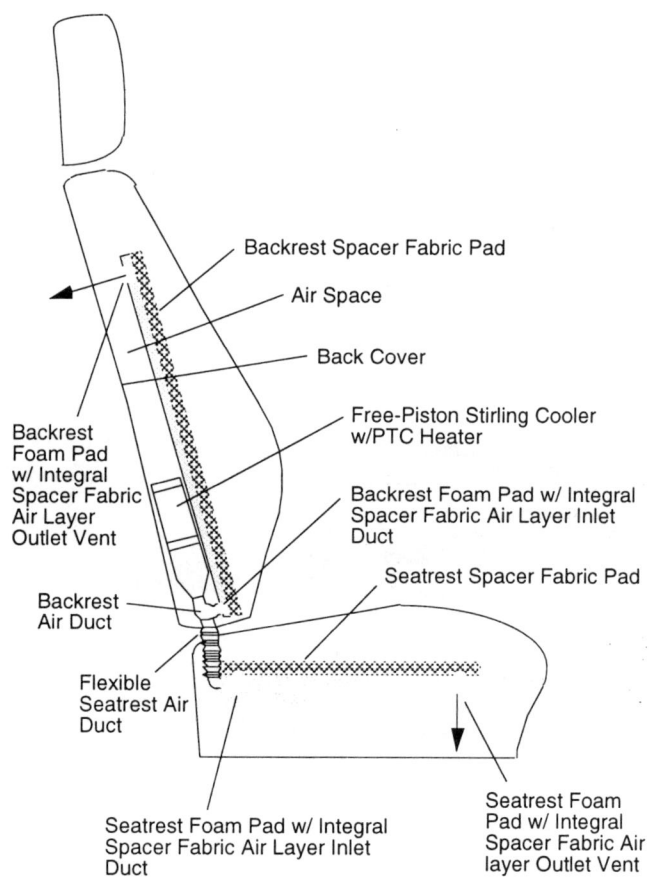

Fig. 11

TABLE 1
COOLING MODE RESISTIVE LOAD TESTS

Tamb. °F/C	Tcold °F/C	dTcold °F/C	Twarm °F/C	dTwarm °F/C	Total dT °F/C	Input W	Lift W	Sens. COP
66.2/19.2	43.3/6.3	22.9/12.6	80.4/26.9	14.2/7.8	37.1/20.5	11.6	34.6	3.00
67.6/19.8	44.4/6.9	23.2/13.0	98.9/37.2	31.3/17.3	54.5/30.5	20.6	50.6	2.46
68.2/20.0	43.9/6.6	24.3/13.4	106./41.1	37.8/20.9	62.1/34.2	31.4	70.5	2.25
104./40.0	47.1/8.4	56.9/31.5	119./48.5	15.3/8.4	72.2/40.8	26.3	53.1	2.02
118./47.9	42.3/5.7	75.9/41.9	140./60.1	21.9/12.0	97.9/54.0	26.8	46.5	1.74
121./49.5	44.9/7.2	76.1/42.0	138./58.9	16.9/9.3	93.1/51.0	20.0	36.5	1.83

TABLE 2
COOLING MODE AIR TESTS

Input Volts	Input Watts	Tamb. °F/C	RH	CFM/m3/hr.	dTcold °F/C	Sens. Lift W	Sens. COP	x Peltier
12.1	19.72	75.4/24.1	64.0%	8.0/13.6	18.5/10.2	45.67	2.32	4.6
12.1	23.84	78.5/25.8	64.0%	8.0/13.6	24.5/13.5	60.50	2.54	6.0

TriLock has been is daily use for over 3 years as of September 1997, and is in excellent condition. It is a unique material with low air flow pressure drop and long-term durability combined with a high degree of suppleness and conformability. TriLock 007 is self-extinguishing, so it exceeds both the FMVSS 302 rating of 10.16 cm/min. and Ford Motor Co. rating of 5.1 cm/min.

Trilock may be glued to the seat foam, and the air inlet ducts leading to and from the TriLock layers on both the seatrest and the backrest may be molded into the foam as shown in Figure 11, which also shows Spacer Fabric air layer outlet vents molded into the front and top ends of the seat support foam. Figure 11 also shows a flexible seatrest air duct with a bellows section that allows the backrest to be positioned within it's range of angular movement with respect to the seatrest.

Table 1 shows some test numbers obtained using an adjustable resistive thermal conduction load on the cold end of the free-piston Stirling machine. It can be appreciated that the machine achieved a maximum cold end ΔT below ambient of 76.1° F/42.0° C at a sensible COP of 1.83, and achieved a COP of 3.0 at a cold end ΔT of 22.9° F, or 12.6° C below ambient.

This means it is possible to obtain seat air at 74-75.0° F/23.3-23.8° C, at a vehicle interior air temperature of 150.0° F/65.6° C.

Table 2 shows test numbers obtained with the radial finned main heat exchanger and convective thermal load, (air), flowing at the indicated rates in cubic feet per minute. Sensible lift and COP simply mean that the condensing rate, under the test conditions, including relative humidity and air ΔT, was not significant enough to be readily considered in calculating total cooling power. It is sufficient at this point to know that, if any condensing rate were to be considered in calculating total cooling power, it would result in a latent COP greater than the already impressive sensible COPs shown in the charts.

Table 2 also shows the difference between SVTS Sensible COP and VTS Sensible COP, (Peltier), under similar conditions of ambient temperature and cold side ΔT. It can be seen that the SVTS is much more efficient than the VTS in cooling mode.

Tests shown in Tables 1 and 2 were done with a closed loop driver controller with a cold end temperature setting of 43-47° F/6.3-8.4° C, measured with a thermistor mounted on the cold end.

Heating mode is not charted because it is relatively straightforward. The COP in heating mode is very close to 1.0. It is not 1.0 exactly because there are some thermal impedances in the heating mode assembly, such as r-theta from the PTC heater to the tips, or crests of the main heat exchanger fins. I2R in the wiring also diverts a minute amount of power that would otherwise be used in raising the temperature of the seat air. However, the COP in heating mode is near 1.0 because the thermal impedances are not very large, and most of them are in the vicinity of seat air flow anyway.

Thermal impedances that reduce COP in heating mode are those which do not contribute to raising the temperature of seat air in heating mode.

CONCLUSIONS

Consequences of the much higher efficiency of the SVTS over the VTS include:

1- Because the SVTS is so much more efficient than the VTS, there is much less concern about additional alternator cost, weight, and energy consumption. Even multiple Stirling Variable Temperature Seats, installed in minivans and luxury cars, for example, will have an almost negligible effect upon fuel economy in conventional vehicles, or on battery range in electric vehicles.

2- Because the SVTS is more efficient than the VTS at substantially greater ΔT's, cooling mode performance is greatly improved over the VTS under high vehicle interior air temperature hot soak conditions. It is now possible to cool leather seats almost as quickly and powerfully as cloth seats.

3- The noise level of the SVTS is lower because it does not need to move as much air over the auxiliary, or rejector, heat exchanger in cooling mode, in order to do an even greater amount of cooling.

4- Because the SVTS is so much more efficient than the VTS in cooling mode, the controls are smaller, lighter, and less expensive, because less power moves much more heat. An SVTS machine with a maximum cooling power rating of 40.0 watts, and with a PTC auxiliary heat exchanger, will require a maximum power input of approximately 13.0 watts, or ~1.1 amp at 12.0 VDC, at maximum power. The main blower need not use more than .6-.7 watt, and the auxiliary fan need not use more than ~3.0 watts. Total wattage: 16.6-16.7. The controller, however, need not control more than about 13.0 watts. The main blower, at a rating of .6-.7 watts, may be controlled with a simple potentiometer, or with a very inexpensive logic-switched resistor network for a more automated control approach.

Other important considerations are:

1- Because the SVTS uses helium as the working fluid, it is as harmless to the environment as the thermoelectric VTS.

2- Manufacturing costs are expected to be approximately equal to optimized thermoelectric systems at similar production levels.

3- From the standpoint of reliability and durability, although Peltier thermoelectric systems are extremely reliable and durable when properly designed and engineered, the basic free-piston Stirling machine, which has been in development for over twenty years, although very sophisticated from a design and engineering standpoint, is mechanically simple and robust.

4- The Spacer Fabric air flow layer material is the major component of the dramatic improvement in orthopedic support and comfort, as well as uniformity and efficiency of thermal transfer, of the SVTS over the VTS. Thin layers of TriLock Spacer Fabric are supple enough to be used with all types of adjustable lumbar supports, including automatic cycling lumbar units, and may also be extended to the seat side bolsters as well. TriLock has also been made with self-extinguishing fibers.

For all practical intents and purposes, there is no difference in environmental compatibility, durability, and reliability between the SVTS and the VTS, however, the SVTS dramatically outperforms the VTS in cooling mode in terms of efficiency, ΔT, and perceived cool-down time, while offering enhanced control simplicity in heating mode with a relatively very small reduction in heating mode efficiency, resulting in a heating mode efficiency that is virtually the same as conventional resistance wire heated seats.

Important advantages of convective, (air), heating over resistance wire heating are:

1- It is virtually impossible for a vehicle occupant to be burned by warm air that is temperature limited by the PTC heating element. Even if the input voltage to the PTC heating element were to exceed specification, the PTC heater can be specified taking potential overvoltage into account because the total convective heating power is spread out over a larger area than with resistance wire seat heaters, which are limited to producing heat at the actual wire grid itself, which is a relatively small area. This means that the resistance wires have to get hotter in order to convey the same effective occupant heating wattage as convective heating can convey at lower temperatures because of the larger effective thermal transfer area percentage of the seat contact surface.

2- Because the total heating power is so relatively evenly spread out over the entire seat contact area, the heating intensity at any given point is lower than than for resistance wire heated seats, resulting in gentle yet effective heating without hot spots while maintaining the same, or even greater, total heating capacity.

Orthopedic support and comfort are significantly improved because the new spacer fabric air flow layer structure is transparent to the user, while preserving high air flow efficiency and very even thermal distribution over the entire seat contact surface.

980916

A Field Study of Distance Perception with Large-Radius Convex Rearview Mirrors

Michael J. Flannagan, Michael Sivak, Shinichi Kojima
and Eric C. Traube
The University of Michigan

Copyright © 1988 Society of Automotive Engineers, Inc.

Abstract

One of the primary reasons that FMVSS 111 currently requires flat rearview mirrors as original equipment on the driver's side of passenger cars is a concern that convex mirrors might reduce safety by causing drivers to overestimate the distances to following vehicles. Several previous studies of the effects of convex rearview mirrors have indicated that they do cause overestimations of distance, but of much lower magnitude than would be expected based on the mirrors' levels of image minification and the resulting visual angles experienced by drivers. Previous studies have investigated mirrors with radiuses of curvature up to 2000 mm. The present empirical study was designed to investigate the effects of mirrors with larger radiuses (up to 8900 mm). Such results are of interest because of the possible use of large radiuses in some aspheric mirror designs, and because of the information they provide about the basic mechanisms by which convex mirrors affect distance perception.

Subjects' distance perceptions for objects seen in large-radius rearview mirrors were measured by magnitude estimation in a static field setting. The results indicate that overestimation of distance continues to decrease as mirror radius increases beyond 2000 mm, and that the overestimation continues to be substantially lower than would be predicted from a model based on image minification and reduction of visual angle. However, even at the longest radius examined in this experiment (8900 mm) the overestimation of distance (8%) is not small enough to be dismissed definitively as trivial. Because various learning effects and changes in driver strategy may compensate for the distortion of distance perception, this does not necessarily mean that convex mirrors of any radius are unsafe. But it suggests that, even for convex mirrors with very long radiuses, the gain in quantity of field of view provided by the convexity comes with a nontrivial cost in quality of field of view, and that the tradeoff between these two characteristics must still be considered in designing optimal mirrors.

Introduction

The purpose of this study was to make empirical measurements of the effects of large-radius (i.e., over 2000 mm) convex rearview mirrors on driver distance perception. The possible perceptual effects of large-radius mirrors have recently become of interest because of new proposals for the design of nonplanar rearview mirrors for use in the exterior positions (driver and passenger side) on cars, light trucks, and vans in the U.S.

Several formal studies have demonstrated that convex rearview mirrors cause drivers to overestimate the distances to vehicles behind them (for a review see Flannagan, Sivak, & Traube, 1997). This effect has raised concern about safety because, to the extent that drivers use convex rearview mirrors to judge gaps in traffic, this misperception of distance could cause drivers to accept smaller gaps than they believe they are accepting in maneuvers such as merges and lane changes. Various forms of driver adaptation, including a strategy of making judgments of speed and distance only by means of flat mirrors (assuming that at least one flat mirror is available), may limit the extent to which this misperception affects actual driving behavior (e.g., Mortimer, 1971).

Previous studies of distance perception have used mirrors with radiuses of curvature ranging from about 300 to 2000 mm. That range covered all the mirror designs that were of major interest in the past. However, recently there has been interest in using mirror surfaces with radiuses of curvature over 2000 mm. Such gentle curvatures are not likely to be used for the entire surfaces of single-radius, spherically convex mirrors, but they might be used for portions of the relatively complex surfaces of aspheric mirrors, especially those designed for use in the U.S.

Aspheric rearview mirrors typically combine an inboard section that is relatively gently curved with an outboard section that is progressively more strongly curved (e.g., Pilhall, 1981). They are thereby able to provide a large total field of view, as well as a relatively undistorted view of more distant traffic (which will be imaged in the inboard, longer-radius portion). The images of vehicles seen in the outboard, shorter-radius section will be strongly distorted, but it can be argued that those vehicles will always be close enough to the observer's vehicle that merely detecting their presence is enough to indicate that a maneuver such as a lane change is not safe. According to this argument, quantitative judgments about their speeds and distances are moot.

The possibility of using surfaces with radiuses of curvature longer than 2000 mm is much more of an issue in the U.S. than in certain other parts of the world, including Europe and Japan. In Europe and Japan, single-radius spherical convex mirrors have been used extensively for many years. These mirrors have typically had radiuses shorter than 2000 mm, partly because the field of view of a single-radius mirror will not be substantially larger than that of a flat mirror if the radius is much longer than 2000 mm. When aspheric mirrors were introduced in Europe, even the most gently curved portions of the aspheric mirrors had radiuses of about 2000 mm. In contrast, drivers in the U.S. did not have much experience with convex mirrors until about 1982, when Federal Motor Vehicle Safety Standard (FMVSS) 111 was amended to allow greater use of convex rearview mirrors on the passenger sides of cars and light trucks ("Preamble," 1982). Consequently, some people have suggested that if aspheric mirrors are to be widely used in the U.S., the inboard portions should be similar to the flat mirrors that U.S. drivers are currently accustomed to, and which are currently required by U.S. law. Thus, the inboard portions would be either flat or have such long radiuses of curvature that they would be virtually flat.

The question of when a convex mirror becomes "virtually flat" for the purposes of driver visual perception is especially of interest because using a completely flat surface as part of an aspheric mirror introduces difficulties for the manufacturing processes that are conventional for automotive mirrors. Models of driver visual perception do not provide a satisfactory answer. The simplest model, based on the visual angles subtended by the minified images seen in convex mirrors, predicts that even mirrors with very long radiuses would cause distortions of distance information that are not small enough to be considered clearly trivial. According to such a model, a mirror with a radius of 10,000 mm would cause distances to be overestimated by about 13% (Flannagan et al., 1997). However, all previous studies of distance perception in convex mirrors have found that overestimations of distance are much less severe than predicted by the visual-angle model (Flannagan et al., 1997).

Given the current practical interest in mirrors with radiuses above 2000 mm, the lack of empirical evidence above 2000 mm, and the theoretical uncertainty caused by the substantial violations of the visual-angle model, we undertook the present empirical study of mirrors with radiuses over 2000 mm. The conditions selected for the experiment were intended to represent something of a worst case for distance overestimation: subjects were not given time to adapt to the different curvatures and they were not given feedback about the accuracy of their distance estimates. This is consistent with the idea that, in order to be considered virtually flat, mirrors should have no substantial effects on perception even under worst-case conditions.

Method

Subjects

Twelve subjects participated in the experiment. There were six subjects in each of two age groups, a younger group ranging from 18 to 35 with an average age of 25.8, and an older group ranging from 63 to 76 with an average age of 71.7. Each age group had three males and three females. All subjects were active, licensed drivers.

Test site and vehicle positions

The experiment was conducted in a paved parking lot. The surface was new asphalt with no markings. There were three cars involved in the study, positioned as shown in Figure 1. All three cars faced in the same direction, simulating the spatial relationships that might occur for three vehicles traveling in the same direction in two adjacent lanes. The car that the subjects were seated in was stationary throughout the experiment. A second car (referred to as the anchor car) was also always stationary, 20 m in front of the subject's eye position. On each of a series of trials, a third car (referred to as the stimulus car) was positioned at one of four distances behind the subject's car (20, 30, 40, or 50 m from the exterior rearview mirror to the front of the rearward car). The stimulus car was offset 3.7 m to the left of the subject's car (the width of a standard lane).

Mirrors and fields of view

Five different driver-side, exterior rearview mirrors were used. Each was 130 mm wide by 70 mm high. Four of the mirrors were spherical convex with radiuses of curvature 2100, 3300, 5400, and 8900 mm. (The radius values used here are averages of six measurements for each mirror, made in pairs at three different points across the mirrors as specified in the Economic Commission for Europe Regulation No. 46 [ECE, 1988]. The individual measurements for all mirrors met the 15% consistency limits specified for mirrors under 3000 mm in that rule.) The fifth mirror was flat (and thus could be described as having an infinite radius of curvature).

The mirrors were fitted with brackets that allowed them to be changed quickly during the experiment, without changing aim. Each of the five mirrors was aimed by the subjects at the beginning of their individual experimental sessions. They were instructed to aim each mirror so that the side of their own vehicle would be just visible at the inboard edge of the mirror. The horizontal field of view was measured for each subject. Average horizontal fields of view are shown in Table 1, along with a summary of the radiuses of curvature. The rearward stimulus car was always entirely visible in the mirror, even at the nearest position (20 m) with the smallest field of view (which occurred with the flat mirror).

The center and passenger-side rearview mirrors of the subject's car were covered during the experiment.

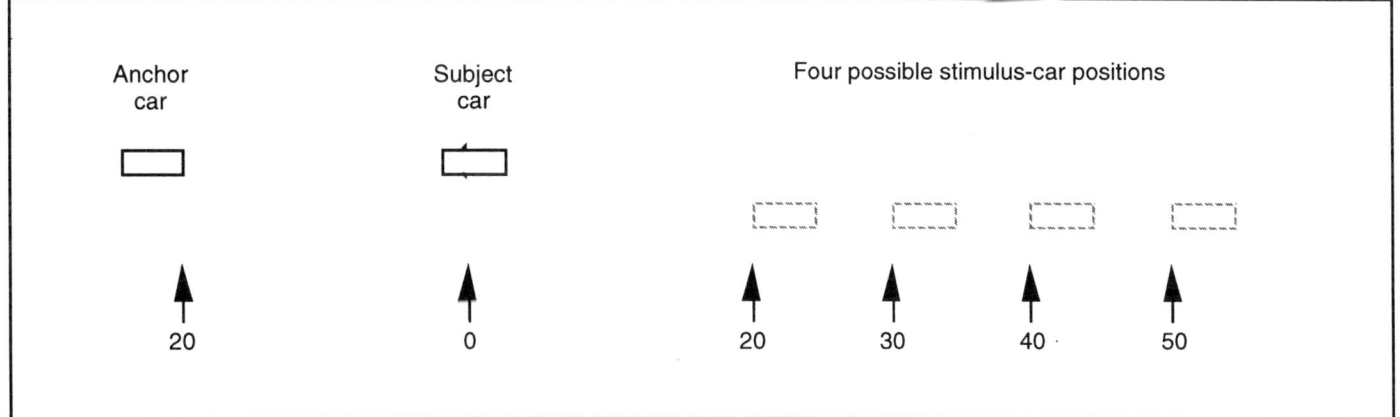

Figure 1. An overhead view of the experimental setup. The fronts of all three vehicles are to the left in this diagram. The anchor car was stationary, 20 m in front of the subject's eye position. On each trial, the rearward stimulus vehicle was positioned at one of four distances from the exterior rearview mirror. The arrows show distances in meters. The rearward stimulus vehicle was offset 3.7 m laterally from the subject's car (the width of a standard lane).

Table 1
Radius of curvature and horizontal field of view for each of the five mirrors.

Radius (mm)	Field of view (degrees)
2100	18.6
3300	17.7
5400	15.2
8900	13.9
∞ (flat)	13.0

Procedure

Each subject participated in a single, individual session that lasted about one hour, including instructions and debriefing. All sessions were conducted in daylight.

One experimenter sat in the rear seat of the subject's car, giving instructions and recording responses. A second experimenter drove the rear stimulus car, moving it among the four rearward positions between trials. A third experimenter stood outside the subject's car, just in front of the driver-side mirror position, and changed mirrors between trials.

At the beginning of each session, the subject sat in the driver's seat of the subject's car and listened to a set of instructions. He or she was informed that the study concerned new rearview mirror designs, but otherwise was not told about the purposes of the study or the nature of the mirrors that would be used.

The anchor car, parked in front of the subject's car, was pointed out, and the subject was told to regard the distance to the rear of that car as 100 units on an otherwise arbitrary scale of distance. The subject was told that he or she would be estimating distances to the rearward car by selecting any positive real numbers that seemed appropriate to represent that distance proportionately, given that distance to the forward vehicle was designated 100. (This is a standard technique in the study of perception, generally referred to as magnitude estimation; see for example Marks, 1974.)

On each of a series of trials, the rearward vehicle was moved into position while the subject looked forward at the reference vehicle, and held up a card that blocked his or her view of the left exterior mirror. When the rearward vehicle was in the proper position, the experimenter in the subject's car asked the subject to lower the card, look toward the left exterior mirror, and view the image of the rearward vehicle. The subject was then to make a numerical estimate of the distance to the rearward vehicle by saying an appropriate number. The subject was not permitted to make a direct look to the rear at any time.

For each subject there was a total of 40 trials, presented in two blocks of 20. Each block consisted of all combinations of the 5 rearview mirrors and the 4 distances to the rear vehicle. Within each block, the order of trials was randomized. Thus, mirrors and distances changed from trial to trial in a way that was unpredictable to the subject.

Results and Discussion

We performed an analysis of variance for distance estimates, using age and sex as between-subjects variables, and mirror type and actual distance to the rear car as within-subject variables. The effects that were statistically significant are discussed below. No interactions were significant.

There was a significant effect of age group, with older subjects giving lower distance estimates than younger subjects, $F(1,8) = 8.98$, $p = .017$. There was a tendency for females to give lower estimates than men, although this effect was not significant, $F(1,8) = 1.24$, $p = .298$. These effects can be seen in Figure 2. The horizontal dashed line in that figure shows where the averaged data points would fall if the subjects were perfectly calibrated to the distance standard provided by the forward anchor car. By that standard, an actual distance of 20 m should correspond to a numerical estimate of 100. The average of the four actual distances presented (20, 30, 40 and 50 m) was 35 m, which should correspond to an estimate of 175.

The statistically significant effect of age and the possible effect of sex on distance estimates are in a sense consistent with the typical relationships of those variables to gap acceptance. Younger drivers and males typically accept smaller gaps for traffic maneuvers, possibly consistent with a greater tendency to takes risks (e.g., Sivak, Soler, & Tränkle, 1989). The present results indicate that older drivers, and possibly females, gave lower estimates of the distance to the rear vehicle, which could give rise to conservative behavior (larger actual sizes of accepted gaps) in real lane changes or merges. However, in this case the subject's task was not to indicate an acceptable gap size, but simply to estimate distance relative to the forward anchor-car distance, independent of any judgments about risk or the possibility of maneuvering in traffic.

The effect of actual distance to the rear car on estimates of distance is shown in Figure 3. This effect was highly significant, $F(3,24) = 59.22$, $p < .0001$ (taking into account the Greenhouse-Geisser correction for degrees of freedom with repeated measures). In this figure, the diagonal dashed line shows where data would be expected to fall assuming perfect calibration with the forward distance standard. The same general tendency for undercalibration that is evident in Figure 2 (and which is mostly attributable to the older subjects) can be seen in Figure 3. It appears that this effect is greater for the longer actual distances.

The effect of primary interest is shown in Figure 4. Mirror type had a significant effect on distance estimates, $F(4,32) = 6.60$, $p = .0006$ (taking into account the Greenhouse-Geisser correction). As expected, the lowest distance estimates were obtained with the flat mirror, and distance estimates were progressively higher with shorter-radius convex mirrors.

In order to obtain an index of the overestimation caused by the convex mirrors we divided the estimates made with the convex mirrors by those made with the flat mirror. (As pointed out above, for most subjects performance with the flat mirror did not show absolute calibration with the forward distance standard. There are several possible reasons why judgments of the distance to a rearward vehicle, seen in a mirror, may not be calibrated with judgments of the distance to a forward vehicle, seen directly. However, it is nevertheless a reasonable assumption that the differences from one mirror to another are valid estimates of the relative effects of convex and plane mirrors in the real world.) The resulting index will be equal to 1.0 when there is no overestimation (i.e., performance with a convex mirror is the same as performance with the flat mirror). This index is shown in Figure 5, along with the levels of overestimation that are predicted by a model based on how the convex mirrors reduce the visual angles of the images of the rearward car (for details, see Flannagan et al., 1997). Note that although the misestimation index shown in Figure 5 is closely related to the averaged distance estimates shown in Figure 4, it cannot be derived directly from those averaged estimates because it is computed for each subject individually.

Distances were overestimated with all of the convex mirrors, although, in agreement with previous results, the degree of overestimation was considerably less than predicted by the visual-angle model. The 2100-mm mirror caused overestimation of 22% (compared to a prediction of 61%), and the 8900-mm mirror caused overestimation of 8% (compared to a prediction of 14%).

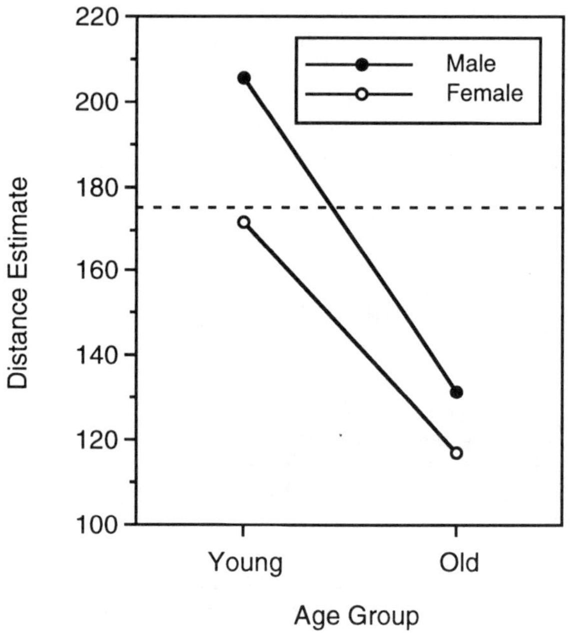

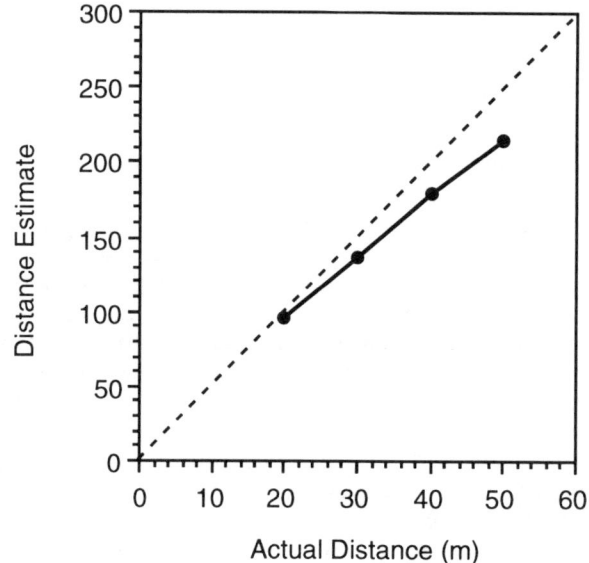

Figure 2. Distance estimates for each combination of age and sex. The horizontal dashed line represents where data would fall if the subjects were perfectly calibrated to the distance standard (by which 20 m should correspond to an estimate of 100).

Figure 3. Distance estimates as a function of actual distance. The diagonal line indicates where data would fall if the subjects were perfectly calibrated to the standard (by which 20 m should correspond to an estimate of 100).

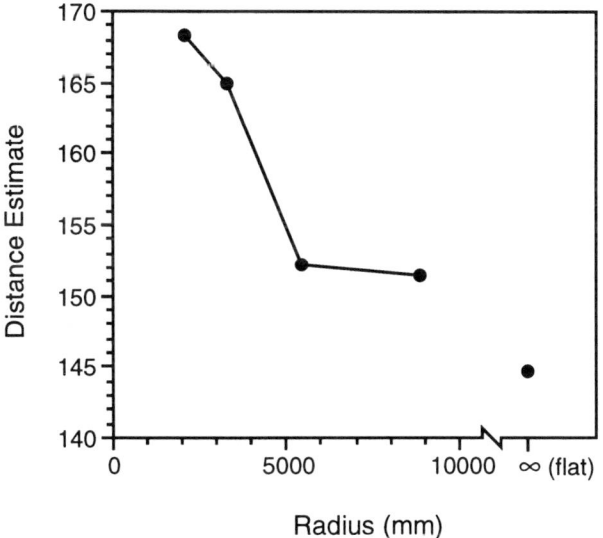

Figure 4. The effect of mirror type on distance estimates.

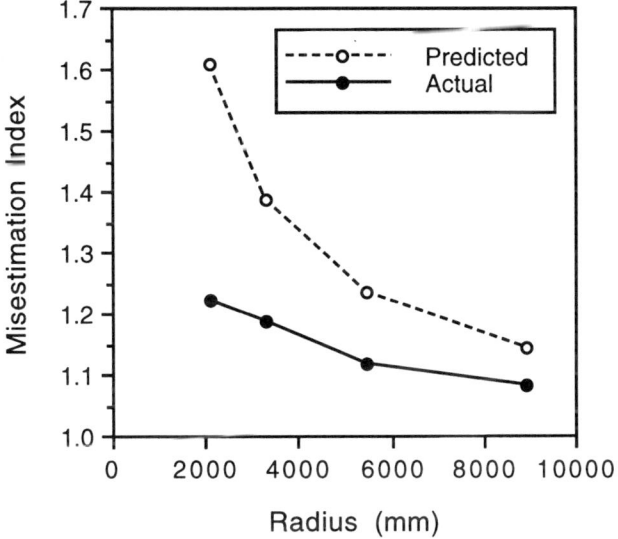

Figure 5. Overestimation of distance (judgments in the convex mirrors relative to judgments in the flat mirror) as a function of radius of curvature. The open symbols show predicted overestimation based on the visual-angle model. The index shown here is closely related to the averaged estimates shown in Figure 4, but cannot be derived directly from them (see text).

Comparison to previous results

Figure 6 shows the data from Figure 5 in comparison to overestimation measurements derived from seven previous studies of the effects of convex mirrors on distance perception (Bowles, 1969; Fisher & Galer, 1984; Flannagan, Sivak, & Traube, 1996; Mortimer, 1971; Smith, Bardales, & Burger, 1978; Sugiura & Kimura, 1978; Walraven & Michon, 1969). Two general types of tasks were used in the previous studies. One type allowed indirect inferences about distance perception by requiring subjects to indicate the last moment at which they would change lanes into the path of an overtaking vehicle seen in a rearview mirror. The other type required subjects to make relatively direct judgments of the distances to vehicles seen in rearview mirrors, but with no connection to a specific driving task. (Note that the method used in the present experiment was of this second type.) The type of task used in each study is designated in Figure 6 by solid lines and filled symbols (the lane-change task), or by dashed lines and open symbols (the distance-judgment task). Figure 6 also shows predicted overestimation based on the visual-angle model (the line with alternating dots and dashes). This is an extended range of the predictions shown in Figure 5. The predictions are tailored to the exact conditions of the present experiment, but they are approximately applicable to the previous studies as well.

The results from the earlier studies that are shown in Figure 6 were selected in some cases to represent, as much as possible, performance prior to any adaptation to the convex mirrors (see Flannagan et al., 1997 for details). This selection might be expected to do several things. First, it should result in maximum measures of overestimation because adaptation should generally cause estimates to become more accurate (Flannagan et al., 1996). Secondly, it should make the previous results reasonably comparable to each other, as well as reasonably comparable to the present data, which were obtained under conditions in which the subjects had little if any opportunity to adapt to the various convex mirrors. Nevertheless, there are large disparities among the previous results, and, as with the current results, all of the previous results are substantially lower than predicted by the visual-angle model (for details see Flannagan et al., 1997). It is not clear how to account for these disparities. One major difference among the studies, the distinction between the two broad classes of methods (a lane-change task or a distance-judgment task) is not very helpful, because, as is shown in Figure 6, studies using the two methods overlap considerably in their results.

Given the large and unexplained range of previous findings, there are perhaps two major points to be made about the current results. First, it is probably a good idea to be cautious about the exact values obtained in any single study. Qualitatively there is a fair bit of consensus in the previous results: most of them indicate that substantial overestimation does occur. But there is little consensus about the exact magnitude of the effect, and there is no satisfactory quantitative model for it. Second, the current results are comfortably within the range that would be obtained by extrapolating the previous results as a group to longer radiuses. Overall, the comparison to previous results indicates that the current results are probably about right, but that there is some uncertainty about their magnitude.

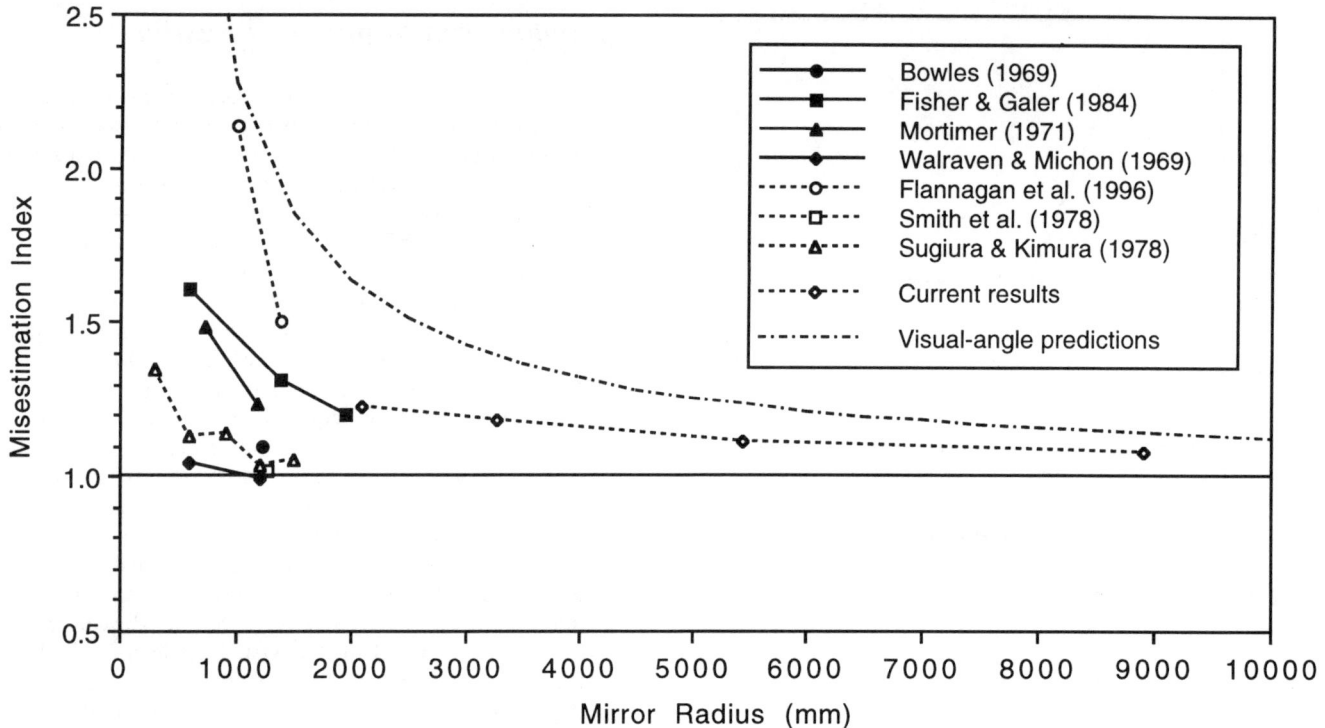

Figure 6. Comparison of the current results to selected data from seven previous studies. In all cases, estimation of distance in convex mirrors is compared to estimation in a flat mirror. The solid lines and filled symbols represent data from studies that used lane-change tasks; the dashed lines and open symbols represent data from studies that used distance-estimation tasks. The line with alternating dots and dashes shows predicted overestimation based on the visual-angle model.

When is a mirror "virtually flat"?

How small would distance distortion have to be to consider a mirror virtually flat with respect to traffic safety concerns? The smallest overestimation observed in this experiment was 8%, which occurred with the 8900-mm mirror. Is this amount, or any amount, of overestimation small enough to establish a consensus among all those concerned with the issue that the effect is small enough to disregard? Any answer to this question must address a number of subtle issues.

To begin with, it is worth noting that to assert that a certain level of perceptual distortion is inconsequential is to take on a heavy burden of proof. Any reasonable uncertainty about whether a certain amount of distortion matters must lead to the conclusion that the mirror in question should not be considered virtually flat. However, even large amounts of perceptual distortion by convex mirrors do not necessarily imply problems with overall safety. People may with experience adapt to minified images, either by recalibrating their judgments (which they had no opportunity to do in this experiment) or by using only flat mirrors to make judgments of speed and distance as suggested by Mortimer (1971). Furthermore, studies of accident data have provided suggestive, although not fully conclusive, evidence that convex mirrors result in better overall safety in comparison to flat mirrors (Luoma, Sivak, & Flannagan, 1995; Schumann, Sivak, & Flannagan, 1998). The finding of a measurable effect of convex mirrors on distance perception in an experiment like the present one should be regarded as necessary but not sufficient to establish that a safety problem exists. If perception is not substantially affected by a certain mirror, it seems reasonable to argue that driver behavior cannot be affected by that mirror. If perception is affected, then behavior may or may not be affected, depending on possible learning effects or changes in driving strategy.

Several standards by which to judge the importance of a given level of overestimation can be suggested. First, one could argue in terms of the geometries of accidents and near misses. If it could be established that all relevant accidents (e.g., side swipes during attempted lane changes) occur because drivers do not see adjacent vehicles at all, and that whenever they do see adjacent vehicles they allow gaps that would be safe even after a certain amount of distance overestimation was factored in, then a relatively clear criterion could be developed for how much overestimation could be tolerated. However, this would require an enormous set of observations of traffic behavior, under a large variety of conditions. To our knowledge, the necessary information is not currently available.

Second, one could rely on expert judgment. However, some would argue that expert judgment should not be regarded as definitive. Furthermore, the expert opinion that is available in perhaps the most well documented form—the seven studies cited in connection with Figure 6—do not offer a clear criterion. Several of these studies (Fisher & Galer, 1984; Mortimer, 1971; Sugiura & Kimura, 1978) have discussions framed primarily in terms of the tradeoff between size of field of view (which is better with smaller radius) and distance distortion (which is better with larger radius). These discussions could be read as implying that any distortion is worth paying attention to, and that the question is not how much distortion matters, but rather how any given level of distortion is balanced by the benefits of a wider field of view. Several of the studies conclude that there is no safety problem with certain mirrors, at least under certain conditions (Mortimer, 1971; Sugiura & Kimura, 1978; Walraven & Michon, 1969), but those conclusions are based on the inability to measure a significant level of distance distortion under the experimental conditions used, rather than on an argument about how important any measurable level of distortion would be.

Third, the amount of distortion could be compared to the normal variability in drivers' judgments of distance. The reasoning here is that judgments of distance must have some natural variability, that drivers may be aware of how bad the variability is, and that they may allow margins of safety to take the known level of variability into account. If a certain level of distortion is very small relative to the natural variability of distance judgments, then it may be reasonable to conclude that it would also be accommodated by the normal margins of safety that drivers use. The data from the present experiment can be used to generate one estimate of normal variability of distance judgments. We did so by calculating, for each subject and at each actual distance, the coefficient of variation (standard deviation divided by the mean) of distance judgments made with the flat mirror. Averaged over subjects and distances, the coefficient of variation is 0.12 (or 12%). By that standard, the 8% overestimation caused by the 8900-mm mirror is not negligible. (Interestingly, the coefficient of variation did not vary substantially with actual distance, meaning that the variability in judgments was always approximately proportional to the distance being judged.)

Fourth, to the extent that there are individual differences among people in how much overestimation is caused by a given mirror, safety problems will always be somewhat worse than implied by the group average. For a safety problem to exist, it is not necessary for all drivers— or even the average driver—to make substantial overestimates of distance; it is only necessary for a few drivers to do so. The misestimation values in Figure 5 are means across subjects, and they are approximately the values that would be expected for a 50th-percentile driver. An experiment of this size cannot reliably measure the overestimation that would be expected for a 95th- or 99th-percentile driver, but it would clearly be at least somewhat higher than the mean values.

In summary, although we cannot conclude that the minimum overestimation observed here (8%) would cause a safety problem, neither can we confidently dismiss the possibility that it might. Given the burden of proof implicit in any claim that a mirror of a certain radius is "virtually flat," it seems clear that the claim cannot be made (at least not without legitimate controversy) for any mirror with a radius in the range studied here (up to 8900 mm). Instead, the same tradeoff between size of field of view and perceptual accuracy that must be considered for convex mirrors below 2000 mm radius must also be considered for mirrors with substantially longer radiuses.

Summary and Conclusions

The present results, together with a number of previous results, indicate that the effects of convex rearview mirrors on distance perception are continuous over a large range of radius of curvature, from well below 2000 mm up to at least 8900 mm. Throughout this range mirrors cause overestimation of distance, although that overestimation is always considerably smaller than predicted by a model of distance perception based on the minified image produced by the mirror and the resulting visual angle experienced by the driver. Whatever mechanisms are responsible for the overestimation of distance, they seem to operate continuously throughout the range of radius examined in this study and in previous studies.

Even at the longest radius examined in this experiment (8900 mm) the overestimation of distance (8%) is not small enough to be dismissed definitively as trivial. Because various learning effects and changes in driver strategy may compensate for the distortion of distance perception, this does not necessarily mean that convex mirrors of any radius are unsafe. But it suggests that, even for convex mirrors with very long radiuses, the gain in quantity of field of view provided by convex mirrors comes with a nontrivial cost in quality of field of view, and that the tradeoff between these two characteristics must still be considered in designing optimal mirrors.

Acknowledgments

We wish to thank Ichikoh Industries, Ltd. for their generous support of this research.

References

Bowles, T. S. (1969). Motorway overtaking with four types of exterior rear view mirror. In *International Symposium on Man-Machine Systems, Volume 2, Transport systems and vehicle control (IEEE Conference Record no. 69C58-MMS)* Institute of Electrical and Electronic Engineers.

ECE (Economic Commission for Europe). (1988). *Uniform provisions concerning the approval of rear-view mirrors, and of motor vehicles with regard to the installation of rear-view mirrors* (ECE Regulation No. 46). Geneva. United Nations.

Fisher, J. A., & Galer, I. A. R. (1984). The effects of decreasing the radius of curvature of convex external rear view mirrors upon drivers' judgements of vehicles approaching in the rearward visual field. *Ergonomics*, 27, 1209-1224.

Flannagan, M. J., Sivak, M., & Traube, E. C. (1996). Driver perceptual adaptation to nonplanar rearview mirrors. In C. Serafin & G. Zobel (Eds.), *Automotive design advancements in human factors: Improving drivers' comfort and performance, SP-1155* (pp. 213-220). Warrendale, Pennsylvania: Society of Automotive Engineers.

Flannagan, M. J., Sivak, M., & Traube, E. C. (1997). Effects of large-radius convex rearview mirrors on driver perception. In J. Jiao & R. Neumann (Eds.), *New concepts in international automotive lighting technology, SP-1249* (pp. 175-181). Warrendale, Pennsylvania: Society of Automotive Engineers.

Luoma, J., Sivak, M., & Flannagan, M. J. (1995). Effects of driver-side mirror type on lane-change accidents. *Ergonomics, 38*, 1973-1978.

Marks, L. E. (1974). *Sensory processes: The new psychophysics.* New York: Academic Press.

Mortimer, R. G. (1971). *The effects of convex exterior mirrors on lane-changing and passing performance of drivers* (SAE Technical Paper Series No. 710543). New York: Society of Automotive Engineers.

Pilhall, S. (1981). *Improved rearward view* (SAE Technical Paper Series No. 810759). Warrendale, Pennsylvania: Society of Automotive Engineers.

Preamble to an amendment to Federal Motor Vehicle Safety Standard No. 111, 47 Fed. Reg. 38698 (1982) (codified at 49 C.F.R. § 571.111).

Schumann, J., Sivak, M., & Flannagan, M. J. (1998). Are driver-side convex mirrors helpful or harmful? *International journal of Vehicle Design, 19*, 29-40.

Sivak, M., Soler, J., & Tränkle, U. (1989). Cross-cultural differences in driver risk-taking. *Accident Analysis & Prevention, 21*, 363-369.

Smith, R. L., Bardales, M. C., & Burger, W. J. (1978). *Perceived importance of zones surrounding a vehicle and learning to use a convex mirror effectively* (DOT HS 803 713). Washington, D.C.: Department of Transportation, National Highway Traffic Safety Administration.

Sugiura, S., & Kimura, K. (1978). *Outside rearview mirror requirements for passenger cars: Curvature, size, and location* (SAE Technical Paper Series No. 780339). Warrendale, Pennsylvania: Society of Automotive Engineers.

Walraven, P. L., & Michon, J. A. (1969). *The influence of some side mirror parameters on the decisions of drivers* (SAE Technical Paper Series No. 690270). New York: Society of Automotive Engineers.

980917

Measuring Curvature of Mirrors using Image Analysis

Dorothy J. Helder

Donnelly Corporation

Copyright © 1998 Society of Automotive Engineers, Inc.

ABSTRACT

This paper describes a method for measuring the radius of curvature of mirrors by measuring the size of the reflected image. The image is generated by a video camera, captured using an image grabber board and analyzed using custom software on a standard computer. The system is shown to be capable to higher resolution than the current SAE defined spherometer and capable of determining radius of curvature over a smaller area. The latter is particularly important in using the system for measuring aspheric mirrors.

INTRODUCTION

Mirrors with a curved surfaces are described by defining the radius of curvature (ROC) at points on the mirror surface. The gage and procedure to make this measurement are described in *SAE J1246 Measuring the Radius of Curvature of Convex Mirrors*.[1]

This gage, shown in Figure 1, is called spherometer. It consists of two outer posts and a center vertically moveable post. When placed on a mirror surface, the contact points of the three posts define three points through which a circle can be drawn. The ROC of the mirror surface is defined as the ROC of that circle.

Because ROC is defined by only those three points, it does not include any small variations which occurs between the posts nor any large variation in ROC which occurs over the whole surface. A standard spherometer with a width of 64 mm (2.5 inches) will average ROC over almost half the width of a typical 150 mm outside mirror. For a uniform convex mirror, this is not a problem. However, for a mirror with purposely varying surfaces (aspheric mirrors), the spherometer cannot accurately report ROC.

This paper describes a system which can determine ROC by measuring the size of a reflected image. The Contour Analysis System (CAS) can measure with higher precision over smaller areas and is, therefore, able to provide better information on aspheric mirrors.

THEORY

The system relies on fundamental paraxial optics. For objects close to the axis of a mirror, the size of the image of an object reflected by the mirror (Size_of_Image), the size of the object (Size_of_Object) and the ROC of the mirror are related by the equation:

$$\frac{2}{ROC} = \frac{1}{Size\_of\_Image} - \frac{1}{Size\_of\_Object} \qquad \text{eq (1)}$$

For these same objects, the location of the reflected image, (Mirror_to_Image), the location of the object, (Mirror_to_Object), and the radius of curvature of the mirror (ROC) are related by the equation:

$$\frac{2}{ROC} = \frac{1}{Mirror\_to\_Image} - \frac{1}{Mirror\_to\_Object} \qquad \text{eq (2)}$$

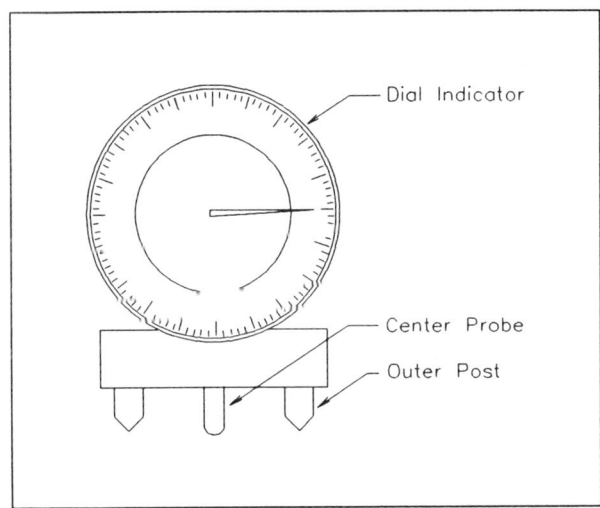

Figure 1. Gage for measuring Convex Mirrors.

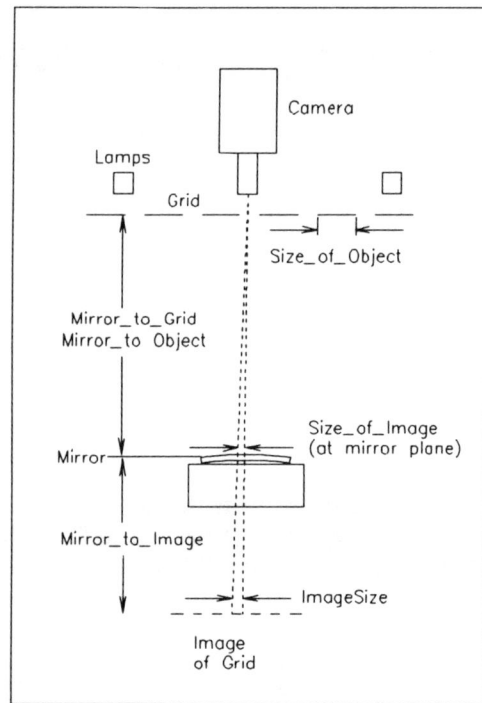

Figure 2. Schematic of VCS System

Equation (1) and (2) are used to derive a formula for ROC based on image size.

A schematic of the CAS unit is shown in Figure 2. The camera is located directly behind the grid and views the image of the grid reflected by the mirror through a hole in the grid. This keeps the object and image as close as possible to the axis of the optical system to allow the use of paraxial optics formula. The object is a grid of dots and the size of the object is the distance between the centers of any two dots. The image size is the distance between two imaged dots. For ease, the system measures this distance in the plane of the mirror. Knowing the distances from the mirror to the grid (Mirror_to_Grid), from the mirror to the camera (Mirror_to_Camera), and the distance between two dots on the grid (Size_of_Object), the ROC can be determined by measuring the distance between the imaged dots in the plane of the mirror (Size_of_Image).

$$ROC = \frac{2 * Mirror\_to\_Grid}{\dfrac{Size\_of\_Object}{Size\_of\_Image} - \dfrac{Mirror\_to\_Grid}{Mirror\_to\_Camera} - 1} \qquad eq\ (3)$$

Although theoretically possible to use this equation to determine ROC, actual execution is difficult and highly subject to error. To get accurate results requires that all distances be precisely known and maintained and that the grid be extremely accurate and be perfectly located. Such a system would not be sufficiently robust for the production floor. Therefore, instead of using absolute readings, measured values were normalized using two standard mirrors: a perfectly flat mirror and a convex mirror of known ROC. Each of the standards was measured and the resulting image sizes used to determine the ROC of the sample mirror.

$$ROC = \frac{C}{\dfrac{Size\_of\_Flat\_Image}{Size\_of\_Sample\_Image} - 1} \qquad eq(4)$$

where:

$$C = ROC(standard) * \left(\frac{Size\_of\_Flat\_Image}{Size\_of\_Standard\_Image} - 1 \right)$$

By normalizing to known standards, errors in distance and alignment were eliminated as issues. Only the error in the standards themselves and those in determining the image size needed to be considered.

NON-PARAXIAL ISSUES

The use of paraxial optics can give results to within 1% of true value only for points within about three degrees of the system axis. The error is caused by the approximations used to develop the paraxial equations and by the fact that the video system is ascribing two dimensional locations to three dimensional points.

If the surface contour of the mirror to be measured is known, the error may be easily corrected. The two dimensional points are ascribed to the proper three dimensional location using the equation of the surface and ROC values determined from iterative ray tracing. Because of the symmetry of spherical convex mirrors, the equations for this are fairly simple and have given good results. If the surface equation is not known, a surface can be approximated point by point over the mirror surface.

Correction can also be obtained by normalizing the distances between the dots using the sample mirror to distances measured using a mirror with known ROC. The standard mirror used for calibration must have the ROC mapped over its entire surface. At calibration, a matrix or table is created within the computer, listing the inter-dot distance corresponding to the known ROC for each area of the mirror. These values are used to calculate ROC for the corresponding areas of the sample mirror.

PROTOTYPE UNIT

DESIGN CONSIDERATIONS

Precision was a primary design consideration. Precision in determining ROC is primarily a function of the precision with which the size of the image can be measured and this in turn is a function of the size of the image on the screen and the distance between the grid and the mirror.

The first effect is obvious. Precision is higher when the image covers a larger area of the video screen. An image which is 100 pixels wide can be resolved to 1 part in 100 whereas an image 30 pixels wide can only be resolved to 1 part in 30. The required width can be obtained by selecting a camera or multiple cameras with sufficient resolution and by choosing an appropriate distance between the dots on the grid.

The second effect on precision, that of distance between the grid and the mirror, is not as obvious. This effect is

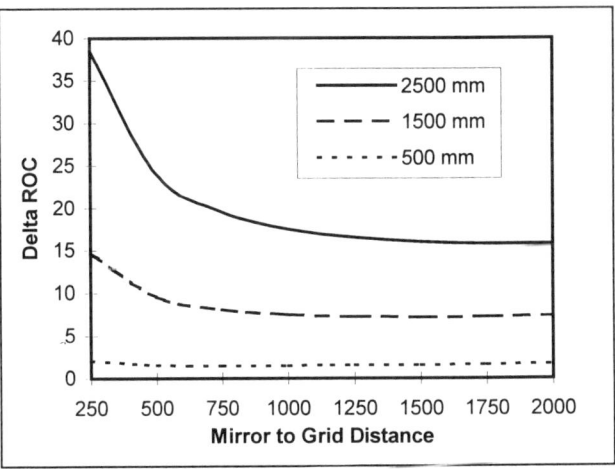

Figure 3 The smallest detectable ROC as a function of the mirror-to-grid distance. The three lines give results for three ROC's.

due to the fact that the ROC does not have as much an effect on image size when the object is closer to the mirror as when it is farther away. The overall effect can be seen in Figure 3 which shows the smallest change in ROC that can be detected as a function of mirror-to-grid distance for a unit with a specific image size and image resolution. At infinitely small mirror-to-grid distances, all three curves go to infinity and it becomes impossible to resolve any differences in ROC. At 250 mm, the system could resolve ROC of 2500 mm to within about ±40 mm. At 750 mm, the ROC can be resolved to within about ±20 mm. Overall, large mirror-to-grid distances give better resolution or precision. However, the marginal improvement diminishes as distance increases.

DESIGN PARAMETERS

The CAS system was designed to be of reasonable cost, capable of duplicating measurements made by the J1246 spherometer, able to measure variable ROC mirrors, and small enough to be moved within the plant. Photographs of the development unit are shown in Figures 4 and 5. The overall unit, shown in Figure 4, consists of a mirror platform, a grid located above the platform and a video camera above the grid. The latter two are hidden by the cover at the top of the unit. They can be more clearly seen in Figure 5. The camera is connected to a standard Pentium computer.

Figure 4. Photograph of the CAS Unit. The mirror is on the stand about 1/3 of the way up. The grid and camera are hidden from view by the black cover at the top.

A video camera and lens were selected which could fill the 640 X 480 pixel video screen with a 150 X 200 mm (6 X 8 inch) convex mirror. The image analysis software allowed for subpixel resolution of 0.1 pixel. The grid was designed to give an average image size of 15 mm at the mirror surface or 45 pixels at the screen.

A graph similar to that in Figure 3 was developed using the image size and resolution requirements and used to select a mirror-to-grid spacing of 850 mm. This distance provided good resolution over all ROC, a workable scale for the grid dimensions and an overall system size that

Figure 5. The camera is located directly above the grid. This unit is within the black enclosure at the top of the unit shown in figure 4.

Table 1. Range of Repeated Measures (mm)			
	Range Containing 99% of Readings		Theoretical Max Range
ROC (mm)	Range	% of ROC	
2109	±15.8	±0.75	±17.2
1879	±12.1	±0.64	±15.3
1643	±13.3	±0.80	±13.6
1249	±9.7	±0.78	±11.1
1110	±9.2	±0.82	±10.2
901	±4.2	±0.47	±9.2
843	±3.9	±0.46	±8.9
794	±3.8	±0.48	±8.6

could be easily moved. The dot spacing for the grid was set at 50 mm along both the x and y directions.

ACCURACY AND REPEATABILITY

The final system used the same grid dot spacing for all measurements. This gave an image size in the plane of the mirror of 9 to 19 mm over an ROC range of 250 to 2500 mm and was able to resolve ROC ±1.7 to ±15.8 mm over the same range.

Table 1 gives the results for repeated measurements made on mirrors within an ROC range of 250 to 2500 mm. The second column of the table gives the ROC range that would statistically contain 99% of the measurements taken. The third column gives the error in terms of percent ROC. Current regulations have ROC tolerance ranges of ±12.5% and ±15.0%. The unit is certainly capable of those measurements. The last column gives the equivalent theoretical ranges based on error analysis. The unit does as expected at higher ROC and much better than expected at lower ROC.

The accuracy and linearity of the system was checked by measuring a series of ROC standards. The results are shown in Table 2. The readings are the average of 20 readings of each standard. The averages are all within the resolution range and show that the system is linear over the range tested.

Table 2. Measurement of Known Mirror Standards All readings in mm		
Known ROC Value	Measured ROC Value	Difference
1295	1294	-1
1168	1168	0
709	710	+1
508	506	-2

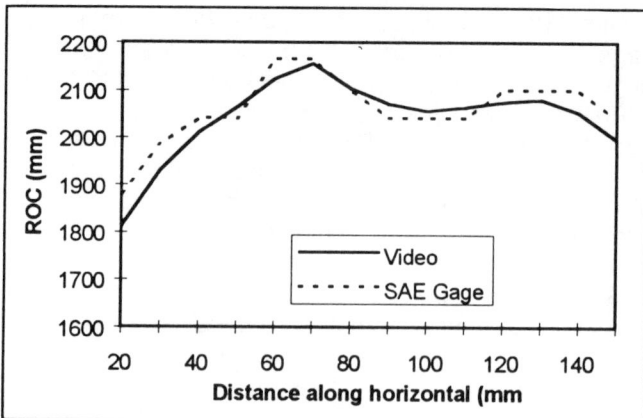

Figure 6. Comparison of ROC readings using the CAS and the spherometer described by SAE..

COMPARISON OF SPHEROMETER AND CAS

The CAS video system was compared to the Spherometer specified in SAE J1246 by comparing measurements taken of a 2000 mm convex mirror. Measurements were taken at 10 mm intervals across a horizontal line through the vertical center of the mirror. ROC was attributed to the location of the center post on the spherometer. ROC was extrapolated to the same point using data from the CAS. The width of the spherometer was 38 mm. The width of the area measured by the CAS varied from 9 to 19 mm depending on ROC.

The results are shown in Figure 6. The two units give equivalent results. The primary difference in the graph is that the curve for the CAS is smoother than that for the spherometer. This is because at this ROC, the spherometer is only able to resolve ±56 mm, whereas the CAS video system can resolve ±12 mm.

It is particularly worth noting that it took about 15 minutes to gather and analyze the date from the Spherometer. The CAS system was able give data for this line, as well as the entire surface of the mirror, in seconds.

USER INTERFACE

Custom software was designed to present a friendly interface to the user. The main screen of the program is shown in Figure 7. This screen allows the choice of viewing the image directly so the mirror can be aligned or initiating the image analysis routine. The program locates the position of the imaged dots, identifies its neighbors in the x and y direction and then calculates the ROC based on the distance between the dot and its neighbors. The program is initialized using flat and convex standards. Samples can then be placed on the platform, aligned and the image analysis routine initiated.

Figure 7 shows the imaged dot matrix for a 1000 mm uniform convex mirror. The dots are evenly spaced in

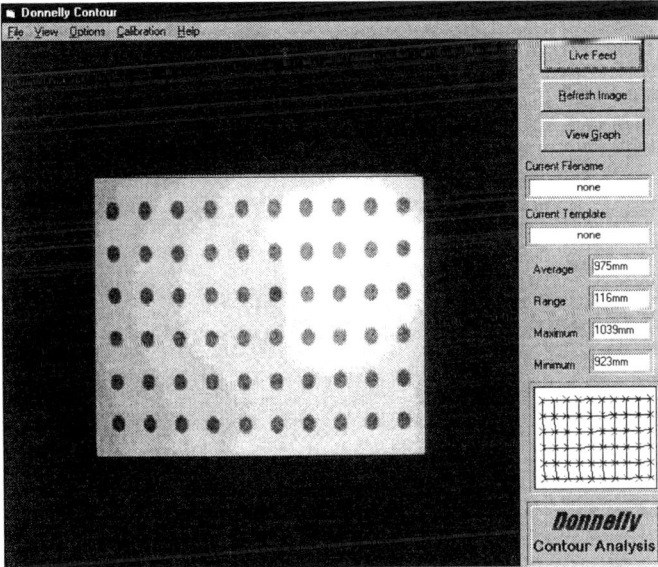

Figure 7. View of the Screen with a Uniform Convex Mirror

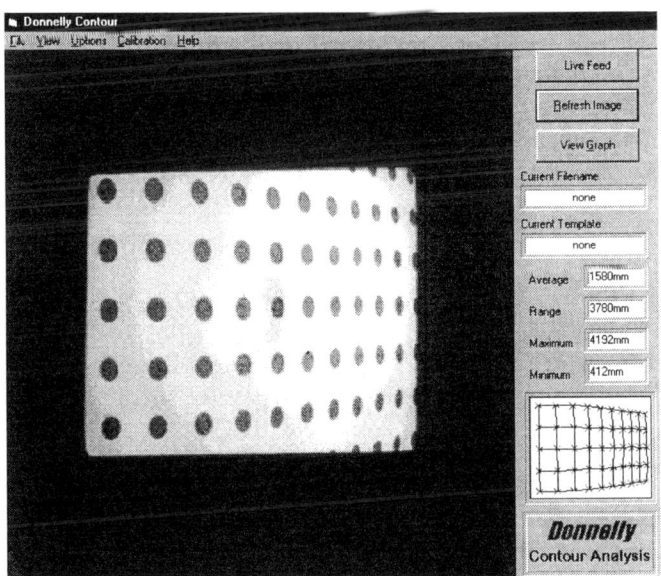

Figure 8. View of the Screen with an aspheric mirror.

spherometer placed along the horizontal and vertical axes respectively. The average and high and low ROC are given in the small boxes on the right.

Figure 8 shows the imaged dot matrix for an aspheric mirror. The changing spacing between the dots shows clearly the changing ROC. As before, vertical and horizontal ROC is calculated between each set of horizontal and vertical dots. The maximum and minimum readings in the boxes to the right show that the mirror ROC ranges from 4192 mm to 412 mm.

The ROC values can be viewed, saved to a file or the user may see a plot of the horizontal, vertical or average ROC at points across the surface. Figure 9 shows such an "iso-ROC" plot for the horizontal ROC of the mirror shown in Figure 8. The ROC along the horizontal axis changes from about 4000 mm at the left to an ROC of about 400mm at the right. The ROC changes more quickly alt the left and more slowly toward the right. The change is the same over the full height of the mirror. A similar plot could be made for the ROC along the vertical axis, average ROC and magnification.

Figure 10 shows an iso ROC plot of a poor quality variable radius mirror. This graph shows uneven changing ROC and an island of higher ROC. This type graph gives production and quality quick feedback on the quality of the mirror.

DISCUSSION

The CAS video system is able to quickly and accurately determine ROC of curved mirror surfaces. Measurement can be make over short distances, making it particularly useful in measuring mirrors with designed-in variable ROC. All results are at least as precise as can be obtained using the Spherometer currently specified by SAE.

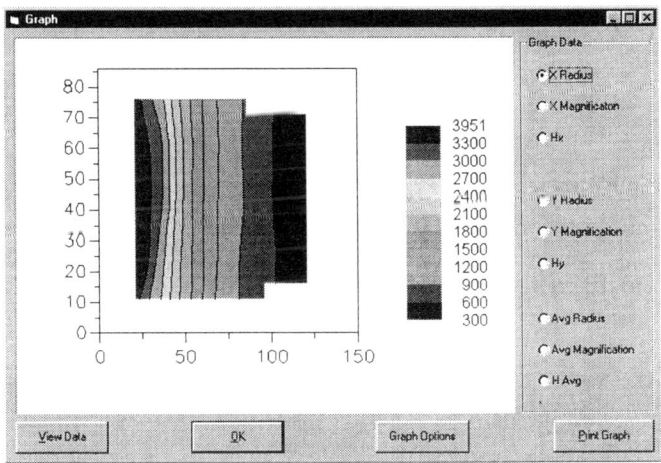

Figure 9. Graph of ROC for the mirror in Figure 7 as measured along the horizontal, (X) axis.

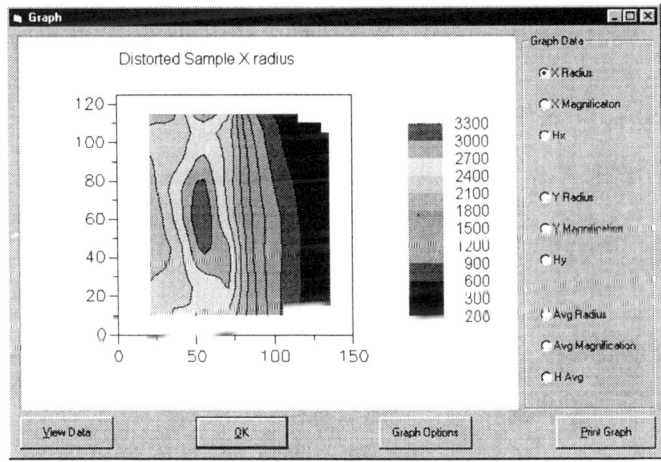

Figure 10. Graph of ROC along the horizontal axis for a distorted mirror.

precise as can be obtained using the Spherometer currently specified by SAE.

The simplified equations are only accurate over a limited area. Outside this area, the user may either normalize readings to known standards or mathematically account for the error.

ACKNOWLEDGEMENTS

The author wishes to acknowledge the efforts of Greg Caskey of Lakeshore Analytical Inc., who designed and built the CAS unit and Nate Oostendorp who developed the CAS software.

REFERENCES

1. *SAE J1246 Measuring the Radius of Curvature of Convex Mirrors.* 1997 SAE Handbook.

980918

Binocular Disparity in Aspherical Mirrors

Stephen M. O'Day
Ford Motor Company

Copyright © 1988 Society of Automotive Engineers, Inc.

ABSTRACT

An aspherical mirror is a convex spherical mirror whose radius of curvature decreases as the line of sight moves horizontally on the mirror from inboard to outboard. This differs from a regular spherical convex mirror which has the same radius of curvature everywhere on the mirror. Aspherical mirrors provide an increased field of view and larger image sizes than would be possible with a traditional spherical convex mirror. One potential concern with aspherical mirrors is binocular image disparity. Binocular image disparity in an aspherical mirror results from the situation where one eye sees an image on a portion of the mirror with a larger radius of curvature than the other eye sees. The difference in image sizes can cause discomfort to the person using the mirror and, if the difference is large enough, the person sees a double image. This paper describes a method for quantifying the binocular image disparity in aspherical mirrors. The method involves measuring images in the mirror in angular degrees with respect to the driver's eye. Binocular Image Disparity is defined as the percent change in angular image size seen by the driver's right eye versus left eye. Four types of targets were used to measure disparity: vertical linear and horizontal linear (one dimensional) targets, area (two dimensional) targets and volume (three dimensional) targets. Three eye to target distances were analyzed: 2000mm, 4000mm, and 6000mm. Four aspherical mirrors were analyzed. This method characterizes aspherical mirrors in terms of the maximum binocular image disparity and the maximum rate of change in binocular image disparity. Binocular image disparity was largely unaffected by target distance. The square area target was recommended over the other target types.

BACKGROUND

Indirect driver vision in the U.S. has mainly been limited to planar mirrors on passenger vehicles and spherical convex mirrors on the passenger side of passenger vehicles and on both sides of larger vehicles. Large indirect fields of view with spherical convex mirrors are possible only with relatively small radii of curvature. This reduces the image size making target detection and depth perception more difficult for the driver. To address this issue of image size versus field of view, aspherical mirrors were developed. An aspherical mirror is a convex spherical mirror whose radius of curvature decreases as the line of sight moves horizontally on the mirror from inboard to outboard. This differs from a regular spherical convex mirror with the same constant radius of curvature everywhere on the mirror. By changing the radius of curvature in this way, an aspherical mirror can provide larger image sizes than would be possible in a spherical convex mirror having the same field of view.

Aspherical mirrors have been in use for a number of years in Europe and in the U.S. in specialty applications (e.g. automobile racing). Their increased field of view and larger image sizes (on inboard portions of the mirror) make them an attractive alternative to large planar mirrors or small radii spherical convex mirrors. However, aspherical mirrors have a deficit not present in planar or spherical convex mirrors, binocular image disparity. In a spherical convex mirror both eyes are looking at portions of the mirror that have the same radius of curvature (since all portions of the mirror have the same radius of curvature), thus both eyes see an image of approximately the same size. With aspherical mirrors each eye looks at portions of the mirror with different radii of curvature. This results in a disparity between the image sizes seen by the driver's right eye versus left eye. This can cause discomfort to the driver using the mirror if the disparity is above a certain threshold and, if the disparity is high enough, can cause the driver to see a double image. However, this paper focuses on the measurement of binocular image disparity.

TARGET PLACEMENT AND MIRROR AIMING PROCEDURE

To measure binocular image disparity a driving scenario was modeled on the computer, as shown in figure 1. Four types of targets were placed in the middle of the adjoining lane at three distances rearward of the driver's eye. It was assumed that the driver's vehicle was in the middle of his/her lane, so that the target was centered 12 feet to the left of the vehicle's center line. The vertical placement of each target was 914.4mm (3 feet) off the ground; this is approximately hood high. Each target was placed at 3 different locations rearward of the driver's middle eye centroid; 2000mm, 4000mm, and 6000mm. For all conditions the distance from the driver's middle eye centroid to the mirror center was held constant at 786.37mm.

Figure 1 - Experimental Layout

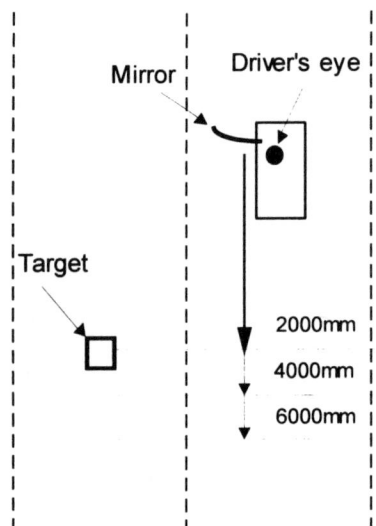

With the target fixed, the mirror was then aimed horizontally at .5 degree intervals such that the image of the target traversed the entire width of the mirror from inboard to outboard. The mirror was aimed vertically so as to keep the target image in the center of the mirror.

Four aspherical mirrors were analyzed. In all cases the mirror size was the same: 190.5mm long by 127mm wide. The mirror was originally designed for a light truck. The distance from middle eye centroid to the mirror center was held constant at 786.37mm.

The four aspherical mirrors that were tested and their characteristics were as follows:

1. E1 mirror based on the Volvo Equations [1] with a 2000mm radius of curvature (ROC) for the spherical portion of the mirror. The spherical portion comprised the inboard 145mm of the mirror. The ROC on the aspherical portion of the mirror changed from 2000mm to 58.5mm. See figure 2.

Figure 2 - E1 Aspherical Mirror

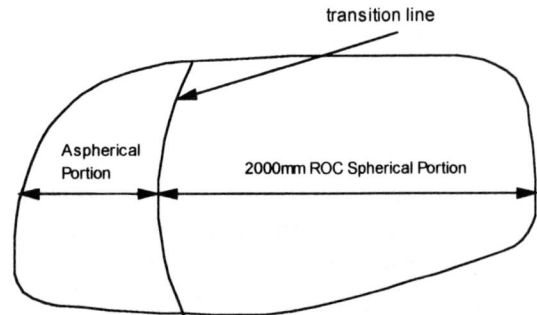

2. E2 mirror based on the Volvo Equations [1] with a 1400mm ROC for the spherical portion of the mirror. The spherical portion comprised the inboard 145mm of the mirror. The ROC on the aspherical portion of the mirror changed from 1400mm to 60.5mm. See figure 3.

Figure 3 - E2 Aspherical Mirror

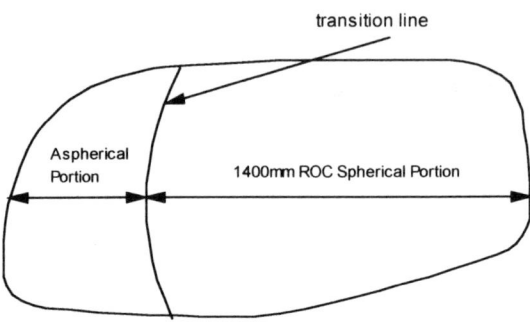

3. M1 experimental mirror with a 2540mm ROC for the spherical portion of the mirror. The spherical portion comprised the inboard 112mm of the mirror. The ROC on the aspherical portion of the mirror changed from 2540mm to 171.7mm. See figure 4.

Figure 4 - M1 Aspherical Mirror

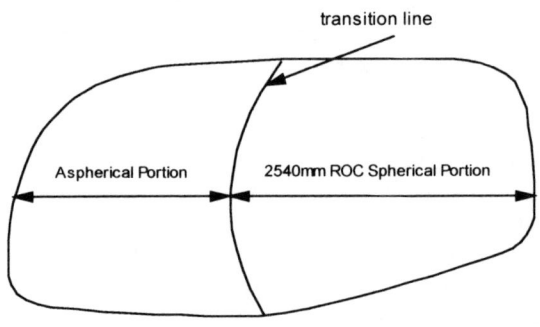

4. M2 experimental mirror with a 3048mm ROC for the spherical portion of the mirror. The spherical portion comprised the inboard 33mm of the mirror. The ROC on the aspherical portion of the mirror changed from 3048mm to 376mm. See figure 5.

Figure 5 - M2 Aspherical Mirror

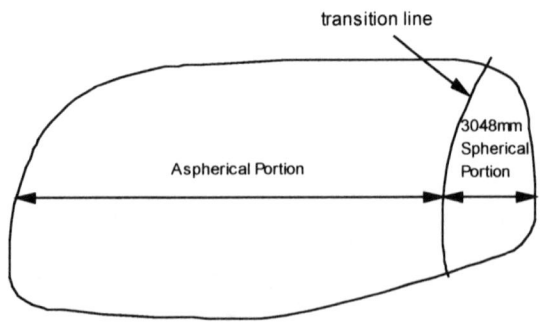

MEASURING BINOCULAR IMAGE DISPARITY

Binocular image disparity was calculated by measuring images in the mirror as an angular displacement relative to the driver's eye. A polar plot [2] uses this principle of showing objects relative to the driver's eye. Figure 9 shows a polar plot of the view through the E1 mirror with a 100mm vertical linear target. An object is shown in this polar plot according to the number of degrees the driver would need to turn his/her eyes right or left and up or down to view the object. Two mirrors are shown since the view of the mirror and target is slightly different when seen through the left eye (solid line) and the right eye (dashed line). The polar plot shows both views as separate images. This can seem confusing since in the real world our brain has learned to combine these two views into one cogent whole. However, it is useful when measuring binocular disparity to keep these two views separate.

Binocular image disparity was defined as follows:

$$\text{Binocular Image Disparity} = \frac{(I^R - I^L)}{I^R} \quad (1)$$

where
  I^R = the image size (in degrees) of the target seen through the driver's right eye,
  I^L = the image size (in degrees) of the target seen through the driver's left eye.

Image disparity measures the percent change between the angular image sizes of the right and left eye. The polar plotting technique provides the angular image sizes required in these calculations.

LINEAR IMAGE DISPARITY. Two linear targets were used, a 100mm vertical and a 100mm horizontal line. In both cases the target was comprised of two points, a and b, connected by a line (see figure 6).

Figure 6 - Vertical linear targets showing point notation

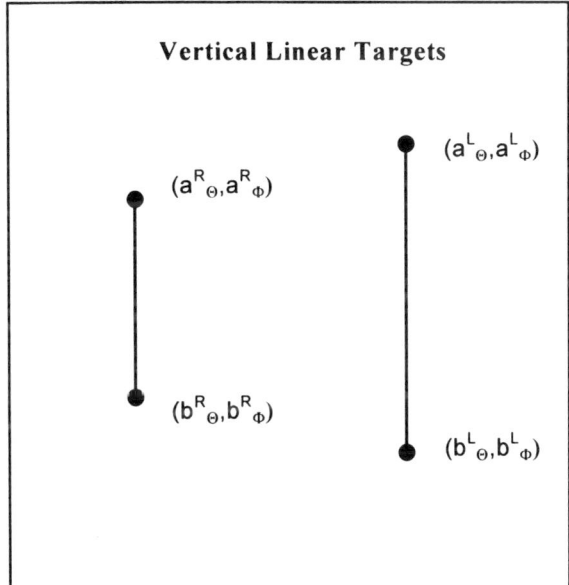

Θ was used to denote the azimuth angle (the horizontal axis on the polar plot) and Φ was used to denote the elevation angle (the vertical axis on the polar plot). Therefore, point a, when seen through the right eye, was defined as (a_Θ^R, a_Φ^R) and (a_Θ^L, a_Φ^L) when seen through the left eye. Point B, when seen through the right eye, was defined as (b_Θ^R, b_Φ^R) and (b_Θ^L, b_Φ^L) when seen through the left eye. Then the binocular image disparity of the linear targets was defined as:

Linear Binocular Image Disparity =

$$\frac{\left(\sqrt{(a_\Theta^R - b_\Theta^R)^2 + (a_\Phi^R - b_\Phi^R)^2} - \sqrt{(a_\Theta^L - b_\Theta^L)^2 + (a_\Phi^L - b_\Phi^L)^2}\right)}{\sqrt{(a_\Theta^R - b_\Theta^R)^2 + (a_\Phi^R - b_\Phi^R)^2}} \quad (2)$$

AREA IMAGE DISPARITY. The area target was a 100mm square comprised of four points, a, b, c, d. The changing radius of curvature of aspherical mirrors distorts the target into the shape of a trapezoid. Therefore, the formula for the area of a trapezoid was used to calculate the area of this target's image:

$$\text{Area of a trapezoid} = \frac{1}{2}(f + g)h \quad (3)$$

Figure 7 - Trapezoid

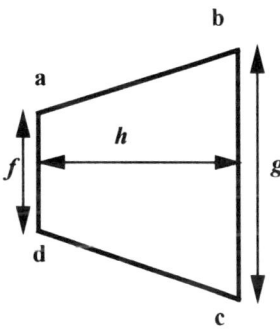

Using the same nomenclature as above, the binocular image disparity of the area target was defined as:

$$\text{Area Binocular Image Disparity} = \frac{(T^R - T^L)}{T^R} \quad (4)$$

where

$$T^R = \frac{1}{2}\left[\sqrt{(a_\Theta^R - d_\Theta^R)^2 + (a_\Phi^R - d_\Phi^R)^2} + \sqrt{(b_\Theta^R - c_\Theta^R)^2 + (b_\Phi^R - c_\Phi^R)^2}\right] * |a_\Theta^R - b_\Theta^R| \quad (5)$$

$$T^L = \frac{1}{2}\left[\sqrt{(a_\Theta^L - d_\Theta^L)^2 + (a_\Phi^L - d_\Phi^L)^2} + \sqrt{(b_\Theta^L - c_\Theta^L)^2 + (b_\Phi^L - c_\Phi^L)^2}\right] * |a_\Theta^L - b_\Theta^L| \quad (6)$$

VOLUME IMAGE DISPARITY. The volume target was a 100mm cube comprised of fourteen points: a, b, c, d defined the front face of the cube, e, f, g, h defined the rear face of the cube, k, l, m, n defined a slice through the middle of the cube, and o, p defined points in the middle of the front face and the rear face of the cube respectively (see figure 8). Due to the changing radius of curvature of aspherical mirrors the target was distorted into the shape of a prismoid. Strictly speaking, the polar plot of the image of the cube in the mirror is two-dimensional. However, the two dimensional image in the polar plot can be treated as a three dimensional image, as can be seen in the formula (7) below. Therefore, the formula for the volume of a prismoid was used to calculate the volume of this target's image:

$$\text{Volume of a Prismoid} = \frac{1}{6} H (T_0 + 4T_1 + T_2) \quad (7)$$

T_0 = the trapezoidal area of (a, b, c, d),
T_1 = the trapezoidal area of (k, l, m, n),
T_2 = the trapezoidal area of (e, f, g, h),
H = the length of (o, p).

Figure 8 - Prismoid

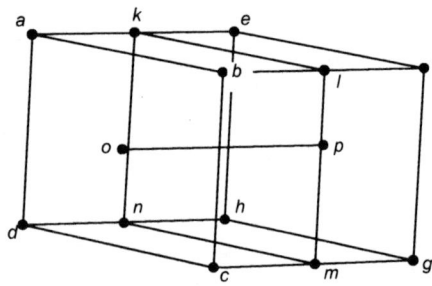

Therefore, the binocular image disparity of the cubic target was defined as:

$$\text{Volume Binocular Image Disparity} = \frac{(P^R - P^L)}{P^R} \quad (8)$$

where
P^R = volume of the prismoid seen through the right eye,
P^L = volume of the prismoid seen through the left eye,

IMAGE DISPARITY RESULTS

The binocular image disparity graphs showed little difference between the three target distances tested. For this reason only the curves for the 4000mm target distance are shown. Figure 10 shows the disparity curves for the vertical linear target at 4000mm. Figure 12 shows the disparity curves for the horizontal linear target at 4000mm. Figure 14 shows the disparity curves for the square area target at 4000mm. Figure 16 shows the disparity curves for the cubic volume target at 4000mm.

The maximum binocular image disparities for all conditions are shown in Table 1. For any given mirror the maximum image disparity changed only slightly when the target distance increased from 2000mm to 4000mm to 6000mm. The general trend was a less than 9% change in maximum binocular image disparity when the target distance increased from 2000mm to 4000mm to 6000mm. The M1 mirror had the greatest maximum binocular image disparity for the vertical linear target, followed by the E1, E2, and M2 mirrors. For all other target types the E1 mirror had the greatest maximum binocular image disparity, followed by the E2, M1, and M2 mirrors.

The overall shape of the disparity curve for each mirror remained the same regardless of target distance. For each mirror the disparity curve seemed very similar across all the targets except for the vertical linear target. The vertical linear target measured the disparity caused by the vertical aspherical properties of the mirror only. All other targets measured the disparity due to either the horizontal aspherical properties or both the horizontal and vertical aspherical properties of the mirror. This made the vertical linear target unique which is reflected in the unique shapes of its binocular image disparity curves (figure 10).

A comparison of the shape of the binocular image disparity curves between the four mirrors showed distinct differences. The disparity curve for the E1 and the E2 mirrors were very similar with a sharp peak at the transition line (point "T" in figures 10, 12, 14, 16) that quickly dropped off. The disparity curve for the M1 mirror increased more slowly to a peak at the transition and then slowly dropped off. For the M2 mirror the curve gradually rose to a maximum near the middle of the mirror and then slowly dropped off.

Table 2 shows the average binocular image disparity once both the right and left eyes were on the aspherical portion of the mirror. These values were calculated by averaging all points on the binocular image disparity curve after and including the "B" point (corresponding to when both right eye and left eye images were fully on the aspheric portion of the mirror), see figures 10, 12, 14, 16. These numbers follow the same trend shown in the maximum binocular image disparity results (table 1).

Table 3 shows the average image disparity as the eyes move across the transition line. These values were calculated by averaging all points on the binocular image disparity curve between the "L" point (corresponding to when the image seen through the left eye is fully on the aspheric portion of the mirror) and the "B" point, inclusive (see figures 10, 12, 14, 16). These values show a general trend toward smaller image disparities at greater target distances.

Table 4 shows the maximum slope angle of the binocular image disparity curve for each mirror in each condition. Again, the E1 and E2 mirrors had the steepest angle, followed by the M1 mirror, and the M2 mirror had the lowest angle.

CONCLUSIONS

Target distance to eye point seemed to have little effect on the binocular image disparity measured, both maximum and averages, as well as, the shape of the disparity curve.

The target type did make a difference in the binocular image disparity measured. The largest disparity was seen with the cubic target, followed by the square target, the horizontal linear target, and the vertical linear target showed the smallest disparity. This effect is, at least partially, due to the fact that the cubic and square image sizes are affected by both the vertical and horizontal aspherical properties of the mirrors.

Of the four target types tested the square target is recommended for measuring binocular image distortion. The vertical linear target only allows the vertical aspherical characteristics of a mirror to be measured. Many aspherical mirrors have greater changes in radius of curvature in the horizontal dimension, therefore, this target is of limited use. The horizontal linear target does not allow the vertical aspherical characteristics of the mirror (meager though they may be) to be included in the disparity measurements. The square target allows for the measurement of the disparity caused by both the vertical and horizontal aspherical characteristics of the mirror. The cubic target allows for this also, though the increased complexity of calculating its binocular image disparity over the square target did not seem warranted.

Measuring binocular image disparity using driver's perspective polar plots, a fixed target and varying the mirror aim provides a method to quantify the binocular image disparity of aspherical mirrors. Three important measurables come out of this method:

1. the maximum binocular image disparity quantifies the maximum difference between the image sizes seen by the right versus left eye, as shown in Table 1.
2. the maximum slope angle of the binocular image disparity curve quantifies the rate at which image sizes change, as shown in Table 4.
3. the shape of the binocular image disparity curve provides a graphic depiction of the way binocular image disparity changes across the face of the mirror.

FUTURE RESEARCH

This paper has suggested a way of measuring the binocular image disparity of aspherical mirrors. It remains to be determined how much disparity is tolerable or acceptable to the driver and when the image disparity becomes bothersome. Also, the level of image disparity that causes the driver to see double images needs to be determined. Luminance or image contrast may affect acceptable levels of binocular image disparity. The driver's state of adaptation to aspherical mirror images can also play a significant role in determining acceptable levels of binocular image disparity.

Image disparity has been defined in this paper as the difference in length, area, and volume between right and left eye images. Another way to define disparity could include a measure of the deformity of an image. Perhaps a combination of disparity and deformation could be useful.

ACKNOWLEDGEMENTS

The research reported in this paper was supported by the Human Factors Engineering and Ergonomics Department and the Advance Vehicle Technology Department of Ford Motor Company. Technical advice and support was given by Gary Rupp and Vivek Bhise of HFEE and Mary Beth Angotti of AVT. Significant assistance was rendered by Robert McCord of MultiVex Mirror Company in the development of the computer models for aspherical mirrors. The SAE Driver Vision Committee provided technical discussions on aspherical mirrors and image disparity. Special thanks to Dorothy Helder of Donnelly Corporation for her helpful suggestions.

REFERENCES

1. **Pilhall, S.** "Improved Rearward View", SAE paper No. 810759, June, 1981

2. **McIsaac, E.J.; Bhise, V.D.** "Automotive Field of View Analysis Using Polar Plots", SAE paper No. 950602

Figure 9 - Polar Plot of the binocular view of the E1 mirror with a 100mm vertical linear target 2000mm rearward of eye centroid.

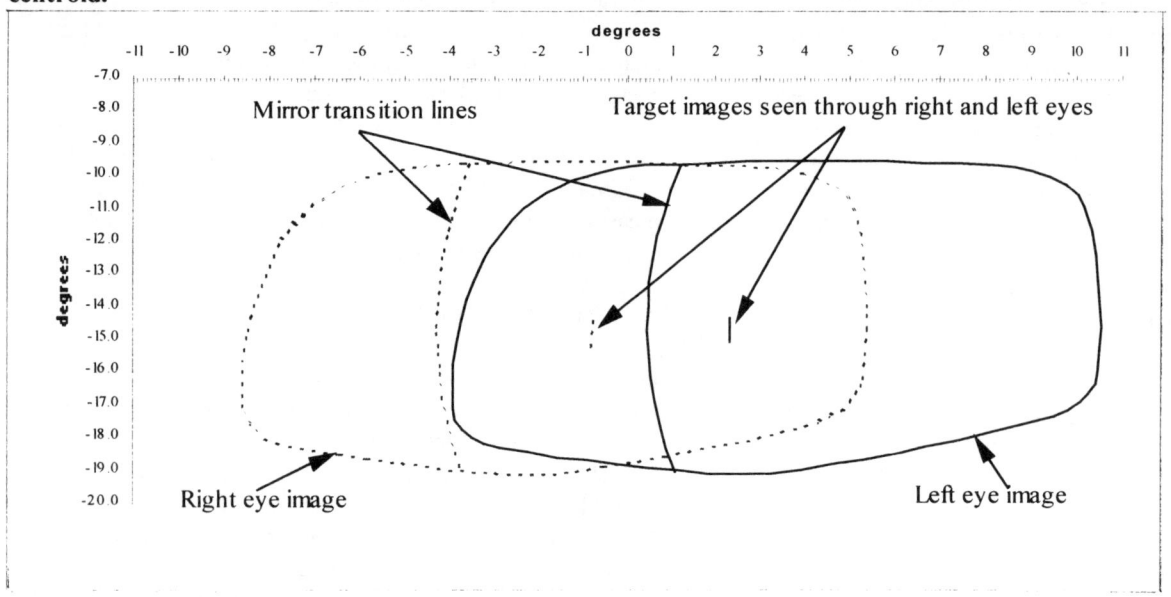

Figure 10 - Binocular Image Disparity of the Vertical Linear Target at 4000mm Rearward of Eye Centroid

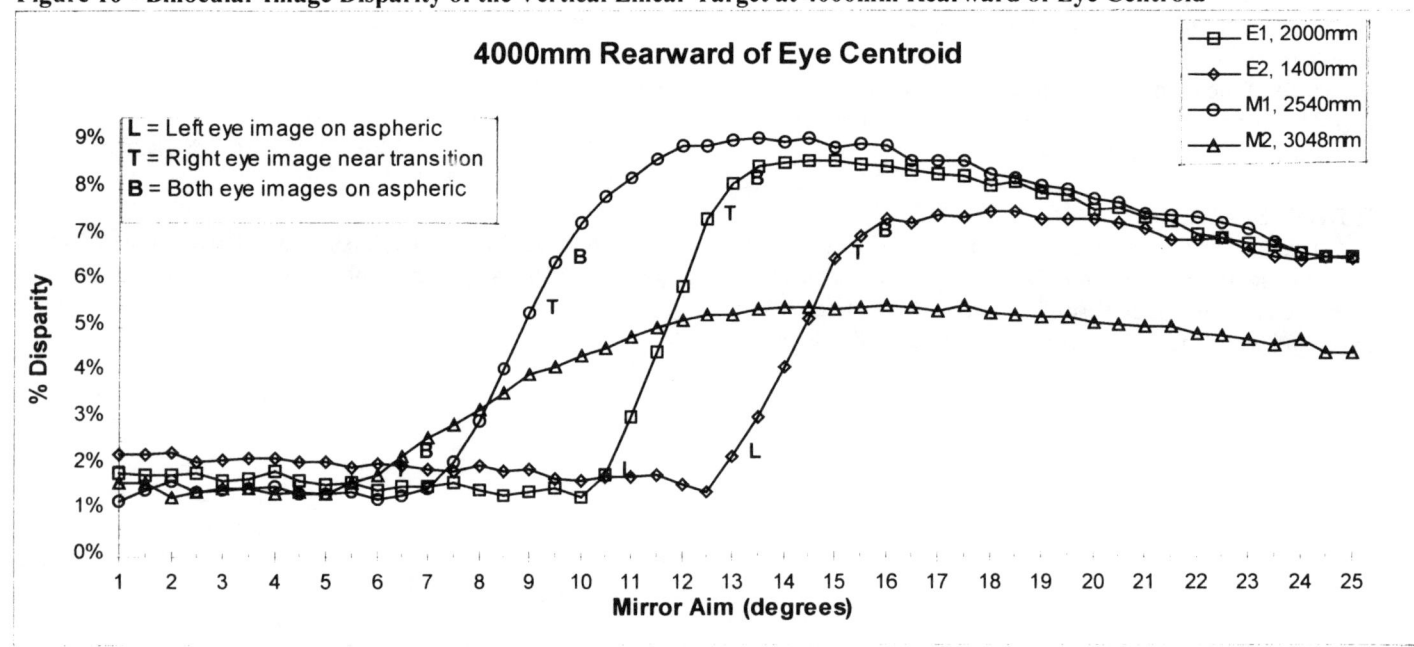

Figure 11 - Polar Plot of the binocular view of the E2 mirror with a 100mm horizontal linear target 2000mm rearward of eye centroid.

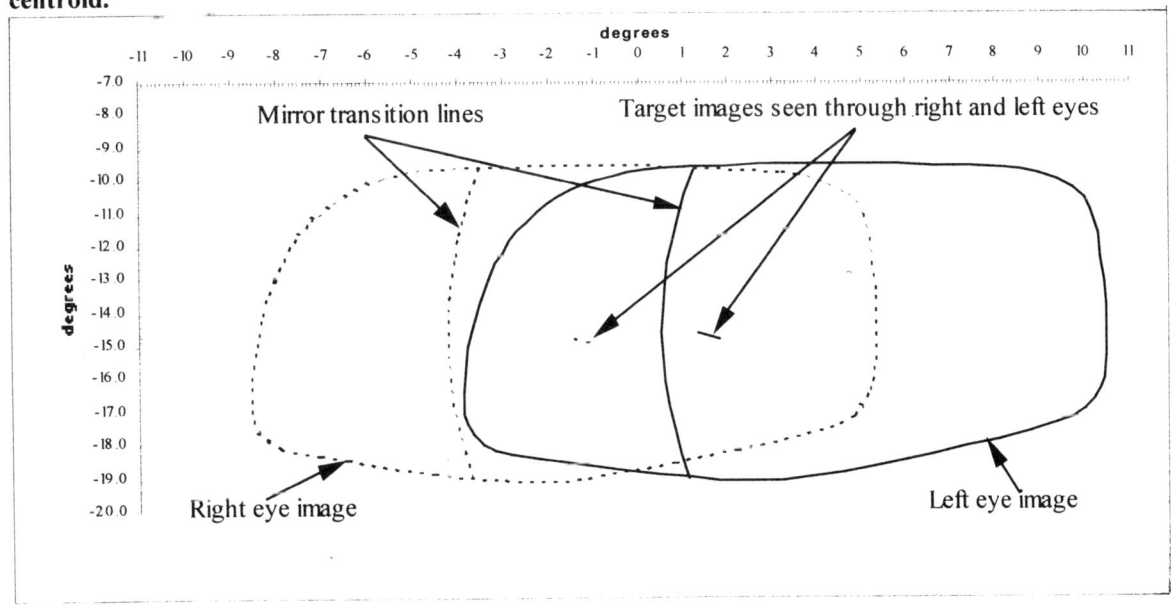

Figure 12 - Binocular Image Disparity of the Horizontal Linear Target at 4000mm Rearward of Eye Centroid

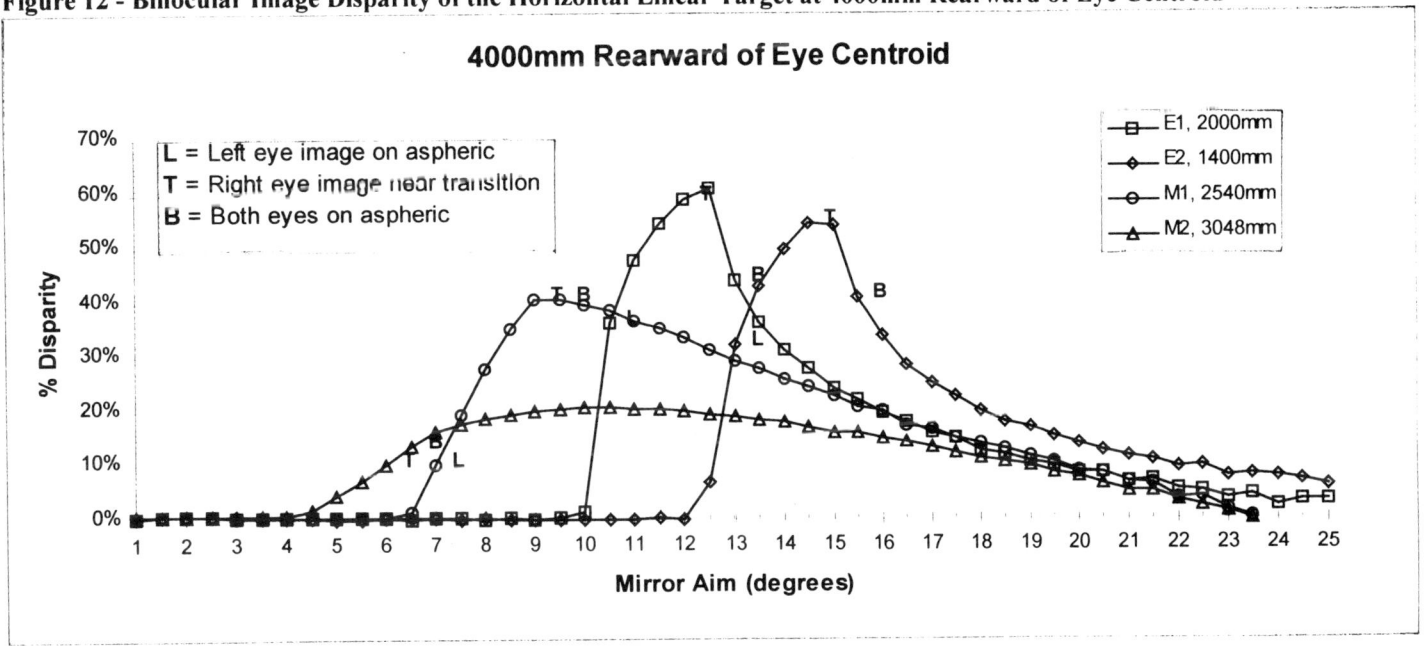

Figure 13 - Polar Plot of the binocular view of the M1 mirror with a 100mm square target 2000mm rearward of eye centroid.

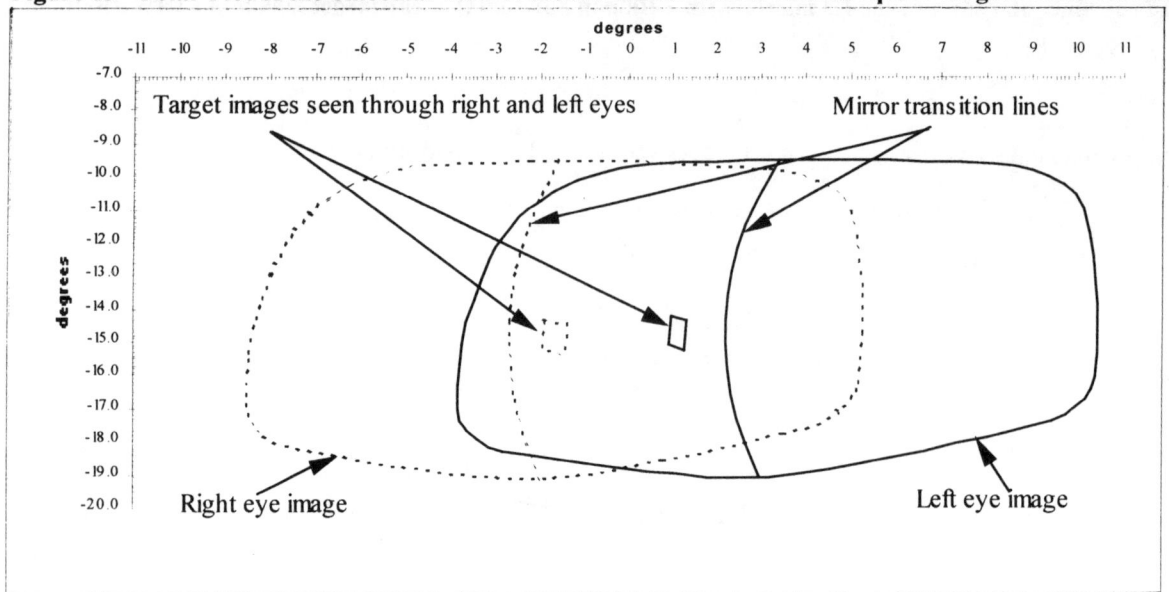

Figure 14 - Binocular Image Disparity of the Square Target at 4000mm Rearward of Eye Centroid

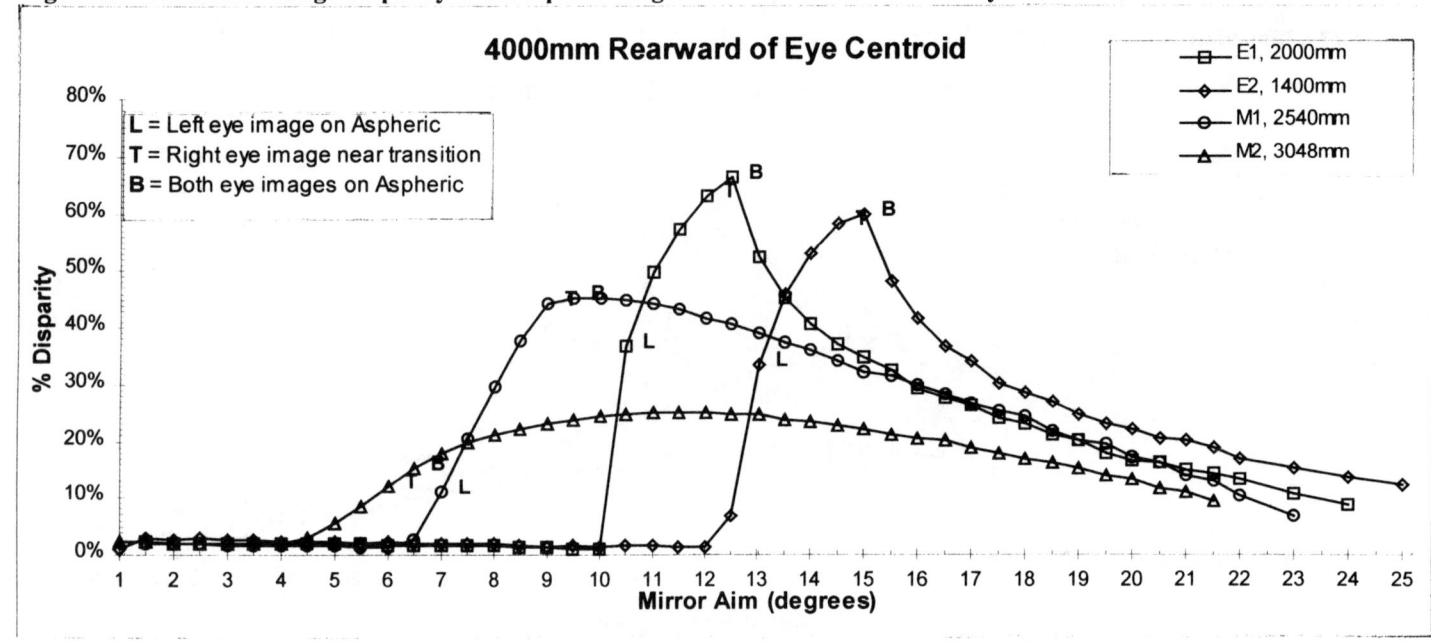

Figure 15 - Polar Plot of the binocular view of the M2 mirror with a 100mm cubic target 2000mm rearward of eye centroid.

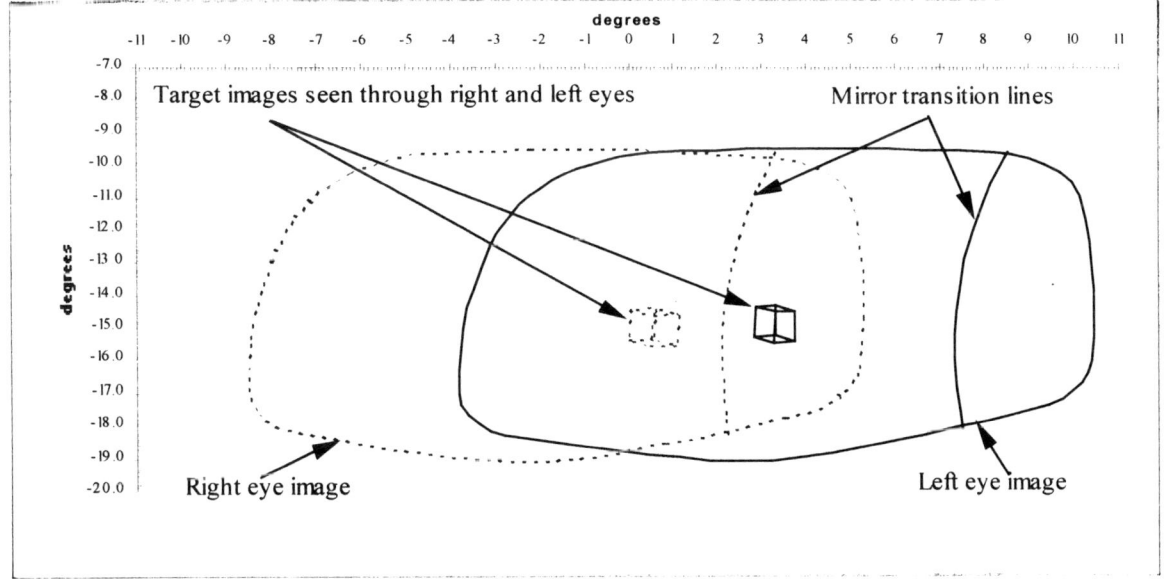

Figure 16 - Binocular Image Disparity of the Cubic Target at 4000mm Rearward of Eye Centroid

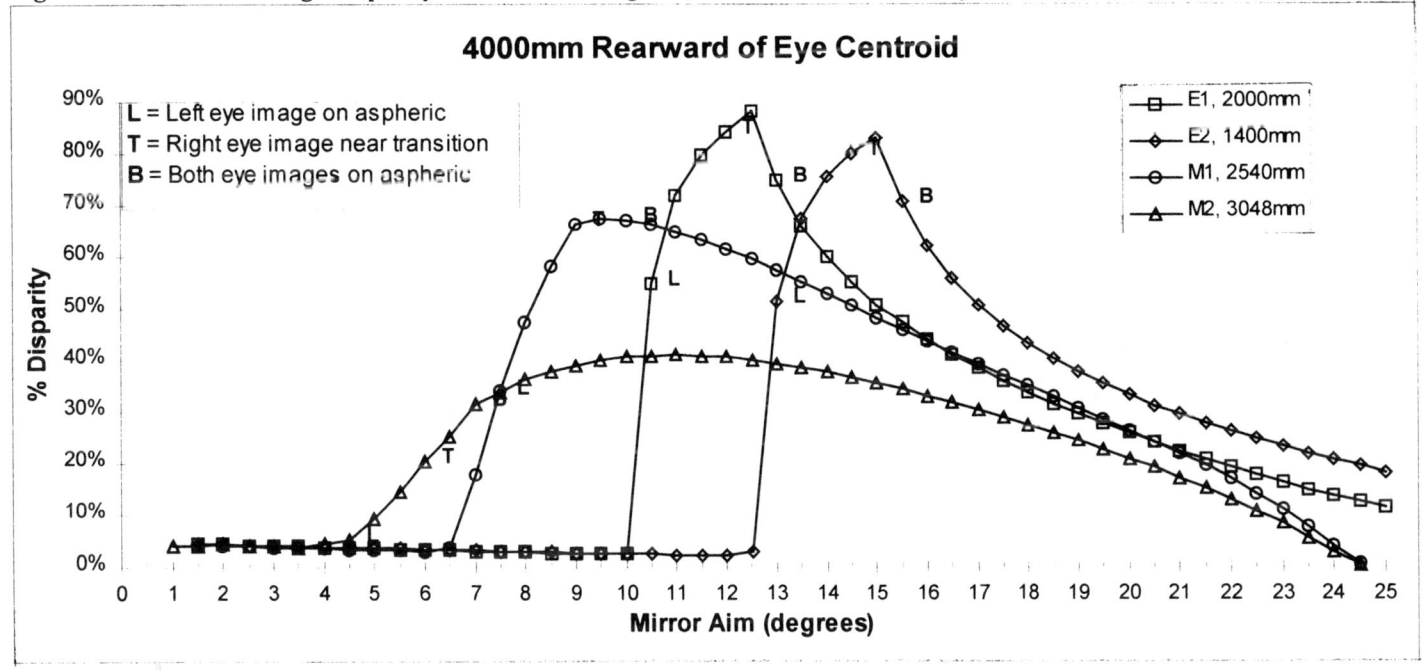

Table 1 - Maximum Binocular Image Disparity

Target	Target Distance	Mirror			
		E1, 2000mm	E2, 1400mm	M1, 2540mm	M2, 3048mm
Vertical Linear	2000mm	8.4%	7.5%	8.8%	5.4%
	4000mm	8.6%	7.5%	9.1%	5.5%
	6000mm	8.7%	7.4%	9.3%	5.6%
Horizontal Linear	2000mm	52.4%	46.8%	36.5%	18.6%
	4000mm	60.8%	54.3%	40.4%	20.5%
	6000mm	60.6%	55.4%	40.7%	20.7%
Square Area	2000mm	68.1%	60.8%	46.6%	25.4%
	4000mm	66.3%	59.7%	45.2%	25.1%
	6000mm	64.3%	58.9%	45.0%	25.0%
Cubic Volume	2000mm	86.7%	82.6%	69.6%	42.0%
	4000mm	88.2%	83.1%	67.5%	41.0%
	6000mm	87.6%	82.8%	66.2%	40.4%

Table 2 - Average Binocular Image Disparity With Both Eyes on the Aspheric Portion of the Mirror

Target	Target Distance	Mirror			
		E1, 2000mm	E2, 1400mm	M1, 2540mm	M2, 3048mm
Vertical Linear	2000mm	7.6%	7.1%	7.8%	4.7%
	4000mm	7.7%	7.1%	7.9%	4.7%
	6000mm	8.0%	7.1%	8.3%	4.7%
Horizontal Linear	2000mm	10.6%	10.7%	13.8%	11.7%
	4000mm	14.3%	16.3%	19.4%	13.6%
	6000mm	20.7%	22.6%	22.3%	16.3%
Square Area	2000mm	18.6%	20.2%	25.5%	17.2%
	4000mm	25.1%	25.6%	29.2%	19.8%
	6000mm	27.8%	29.4%	30.0%	19.5%
Cubic Volume	2000mm	27.8%	30.7%	38.5%	28.8%
	4000mm	33.4%	36.0%	37.6%	28.4%
	6000mm	40.6%	43.7%	43.3%	31.7%

Table 3 - Average Binocular Image Disparity Across the Transition

Target	Target Distance	Mirror			
		E1, 2000mm	E2, 1400mm	M1, 2540mm	M2, 3048mm
Vertical Linear	2000mm	5.6%	5.4%	4.1%	1.8%
	4000mm	5.1%	4.7%	3.7%	1.6%
	6000mm	4.1%	3.7%	3.1%	1.5%
Horizontal Linear	2000mm	41.7%	37.8%	28.3%	8.8%
	4000mm	50.3%	45.7%	28.8%	6.3%
	6000mm	46.0%	41.8%	31.2%	8.0%
Square Area	2000mm	55.8%	50.9%	34.1%	10.1%
	4000mm	54.2%	49.7%	31.4%	7.7%
	6000mm	49.2%	44.9%	28.3%	8.9%
Cubic Volume	2000mm	78.3%	73.4%	57.6%	18.6%
	4000mm	75.6%	71.4%	61.3%	17.6%
	6000mm	77.9%	72.5%	52.5%	16.4%

Table 4 - Maximum Slope Angle (degrees) of the Binocular Image Disparity Curve

Target	Target Distance	Mirror			
		E1, 2000mm	E2, 1400mm	M1, 2540mm	M2, 3048mm
Vertical Linear	2000mm	1.64	1.32	1.40	0.42
	4000mm	1.68	1.47	1.39	0.47
	6000mm	1.99	1.59	1.66	0.58
Horizontal Linear	2000mm	25.34	25.30	10.53	7.06
	4000mm	34.90	26.81	10.39	3.80
	6000mm	32.40	26.13	10.50	6.82
Square Area	2000mm	29.64	28.51	11.70	4.27
	4000mm	35.45	28.10	10.76	3.78
	6000mm	32.01	25.73	10.80	5.05
Cubic Volume	2000mm	47.13	38.00	20.19	8.54
	4000mm	46.27	44.17	17.87	7.06
	6000mm	50.24	42.69	18.29	7.42

980919

Acceptance of Nonplanar Rearview Mirrors by U.S. Drivers

Carol A. C. Flannagan and Michael J. Flannagan
The University of Michigan

Copyright © 1988 Society of Automotive Engineers, Inc.

Abstract

Five different nonplanar mirrors were evaluated as driver-side rearview mirrors in a field test using Ford employees. Two were spherical convex (differing in radius of curvature), and three were aspheric (differing primarily in the proportion of their surfaces over which radius of curvature was variable). Each participant drove for four weeks with one of the nonplanar mirrors. At three times during the test the participants filled out questionnaires concerning their experience with the mirrors. Driver preferences for the experimental mirrors increased moderately between surveys at one week and at four weeks. At four weeks, all five nonplanar mirrors were preferred to the standard flat mirror by at least a small amount. For each of the five mirror designs there was a large range of opinion. Most notably, a small number of people strongly disliked the aspheric design that involved the largest variable-radius area. These results indicate that nonplanar mirrors are likely to be welcomed by a large number of U.S. drivers, but some designs seem more acceptable than others, and for almost any design there may be a small but significant number of drivers who strongly prefer planar mirrors. The sample in this study had a limited range of subject ages, a relatively small number of females, and may have been biased toward people who have generally positive attitudes toward new technologies. The results should therefore be considered preliminary.

Introduction

In the early 1980s, U.S. drivers began to become familiar with nonplanar rearview mirrors as original equipment on passenger cars. Specifically, Federal Motor Vehicle Safety Standard (FMVSS) 111 was revised to explicitly allow spherical convex mirrors at the passenger-side exterior position ("Preamble," 1982). At the interior position and driver-side exterior position, planar mirrors are still required as original equipment, and most drivers probably have little or no experience with anything else. Various convex mirrors are commonly available as aftermarket equipment in the U.S., but a comprehensive study of cars in use as of the mid 1980s found that only 7 percent of them had convex mirrors on the driver side, and virtually all of those were small, short-radius mirrors mounted on or near the standard planar mirrors (Olson & Winkler, 1985).

Two general types of nonplanar mirrors—spherical convex and aspheric mirrors—have been used on the driver side in some areas outside the U.S. for many years. Both of these types of mirrors offer the benefit of wider fields of view than planar mirrors of the same size, and both entail some cost in diminished image quality. Spherical convex mirrors have surfaces that are sections of the surfaces of spheres. The curvature of each such mirror therefore can be described by a single radius value, and they produce images that are approximately uniformly minified, thereby making objects appear farther away than they really are.

Aspheric mirrors designed for use as exterior rearview mirrors are also generally convex, but they typically vary in radius of curvature across their horizontal dimension, being more gently curved (i.e., have longer radius of curvature) on the inboard side and more strongly curved (i.e., have shorter radius of curvature) on the outboard side (e.g., Pilhall, 1981). This somewhat more complex shape is intended to provide a larger field of view while reducing the problems with minification of images (and distance distortion) that spherical convex mirrors produce. Consider a vehicle that is well behind an observer's vehicle, and in the adjacent lane on the side of the mirror in question. The rearward vehicle will be imaged in the inboard, gently curved portion of the mirror, and will therefore be seen in a relatively undistorted form. As it approaches the observer's vehicle from behind, its image will move outward on the mirror, eventually being imaged in the outboard, strongly curved portion, where it will be strongly minified (and also distorted in shape). The intent of this mirror design is that, when vehicles are imaged in the strongly curved portion, they will be so close that a driver will merely have to detect their presence to know that a maneuver such as a lane change is not safe, and that precise judgments of speed or distance, which might require a better quality image, will be moot.

There is suggestive, although not conclusive, evidence that nonplanar mirrors result in improved safety in countries that have a mixture of planar and nonplanar driver-side mirrors (Luoma, Sivak, & Flannagan, 1995; Schumann, Sivak, & Flannagan, 1998). This benefit, if it does in fact exist, presumably results from reducing the blind zone that typically exists on the driver side when a planar mirror is used.

Recently there has been considerable interest in the possibility of using nonplanar driver-side mirrors in the U.S. as a way of addressing the blind zone problem (Flannagan & Sivak, 1996). However, given that such mirrors are relatively unfamiliar to U.S. drivers, there are uncertainties both about how drivers in the U.S. would react to them subjectively, and about the objective safety effects the mirrors would have.

The present study was designed to assess the subjective reactions of U.S. drivers to a variety of both spherical convex and aspheric mirrors (five different designs in all) used on the driver side. Clearly, subjective preferences do not always correlate with performance (e.g., Andre & Wickens, 1995), and we do not mean to suggest that subjective reactions are the whole story, or even the most important part of it. But, when a new system requires a period of adjustment, as seems to be the case for many drivers with nonplanar mirrors, subjective reactions may be particularly important because of their effect on motivation and willingness to give something new an extended try.

In this study, we asked volunteer subjects to have an experimental mirror placed on the vehicle that they were already routinely driving, to drive with it for four weeks, and to fill out questionnaires at various times during that period. The four-week period was chosen because anecdotal evidence had suggested that some drivers seemed to change their opinions about nonplanar mirrors over about that amount of time, with most developing a more favorable opinion. Because we were particularly interested in possible changes in peoples' reactions to mirrors that were initially novel to them, each subject used only one of the five mirror designs being tested.

The sample was not ideal, and was determined partly by availability. It consisted of volunteers from among employees of Ford Motor Company who were participating in a vehicle leasing program. The volunteers were mostly middle aged and male. Given their automotive backgrounds, it is possible that they were more receptive to new technologies than the general population. However, the sample is probably adequate to make preliminary assessments of a number of things, including the time course of changes in subjective reactions, and the relative acceptance of the various mirror designs. (Absolute levels of acceptance by, for example, older drivers might be expected to diverge more from the results obtained with our sample.)

The mirrors used were selected to represent a reasonably full range of designs for both spherical convex and aspheric mirrors. This is easier to accomplish for spherical convex designs because they differ in only one major way: radius of curvature. Two spherical convex designs, differing in radius of curvature, were used. In contrast, aspheric designs in principle could vary in many ways. In this study, the aspect that was manipulated was the proportion of the surface of the mirror over which the radius of curvature was variable. Three aspheric designs were used. Each had an inboard section over which the radius of curvature was constant, and an outboard section over which the radius of curvature diminished, being shortest (most strongly curved) at the outboard edge. The mirror with the smallest variable area also had a dotted line on the mirror surface to indicate the place of transition between constant radius and variable radius.

A single vehicle type was used in the study, a relatively large sport-utility vehicle. This was chosen partly because a large group of drivers was available from which to solicit volunteers, and partly because the vehicle was equipped with mirrors slightly larger than those of a typical car. That allowed somewhat greater flexibility in mirror designs.

Method

Subjects

Subjects were recruited from among Ford employees who leased a Ford Explorer or Mountaineer through the employee lease program. One-hundred twenty subjects were tested, 24 in each of five mirror groups. Of these, four dropped out because either they or a spouse disliked the mirror or because they turned their vehicle in before the end of the study for reasons unrelated to the study. Two more were eliminated because a husband and wife alternately filled out different questionnaires.

Of the remaining 114 subjects, 90 were male and 24 were female. Subjects' ages ranged from 28 to 61, with an average age of 45. Twenty-seven subjects had prior experience with some type of driver-side exterior nonplanar mirror, by self-report.

Because of the recruiting strategy, the sample is not fully representative of the general population of drivers. Most obviously, the members of the sample are mostly male and middle-aged. In addition, because the sample is made up primarily of professional-level employees of an automobile company, it probably includes a number of subjects who are particularly interested in or open to new automotive technology. Consequently, the results should be considered preliminary.

Experimental mirrors

Five different driver-side, exterior rearview mirrors were used. Two were spherical convex, and three were aspheric. Their primary characteristics are listed in Table 1. As noted in the table, on the aspheric mirror with the small proportion of variable area the boundary between the constant-radius and variable-radius portions was marked by a dotted line. On the other two aspheric mirrors there was no explicit demarcation. Thus, across the three aspheric mirrors there was a confounding of the proportion of variable area with presence or absence of explicit demarcation. However, for the mirrors with medium and large variable areas, a demarcation line would intrude more on the overall field of view of the mirror. Therefore, it could be argued that this is a natural confounding, at least somewhat representative of realistic constraints on mirror design.

Table 1
Characteristics of the five experimental mirrors

Spherical convex mirrors (with nominal radius)	Actual radius from a sample of mirrors Mean (n, SD)
Convex, 1500-mm radius	1445 (6, 59)
Convex, 2000-mm radius	2027 (5, 15)
Aspheric mirrors	Percentage of width with variable radius
Aspheric, small variable area[1]	34
Aspheric, medium variable area	40
Aspheric, large variable area	66

[1] The boundary between the constant-radius and variable-radius areas on this mirror was marked by a dotted line.

Procedure

Subjects were contacted by e-mail and phone to set up an initial appointment with the experimenters. Each subject brought his or her vehicle to the test facility, a garage operated by Ford Motor Company. The experimenters first described the study and asked the subject to sign an informed consent form and fill out a driver-information form. The driver-information form asked for basic background information such as year of birth, height, whether the subject wore glasses, and whether the subject had used nonplanar mirrors on the driver's side before. The subjects were instructed that the mirrors caused image distortion, and that they should use caution in interpreting size and distance information. (This instruction was identical for all subjects, regardless of which mirror they were assigned.)

After gathering the initial driver information, the experimenters measured the field of view of the subject's original mirror before installing one of the five nonplanar experimental mirrors. The subject was seated in the driver's seat with his or her head turned to look in the mirror as though checking it during normal driving. The experimenter moved an orange ball, mounted on a post, along a line perpendicular to the long axis of the vehicle and even with the rear bumper, beginning far outboard of the field of view of the mirror and moving it towards the vehicle until the subject said it was visible at the outer edge of the mirror. After the subject indicated that the ball had appeared in the outer edge of the mirror, the experimenter reversed the process, moving the ball away from the vehicle until it disappeared. The process was repeated for the near edge of the field of view. In most cases, the subject could see part of the vehicle in the mirror, making the vehicle the near limit of field of view.

In addition to measuring field of view, the experimenters recorded the vehicle odometer reading on the driver information form. Subjects were given a packet of materials and instructed to fill out and return three different questionnaires at three specific times.

First-Week Questionnaire. The first-week questionnaire contained a few questions about how well the subject had adapted to the mirror and how much he or she liked it relative to the original mirror. The form was filled out after one week of using the mirror and returned by mail.

Fourth-Week Questionnaire. The fourth-week questionnaire contained several pages of questions about how well the subject had adapted to the mirror, how much he or she liked it relative to the original mirror, and how he or she used the experimental mirror when driving. The form was filled out after four weeks of using the mirror. At four weeks the test period was complete, so the subject returned to have the mirror removed at that time. During the return appointment, the experimenters measured the field of view of the experimental mirror before removing it.

Post-Test Questionnaire. Two weeks after removal of the experimental mirror, subjects were asked to fill out the post-test questionnaire, which contained a few questions about the experience of returning to the original mirror and about the subject's preference for the original or the experimental mirror.

Results and Discussion

Fields of view

Total field of view for both the planar and nonplanar mirrors was calculated by first taking the average of the two measurements (ingoing and outgoing) of the outer limit of the field of view. For those who could see the vehicle in their mirror, the outer edge of the rear bumper was used as the inner limit of the field of view. For those who could not see the vehicle in the mirror, the average of the two inner-limit measurements was used. Using these two limits, the total angle of the field of view was calculated.

An analysis of variance (ANOVA) was conducted on the fields of view to determine whether they differed for different mirror groups. There were no significant differences between groups in field of view of the planar mirror, indicating that there were no unexpected systematic differences between subject groups in mirror-aiming habits. For the nonplanar mirrors, however, fields of view differed significantly across groups. The results are given in Table 2.

Eighty-one percent of subjects aimed their original mirrors so that the vehicle was visible. Of these, 93% also aimed the nonplanar test mirror so that the vehicle was visible. In contrast, half of the minority who could not see the vehicle in their planar mirror also could not see their vehicle in the nonplanar mirror. In addition, within mirror groups, planar and nonplanar fields of view were correlated at $r = 0.461$, on average.

Subjective mirror ratings

After one week of driving, 86% of subjects reported that they were comfortable with the mirror. After four weeks of driving, 93% of subjects reported that they had gotten used to the mirror. Table 3 gives the percent of subjects in each mirror group who reported that they were comfortable with the mirror after one and four weeks of use. Subjects were significantly less likely to be used to the aspheric mirror with large variable area after both one and four weeks.

Table 2
Actual fields of view for planar and nonplanar mirrors, measured with mirror aim as set by the subjects

Mirror	Min	Mean	Max	Std. Dev.	N
Planar	11.0	19.9	25.4	2.70	114
Convex, 1500-mm radius	28.2	34.1	39.1	3.05	20
Convex, 2000-mm radius	22.7	29.9	36.1	3.87	21
Aspheric, small variable area	33.9	39.9	46.3	3.23	20
Aspheric, medium variable area	24.5	39.4	47.6	4.89	23
Aspheric, large variable area	26.3	35.9	39.7	3.27	21

Table 3
Subject reports of becoming accustomed to the mirrors

	Percent comfortable with mirror	
Mirror Type	First Week	Fourth Week
Spherical, 1500	90%	95%
Spherical, 2000	91%	100%
Aspheric, small variable area	95%	100%
Aspheric, medium variable area	87%	91%
Aspheric, large variable area	68%	82%
Overall	86%	93%

At each time period, subjects were asked to mark an X on a line indicating the extent to which they preferred either the test mirror or their original mirror over the other. The location of the mark was coded with a number between 1 and 100, with 1 indicating extreme preference for the original mirror, 100 indicating extreme preference for the nonplanar mirror, and 50 indicating no preference.

Figure 1 shows the average response to this question for each mirror over time. At all time periods, the nonplanar mirrors are preferred to the planar mirror (except the large-variable-area mirror after one week). Overall, this preference increased significantly between the first and fourth weeks and held steady two weeks after subjects returned to using planar mirrors. An ANOVA conducted on the preference data showed that the large-variable-area aspheric mirror was preferred to the planar mirror significantly *less* than the other mirrors at all time periods. In contrast, the 2000-radius spherical mirror is preferred more, across time.

Subjects' responses differed enough that the distributions of individual preference ratings are also of interest. Figure 2 contains histograms of preference ratings for each mirror group after one week, and Figure 3 contains analogous histograms of ratings after four weeks. In general, the histograms show that subjects prefer the nonplanar mirrors to their original mirror, though they are still getting used to them after one week. After four weeks, the bars are more concentrated at the right end for all but the large-variable-area aspheric mirror. For the latter, the bars are spread across the spectrum at both one and four weeks. For all mirrors, however, there are some subjects who continue to prefer their original mirror after four weeks.

Preferences and subject characteristics

Preference for nonplanar mirrors might also be related to subject variables, such as age or sex, so a series of analyses were done to test these hypotheses. There was no correlation between age and mirror preference after either one or four weeks ($r = .05$ and $-.19$, respectively), though the age range in this study was somewhat restricted. Similarly, there was no effect of wearing glasses, sensitivity to motion sickness, or prior experience with nonplanar mirrors on mirror preference rating. However, there was a marginally significant difference between males and females after both one and four weeks in mirror preference. Females preferred the nonplanar mirror to their own mirror *more* than did males, on average, after one and four weeks. It is not clear how this result might generalize to the larger female population, given the small sample of females and the non-representative sampling strategy.

Finally, the objectively measured field of view had no relationship either to mirror preference judgments or the subject's perception of the effectiveness of the mirror in providing information for making lane changes. The correlation between field of view of the test mirror and preference was .004 after one week and .038 after four weeks. The correlation between field of view and the reported extent to which subjects had to turn their heads (relative to the original mirror) was -.053 after one week and -.007 after four weeks. Subjects were also asked to rate how well they could see to make a lane change, whether they thought they could change lanes safely using only the mirror, and whether they had used the mirror exclusively to make lane changes. For all three variables, there was no relationship to the objectively measured field of view.

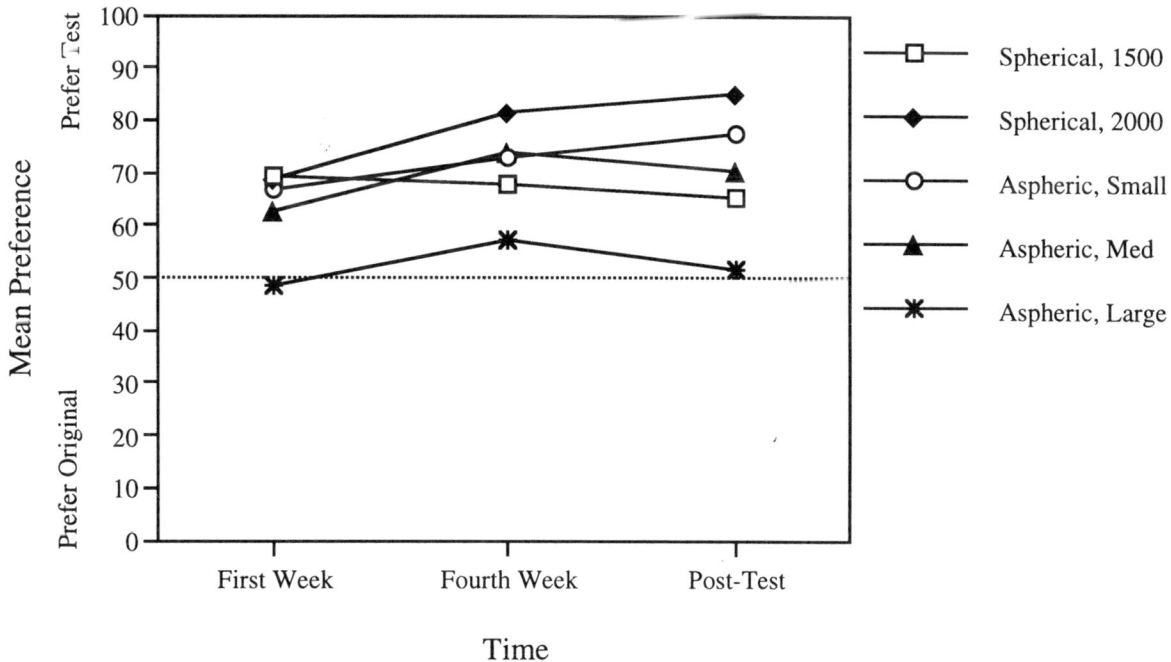

Figure 1. Preference for nonplanar test mirror over original, across time.

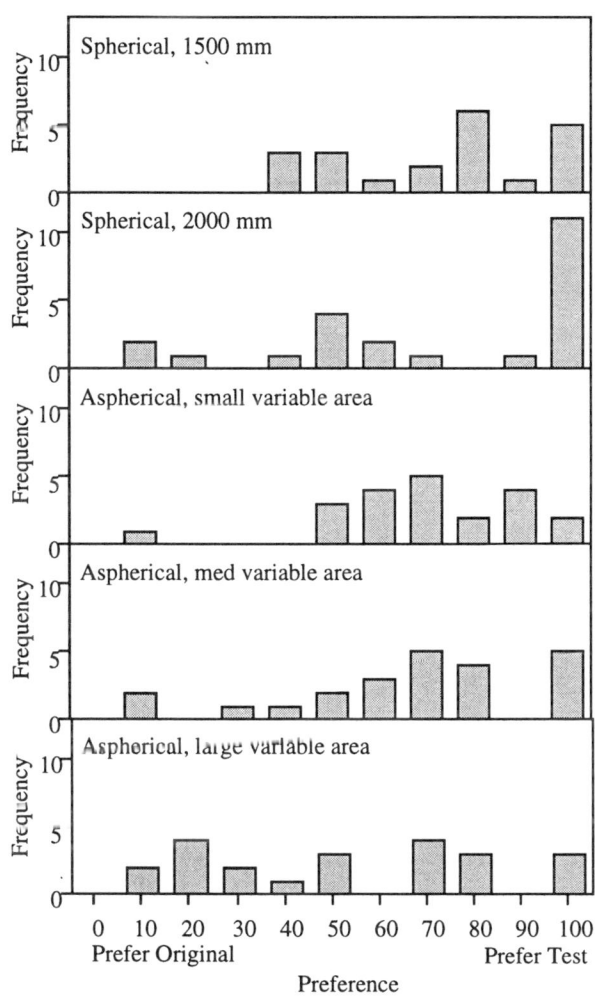

Figure 2. Histograms of preference for nonplanar mirrors after one week.

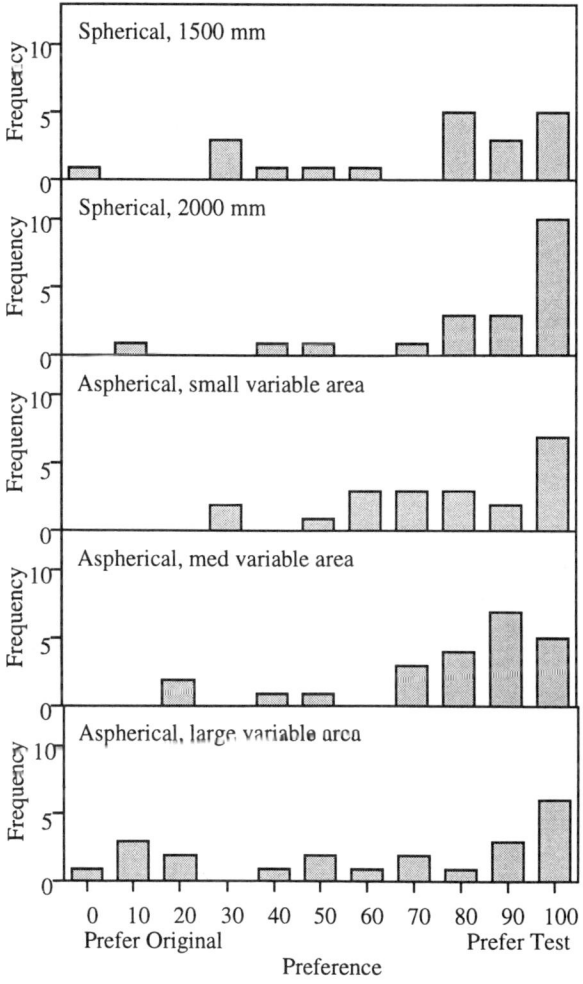

Figure 3. Histograms of preference for nonplanar mirrors after four weeks.

Summary and Conclusions

This study was a preliminary attempt to assess the subjective impressions of U.S. drivers who used nonplanar mirrors over an extended period. Although the sample was not representative of the general driver population, the results provide insights into changes in acceptance over time, and at least the relative levels of preference for the five mirror designs that were tested.

In general, subjects preferred the nonplanar mirrors to their original planar mirrors, and that preference grew over time. However, for the aspheric mirror with a large variable area, subjects appeared to become more extreme in their opinions, creating a strongly bimodal distribution. Most subjects moderately or strongly preferred the experimental mirror, but there were also a small number of subjects who strongly disliked it.

Subjects tended to maintain their aiming practices when they switched mirrors. Although they indicated that the nonplanar mirrors provided them with the necessary information to make lane changes safely, their confidence in the mirror was not related to objective measures of field of view.

Finally, subject variables such as corrected vision, prior experience with the nonplanar mirrors, and age were not related to mirror preference. Females were marginally more positive about the nonplanar mirrors than males, but the female sample was small and may not have been representative.

Future studies might build on these results by repeating the procedure with a larger and more representative sample of drivers. Another important addition to future studies would be information about the effects of nonplanar mirrors on objective driver performance and safety.

Acknowledgments

This work was conducted under the sponsorship of the American Automobile Manufacturers Association. A large number of people contributed to this study, far more than can be acknowledged individually. We wish especially to thank Russell Rockett and John Kuzemka, who conducted the field work. Many people at Ford were extensively involved in numerous ways throughout the planning and conduct of the study, particularly Mary Beth Angotti, Carl Buttermore, Dave Houston, Ric Karbowski, and Pat Spect. Mike Perel provided insightful advice and suggestions concerning the design of the survey forms. George Dalby contributed many valuable thoughts in numerous background discussions. The members of the Vision Standards Committee of the Society of Automotive Engineers, including Chairperson Gary Rupp and Vice Chairperson Dorothy Helder, made many thoughtful contributions to a series of detailed technical discussions as the work progressed.

References

Andre, A. D., & Wickens, C. D. (1995, October). When users want what's not best for them. *Ergonomics in Design*, 10-14.

Flannagan, M. J., & Sivak, M. (1996). *Workshop on rearview mirror human factors research needs: Summary of recommendations* (Report No. UMTRI-96-27). Ann Arbor: The University of Michigan Transportation Research Institute.

Luoma, J., Sivak, M., & Flannagan, M. J. (1995). Effects of driver-side mirror type on lane-change accidents. *Ergonomics, 38*, 1973-1978.

Olson, P. L., & Winkler, C. B. (1985). *Measurement of crash avoidance characteristics of vehicles in use* (Report No. UMTRI-85-20). Ann Arbor: The University of Michigan Transportation Research Institute.

Pilhall, S. (1981). *Improved rearward view* (SAE Technical Paper Series No. 810759). Warrendale, Pennsylvania: Society of Automotive Engineers.

Preamble to an amendment to Federal Motor Vehicle Safety Standard No. 111, 47 Fed. Reg. 38698 (1982) (codified at 49 C.F.R. § 571.111).

Schumann, J., Sivak, M., & Flannagan, M. J. (1998). Are driver-side convex mirrors helpful or harmful? *International journal of Vehicle Design, 19*, 29-40.

980920

The 'Double Objective' Milner Prismatic Exterior Rear View Mirror

Peter J. Milner
de Montfort, UK

Richard E. Berg
de Montfort, US

Copyright © 1998 Society of Automotive Engineers, Inc.

ABSTRACT

The Double Objective variant of the Milner mirror represents the latest evolution of a family of designs developed from the original 1989 Milner prismatic exterior rear view mirror configuration. Papers reviewing this design evolution are referenced.

In Double Objective form, the Milner mirror concept has attained a maturity of packaging and function that further enhances its appeal as an OEM system to both the auto makers and their suppliers. The paper briefly reviews other Milner mirror family members before describing the particular constructional and functional features of the Double Objective system in greater detail.

INTRODUCTION

In 1989 a patent application, now internationally granted [1], was filed by the first author disclosing a design of prismatic exterior rear view mirror having substantially reduced protrusion outside the vehicle body. The exterior optical component of this invention was a refracting 'object' prism, housed in an aerodynamic fairing, that was configured to bend light arriving from behind and to the side of the vehicle inwards, so that it could fall on a reflector of normal size and shape placed inside the vehicle. The key to the invention was the design and location of a second, 'ocular' prism that served to correct the optical aberrations introduced by the object prism; see Figure 1

The use of an exterior refractive surface to 'collect' light from behind and to the side of a vehicle, and then to process the light and present it to the driver in the form of an image is well known [2]. In general, earlier designs suffered from optical problems, not so much at 'collection' as during 'processing', and/or physical ones such as packaging and cost.

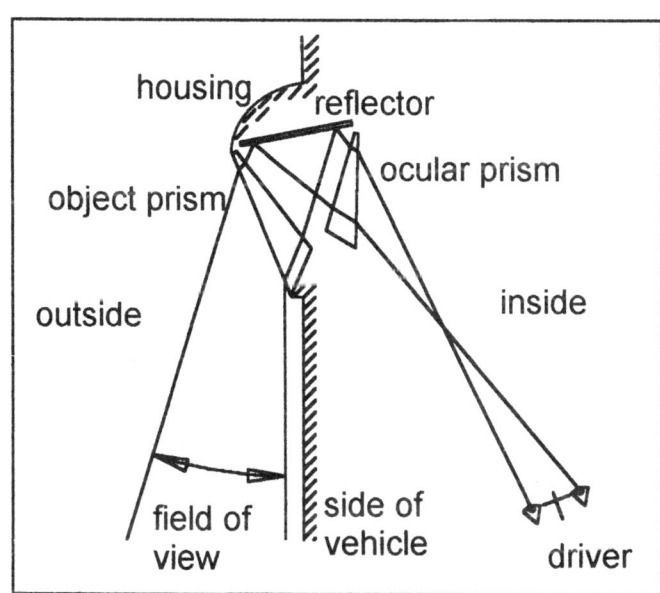

Figure 1. Original Configuration of Milner Mirror

Although careful design had ensured that the Milner configuration possessed no inherent optical problems that might condemn it, during early development it was always recognised that the thick prisms of the original configuration would need to be made thinner in order to reduce weight and aid manufacturability, at the same time reducing cost, before the system could be considered for OEM fitment.

Also during the early years of development, the location of the ocular prism was identified as the cause of an impression of 'tunnel vision' experienced by drivers of evaluation vehicles converted to the system. Details of the optical characteristics responsible for this phenomenon have been described in an earlier paper [3].

Solutions to both the above problems were successfully developed and combined in a new form of the mirror system known as the Wedge configuration [4] illustrated in Figure 2 and first shown in 1995 installed in a small European built coupé, the Opel/Vauxhall Tigra, see Figure 3.

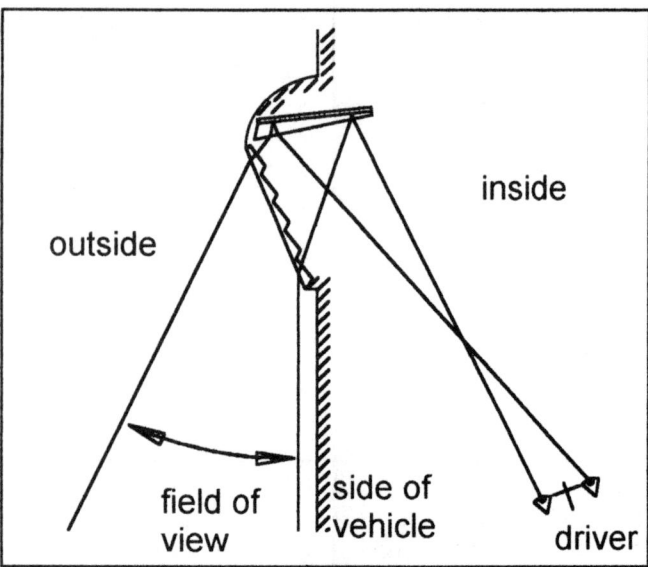

Figure 2. Wedge Type Milner Mirror

Figure 3. Wedge Type Milner Mirror Evaluation Vehicle

The essential changes incorporated in this new form of the mirror were as follows:

- Fresnel type object prism
- Relocation of the ocular prism to lie adjacent to the reflector

The adoption of Fresnel technology for the object prism fulfilled the dual requirements of low weight and manufacturability while the relocation of the ocular prism eliminated the tunnel vision effect. Furthermore, since light forming the image was caused to pass twice through the ocular prism, its apex angle could be halved, and by choosing a material having a much lower Abbe value than that of the object prism it could be halved again. While the dual light transit in different directions prevented the ocular prism from also being converted to the Fresnel type, because of excessive and unavoidable light blockage by the riser (or draft) facet of the Fresnel profile, these measures ensured that the prism would be thin enough to be lightweight and, importantly for manufacturing cost, capable of being injection moulded.

The Tigra evaluation vehicle provided the first means by which the industry was able to assess the results of these developments. Over two years following its early 1995 unveiling, several hundred drivers carried out between them several thousand miles of mirror system evaluation in the Tigra on public roads. As some of these evaluations were unavoidably performed in territories for which the Tigra driving position was on the 'wrong' side, a number of proprietary 'right' side vehicle conversions were commissioned in order to improve assessment realism. The principal conclusion to be drawn from all this evaluation experience was that the new configuration was successful in overcoming the former problems.

However, the new configuration did introduce two new and unwelcome optical phenomena that needed to be addressed, namely:

- Flare from the Fresnel object prism
- Secondary reflections from the ocular prism

Flare is defined here as the appearance of multiple faint images of a source in a line extending either side of the principal image at right angles to the Fresnel groove direction.

Neither of these effects was manifest during normal daytime driving but both were noticeable and distracting at night and under some dawn and dusk conditions. Development programmes were implemented aimed at reducing both effects to levels no lower than those of similar phenomena that can affect conventional mirror systems.

These development programmes resulted in the following conclusions:

- Flare could be extinguished by use of a suitable mask shadowing the riser portion of each Fresnel pitch element
- Identification of two new versions of the mirror system, known as the Lens Prism (Figure 4) and Double Objective (Figure 5) variants, that eliminated problematic secondary reflections by new prism configurations

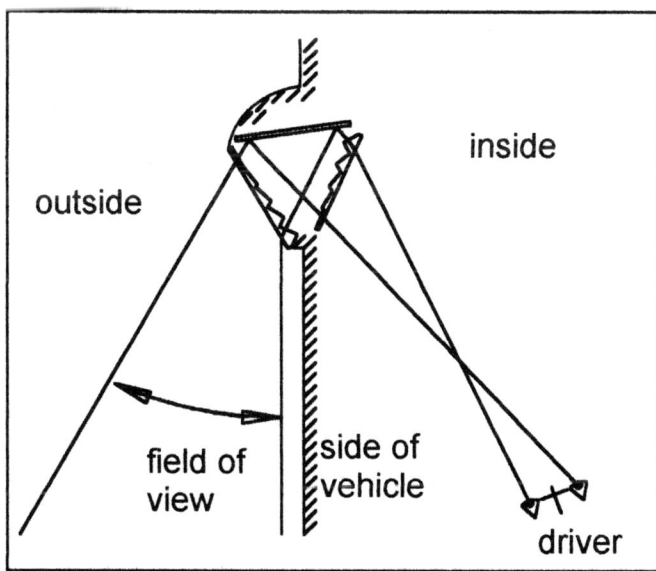

Figure 4. Lens Prism Variant of the Milner System

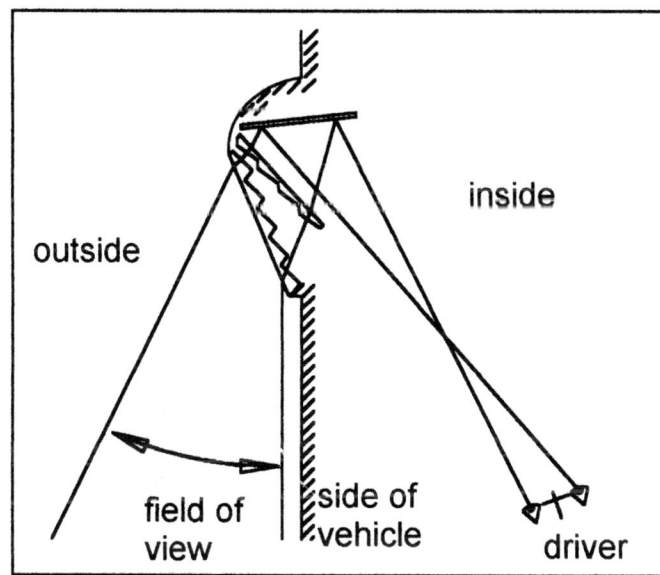

Figure 5. Double Objective Variant of the Milner System

The Lens Prism variant is more lens than prism in so far as its optical behaviour is at variance with a fundamental but invaluable property common to the other members of the Milner system, and indeed to all purely prismatic systems, namely that parallel rays entering the system remain parallel through the system and emerge parallel. This means that the driver eye accommodation requirement is unaffected by the presence of the prisms and that wide tolerances are permissible on the location of the optical elements.

Lens Prism system design is greatly complicated by the fact that acceptable levels of these desirable features have to be specifically engineered, and that in so doing compromises normally have to be made. The reward for this complexity is compactness; the Lens Prism variant offers the prospect of a shorter package length than the other configurations, and for this reason development

work continues.

Unlike the Lens Prism variant, the other new design developed to overcome the secondary image problem, the Double Objective variant, is a purely prismatic device requiring none of the former's optical compromises. Its optical behaviour is already well understood and proven, being similar to that of other purely prismatic members of the Milner system, thus opening the way to faster track production development than would otherwise be the case. It is expected that the present paper will assist this process by providing an appreciation of the essential optical and physical features of the Double Objective form of the Milner prismatic mirror system.

DOUBLE OBJECTIVE MILNER PRISMATIC MIRROR

Mirror hardware comprises two types of component, optical and non-optical, and mirror design is logically undertaken in that order. Furthermore, design of the optical components naturally follows design of optical function, and the following sections are presented in the same order. Before commencing detail design, though, it is important to establish the overall objectives of the design with a view to setting design priorities and boundary conditions which will to some extent constrain the design.

OVERALL DESIGN OBJECTIVES

The most fundamental objective of the design process is the specification of an optical system which incorporates the environmental benefits inherent in the system concept while preserving the look, feel and function of a conventional mirror. These environmental benefits are both optical and non-optical in nature and include the following:

- Better aerodynamic efficiency for economical, quiet and clean operation
- Reduced dynamic and aerodynamic reflector disturbance for enhanced image stability
- Lower exterior protrusion for impact avoidance
- Elimination of effects of cabin glass aberrations and soiling for clearer image

That the prismatic mirror should offer the look, feel and function of a conventional mirror is considered essential in what will be a mixed prismatic/non-prismatic mirror vehicle pool. Drivers must know instinctively where to look for their rear views and be able to extract from the displays all the safety related information they are used to. This means the prismatic mirror should present its display more or less where a conventional mirror would, and the display should possess the same basic characteristics as a conventional mirror. These display requirements may be summarised as a virtual image containing parallax and perspective information presented at a representative distance behind the reflector. In addition, image and object fields of view should be similar to those provided by conventional designs and, of course, all this should be achieved within the most compact package possible.

Regarding physical characteristics, it has been empirically established that a suitably aerodynamic fairing may possess a maximum protrusion of up to about 100mm from the underlying vehicle envelope before any measurable additional aerodynamic drag is incurred. Since a substantially lower protrusion than this incurs optical penalties in terms of mirror function, optimum protrusion is generally considered to be 80-100mm.

OPTICAL CHARACTERISTICS

In the Double Objective design all the prismatic action takes place before the light impinges upon the reflector. The inner prism is therefore better described as a second object prism than as an ocular prism, hence the term Double Objective. In principle this prism may be placed anywhere between the object prism and the path of light leaving the reflector bound for the eye, as is clear from Figure 5. (In practice, as will be explained later, there is normally a preferred position which is determined by performance priorities for the particular application.)

Chromatic aberration

In order to minimise the length of the outer prism for a given reflector width, it is required to maximise the deviation of the light. This is achieved in part by specifying the prism apex angle to be as large as possible, but the passage of light through such a large apex angle prism necessarily causes it to be subject to substantial chromatic dispersion which, if uncorrected, produces an image possessing the undesirable phenomenon of chromatic aberration. Dispersion is the splitting of light into its spectral (rainbow) colours, and chromatic aberration is manifest as coloured fringes in the image as a direct result of this dispersion. The principal task of the correcting prism is, therefore, to nullify the effect of dispersion produced by the object prism. In this regard the prism pair may be likened to the well known achromat design for lenses.

This likeness is particularly useful as it aids appreciation of the choice of materials for the two prisms, which must possess contrasting optical properties, as is the case for the two elements of the lens achromat. In the classic lens achromat the outer element is a crown glass while the inner is a flint; in the Double Objective the materials are plastic equivalents, normally acrylic and polycarbonate, respectively. In both cases the chosen materials permit chromatic correction to be achieved while still preserving a net refractive 'bend' in the light after passage through the pair. This is where the similarity ends, however, as the use to which the bend is put is quite different for the two types.

In the case of the Double Objective the bend is fundamental to the function of the mirror unit in that it is what causes the inboard end of the reflector to be illuminated. As is clear from study of Figure 5, wide spacing of the elements increases the inboard 'throw' of light onto the reflector, due to the fact that the unavoidable 'unbending' effect of the correcting prism occurs later than for closer prism spacing.

In addition to the basic requirements of inward light deviation and dispersion control, the four refractions imposed on light that passes through the two prisms combine to generate some subtle effects that profoundly influence the following critical system characteristics:

- Image magnitude (lateral and vertical)
- Field of view (object and image fields)
- Image distortion
- Binocular alignment

All these effects, which are discussed in greater detail below, have been fully analysed and special tools have been incorporated into the custom optical design software developed for the system in order to permit the designer to maintain the strict control of them that is necessary.

Image magnitude

An understanding of how prisms can create such effects is aided by considering the relationship between the entry and exit properties of typical ray pairs passing through the system, see Figure 6.

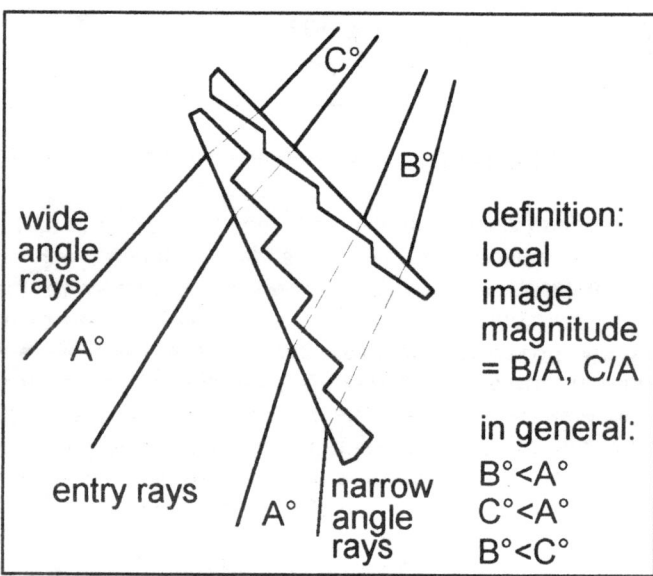

Figure 6. Ray Pair Entry & Exit Relationships

Angles indicated are those between the rays on either side of the respective angle labels, and the term 'local' image magnitude is used because image magnitude generally varies across the field of view. The following observations should be noted:

- Only angular relationships are of interest - ray location is irrelevant
- Local image magnitude is a relationship between entry and exit conditions of a pair of rays - properties of a lone ray are irrelevant
- Local image magnitude is generally less than unity
- Local image magnitude is generally less for narrow angle rays (eg bundle 'B') than for wide angle ones (eg bundle 'C')

The last two observations in the above list refer to properties in the plane of Figure 6, ie the lateral properties of the system. Since prismatic action takes place only in this plane, the 'vertical' properties of the system are unaffected by the prisms and are the same as for a conventional system.

It was mentioned previously that wide spacing of the two prisms produces maximum inboard 'throw' of the light, and this is the configuration normally used when the design priority is to maximise image and object fields of view (see below). If, however, achieving a large lateral image magnitude is the priority, then prism spacing is preferably smaller. The smallest possible separation, when the two prisms are a contacting pair, produces the largest lateral image magnitude of slightly less than unity (when combined with a flat reflector). The additional increment of image magnitude needed to achieve unity, if it is required, may be provided by a small amount of cylindrical curvature applied to one or both prisms.

Field of view

Image field of view, as distinct from object field of view, is the angular size of the image, which in turn is a function of the physical size of the image on the reflector and the driver's distance from it. For a given reflector (assuming it is filled with image) at a given distance from the driver the image field of view is fixed and is the same whether the reflector is being 'served' by the Double Objective's prism system or not.

Prismatic systems are normally designed with a reflector of similar size to that of a conventional system, but sited a little closer to the driver, thereby offering the possibility of a slightly larger image field of view. In practice, however, this advantage may not be fully realised as the effective aperture of the system is often set by the outer prism such that the reflector is not fully filled with image, as indicated in Figure 7.

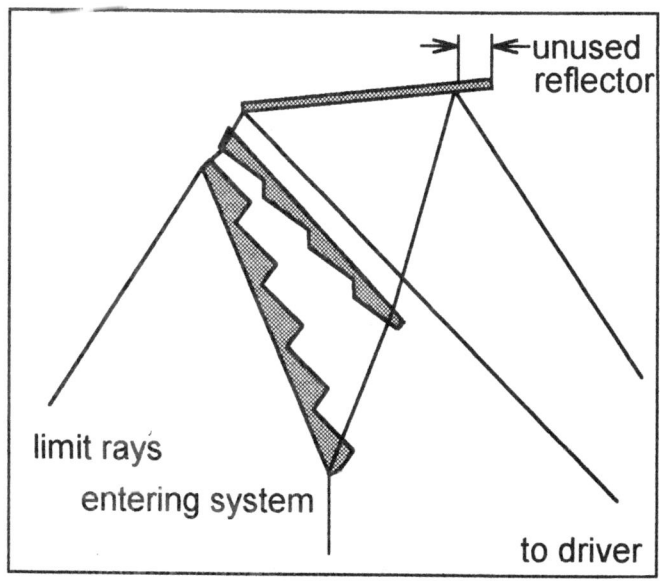

Figure 7. Outer Prism Forming Limiting Aperture Stop

The reason for this deliberate 'mismatch' is to permit a degree of eye articulation (driver head movement) before the image begins to be cut off by misalignment between the aperture stops at the two ends of the system (ie the outer prism and reflector perimeters). The same result can be achieved with the mismatch the other way around, such that the reflector is filled with image and the outer prism is slightly oversized, but this option is normally rejected after packaging and design considerations. Of course, there is also the option to precisely match the apertures and accept that the shape and size of the image frame will change slightly for every movement of the head.

Object field of view is the angular size of the object field and is a function of the image field of view and the mean image magnitude across it. Since the prisms influence image magnitude, the object field of view is similarly influenced by them. Specifically, since the prisms reduce lateral image magnitude they increase lateral field of view.

Convex reflectors are the normal means employed to increase object field of view, but the prismatic alternative has the following advantages:

- The wider field of view is achieved by reducing image magnitude in only the plane necessary, thereby leaving vertical information 'uncorrupted' (for improved distance estimation) and minimising area shrinkage of the image
- The driver's eyes do not need to be refocused when switching to and from the mirror; this is quite different from the convex mirror image which requires eye focus to snap from near infinity to only a few metres distant then back to near infinity again

Image distortion

The only significant form of distortion it is possible for the prismatic system to introduce is known as lozenge distortion and it arises as a direct consequence of the different image magnitudes in the lateral and vertical planes.

If the prisms are configured such that their apex angles are effectively vertical, then the principal axis of prismatic action is lateral and a rectangular object aligned with the horizon also appears as a rectangular image, albeit of a different aspect ratio, as shown schematically in the upper drawing in Figure 8.

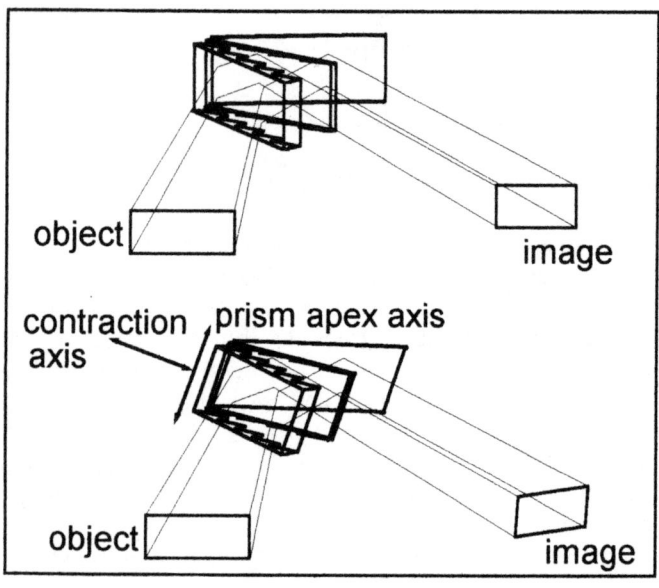

Figure 8. Lozenge Distortion Caused by System 'Roll'

However, if the system is 'rolled' inwards, as is often required by vehicle designers in order to blend the exterior prism surface with the 'tumblehome' of the cabin glass, the result is an image that is a parallelogram, as shown in the lower drawing in the figure. This effect may be understood by noting that the axis of prismatic contraction is now no longer horizontal but, in the example shown, approximately parallel to a diagonal of the object rectangle, and if a rectangle is contracted parallel to a diagonal, a parallelogram results.

Fortunately, methods exists to counter this effect, including adjustments to the roll angle of the correcting prism and the directions of the axes of the Fresnel grooves of both prisms.

Binocular alignment

The width of reflectors used in automotive exterior rear view mirror systems is normally at least twice the driver's inter-ocular distance. Such a width permits the possibility of a large central portion of the image field to be advantageously observed using binocular vision (ie by both eyes simultaneously) but the driver is only able to do this if the directions of the two beams of light entering for the eyes are related in a certain way. A normal pair of eyes is accustomed to being aligned parallel, for viewing distant objects, or convergent for closer objects, but not divergent or vertically misaligned. In general, when presented with an optical system that demands such contortions, human observers have difficulty and report seeing double. (The same phenomenon is perhaps most commonly experienced when viewing through misaligned field binoculars, and it may be noted that it involves both 'hardware' and 'software' adaptation failures; the eyes are unable to transmit registered images to the brain which in turn is unable to identify and overlay the parts common to both.)

Any arrangement of flat reflectors and prisms, either separately or together, is incapable of producing an optical system flawed in this way, but the introduction of a single curved reflector is a fundamental change that puts the entire system at risk. In a typical installation of an orthodox automotive exterior rear view mirror, use of a spherically convex reflector does introduce a small degree of binocular vertical misalignment which is normally comfortably accommodated by the driver. However, the addition of prisms considerably 'destabilises' the system, such that much larger misalignments become possible than would normally be associated with that particular reflector. This situation demands that binocular vertical misalignment must be calculated and closely controlled during the design stage. The same design parameters used to control image distortion are effective here, too, but care must be taken to avoid excessive conflict between the parameter demands of the two phenomena.

Finally in this section, additional information on the means by which the Double Objective configuration avoids the unwanted flare and secondary reflections of earlier designs is presented.

Flare

A simple mask was developed that was confirmed to be effective in reducing flare to an acceptable level. The solution was developed using masks manufactured by ink jet printing parallel black lines onto transparent film. Line spacing was made equal to the pitch of the Fresnel prism, and masks were characterised by line thickness.

It was found that, with optimum registration of the mask such that the mask lines 'shadowed' the Fresnel riser facets, flare was effectively eliminated when the thickness of the lines was about 15% of the pitch, or greater. The preferred location for the mask was established to be in contact with the Fresnel profile, although masks fitted to the other side of the material also worked. It may be noted that this solution is applicable to the Fresnel prisms used in all versions of the Milner mirror.

Other solutions in which flare control is achieved directly within the prism, either by changes to profile design or by post processing, are under development.

Secondary reflections

The problem of secondary reflections was caused solely by the specific location of the ocular prism in the Wedge configuration, specifically its close proximity to the reflector. By relocating the ocular prism to the position it occupies in the Double Objective design the problem is eliminated.

OPTICAL COMPONENT DESIGN

As a result of establishing the overall design objectives and optical characteristics of the system, as described above, most prism design parameters have already been determined, including:

- Plane location
- Perimeter shape
- Material
- Apex angle (but see below)
- Fresnel groove axis orientation

It remains to specify only the following parameters:

- Riser angle
- Elemental pitch

The apex angle of the outer prism was earlier specified to be 'as large as possible' (in order to minimise the length of the outer prism) and is listed above because its value needs to be known (or assumed) in order to calculate all the critical optical characteristics discussed above. However, the practical limit to how large this apex angle may acceptably be is determined by the influence of a feature of Fresnel prisms that is introduced in this section, namely the riser facet. This influence is illustrated in Figures 9, 10 and 11.

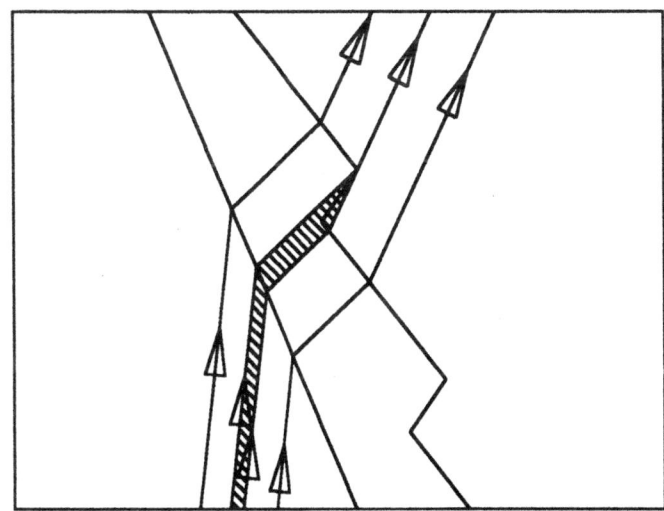

Figure 9. Fresnel Apex/Riser Angle Relationship (1)

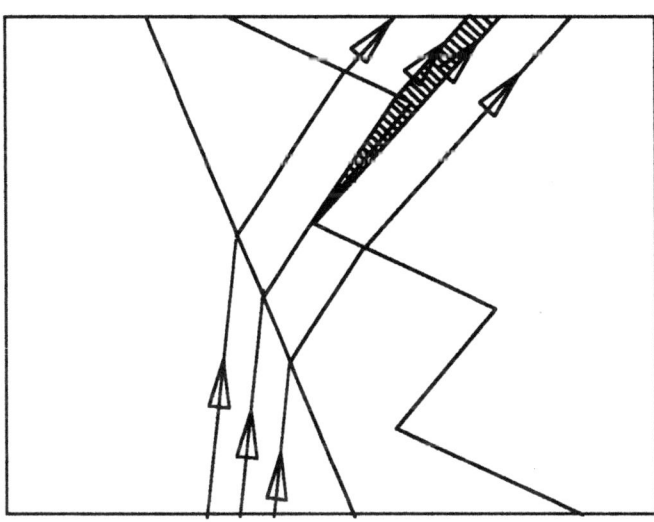

Figure 10. Fresnel Apex/Riser Angle Relationship (2)

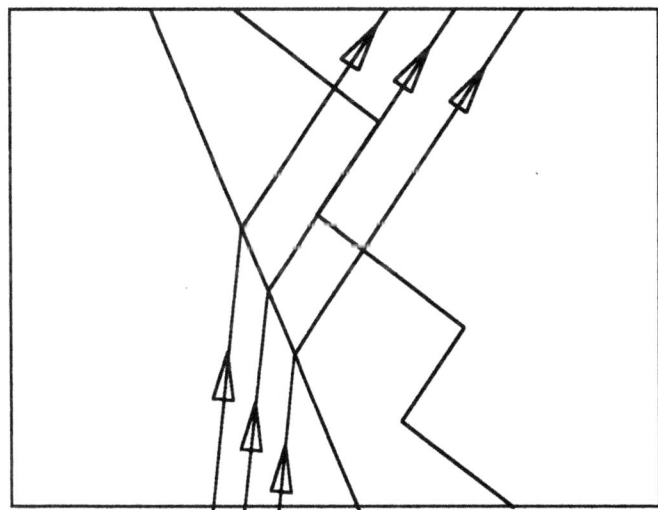

Figure 11. Fresnel Apex/Riser Angle Relationship (3)

In order to appreciate the significance of these figures, it should be appreciated that the observer sees only the output rays when using the mirror. Figure 9 illustrates a condition in which the prism apex angle is smaller than it might usefully be. The hatched area, which includes the riser, is invisible to the observer, but the apex angle may be made substantially larger before this advantageous condition is lost. In Figure 10 the apex angle has been increased too much, and the hatched area, which contains only views of the riser, is visible to the observer. Figure 11 shows the limiting condition between the previous two, which represents the maximum useful apex angle condition, normally about 30° for acrylic. This raytrace also defines the optimum riser angle, which may advantageously vary across the field of view.

Once the apex angle of the outer prism is decided, the optimum apex angle of the inner prism follows, based on the criterion of minimum chromatic aberration. For a 30° apex angle acrylic outer prism the apex angle of a polycarbonate inner is about 15°. The optimum riser angle for the inner prism is determined by raytracing, in similar fashion to the outer prism, and it may also advantageously vary across the field of view.

Both prisms may be manufactured in the form of a raw material that is suitable for all applications for which the same apex and riser angles are appropriate. Typically this might be a thin sheet of prism material which is cut to shape and laminated to a glass carrier. This method is used successfully for prototype applications and is adaptable to volume production.

Elemental pitch of the Fresnel elements is not constrained by any of the forgoing optical criteria, but may be selected on the basis of manufacturing capability, possibly influenced by secondary optical considerations such as possible light scatter from the crests of the elemental prisms. Depending on the manufacturing method favoured, any pitch values in the range 0,1mm to 5mm might be considered appropriate.

As regards reflectors suitable for the Double Objective prismatic mirror, any type appropriate for conventional mirrors may be considered.

NON-OPTICAL COMPONENTS

The non-optical components required to install a Double Objective mirror system in a vehicle comprise the following:

- Prism cell and mounting arrangement
- Adjuster and mounting arrangement
- Heating (if required)

Prism cell and mounting arrangement

The prism cell is an injection moulding that locates the prisms and protects their inner surfaces from damage and soiling. A typical design is illustrated in Figure 12.

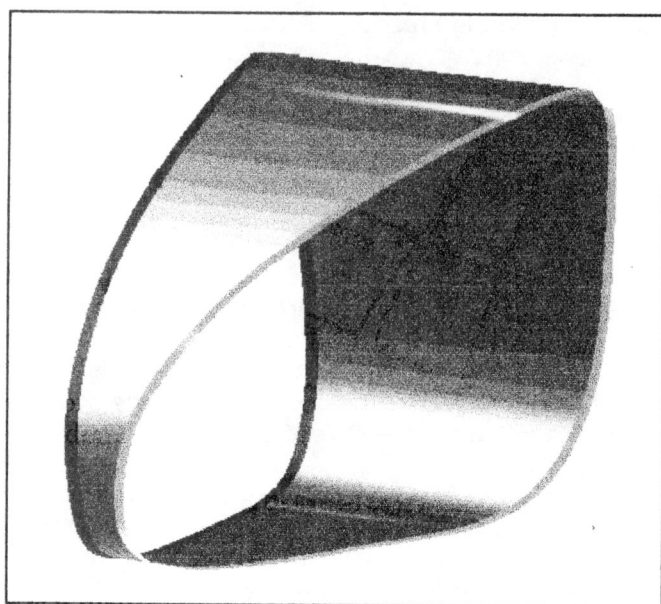

Figure 11. Typical Prism Cell Design

With the prisms fitted, the interior of the cell is sealed and the sub-assembly thereby formed is a robust structure with the durable shell required for the automotive assembly environment and the required life in service.

Since the perimeter of the outer prism needs to be sealed into the outer skin of the vehicle for environmental reasons, the same sealing line may be advantageously used to effect the mounting of the cell sub-assembly in the vehicle. The host component for this mounting may be a specially formed skin panel or a window.

Adjuster and mounting arrangement

The mirror adjuster normally used is an industry standard device mounted on one side to the back of the reflector and on the other side to a mounting bracket affixed either to the vehicle door or to a fixed part of the body structure such as the instrument panel.

Due to its interior location, reflector aerodynamic and vibrational disturbances are reduced. In addition to producing an extremely stable image, as previously noted, this may also permit the use of a smaller, lighter adjuster.

The adjuster itself may be electrically powered or manually operated. Minimum weight and cost are achieved if the same system is used to adjust the two new interior reflectors as is used for the existing one, namely the ball and socket arrangement almost universally adopted for interior mirror adjustment. If the adjuster is electrically operated, however, then it is desirable to specify a low noise type.

Heating

If heating is justified it has to be on the basis of de-icing the outer surface of the outer prism, and that is equivalent to de-icing the outer surface of the side window in a vehicle with conventional mirrors, a task that is normally left to the HVAC system of a vehicle. In order to match such a performance, a dedicated, low-power warm air or infra-red heating system located inside the prism cell is all that is required.

SYSTEM INTEGRATION & COST

The radical low profile silhouette of the system as typically installed is strikingly illustrated in the three photographs of the same vehicle shown in Figure 12.

12.1. Plan view of conventional exterior rear view mirror

12.2. Plan view of same mirror in 'parked' position

12.3. Plan view of Milner Mirror

Figure 12. Three Mirror Configurations on Same Vehicle

The level of integration illustrated in the bottom picture allows considerable weight saving to be achieved. First there is the direct saving associate with deletion of the normal mirror's housing and folding or breakaway mechanism; this more than offsets the weight of the prism cell sub-assembly, which is about 300g. In addition there is a further weight saving available in the door structure which no longer has to provide a stiff mounting for the heavy, cantilevered conventional mirror assembly; this saving is estimated to be typically around 100g.

Manufacturing cost of the Double Objective mirror is likely to be similar to that of an equivalent conventional mirror. This projection is based on indications that the cost of the extra components required for the prismatic system is expected to be no more than the cost of the components no longer required.

SUMMARY

The Double Objective version of the Milner prismatic exterior rear view mirror system was developed to eliminate unwanted optical phenomena sometimes observed under adverse conditions with earlier versions of the system.

The new configuration comprises a cell to house and protect the two prisms, both of which are of the Fresnel type, and a conventional reflector mounted separately from the prism cell which the driver views directly, in the same way that he views the normal interior mirror.

It is in this form that the Milner mirror is currently undergoing production development in support of its targeted first appearance on a volume production vehicle in the year 2000.

REFERENCES

1. US Patent 5,594,593.
2. P J Milner, 'Prismatic Exterior Rear View Mirror Systems', IBEC'96, October 1996, Detroit, MI, USA.
3. P J Milner, 'Evolution of the Milner Prismatic Exterior Rear View Mirror', IBEC'97, October 1997, Stuttgart, Germany.
4. US Patent 5,617,245.

BIOGRAPHY

Peter J. Milner

With a B. Eng. in mechanical engineering, he commenced his career in the automotive industry with Rootes Group (later Chrysler Europe) followed by a spell of several years with Volkswagen in Wolfsburg, Germany. After returning to Chrysler Europe as a senior engineer he became chief engineer of DeLorean Motor Cars Ltd in Northern Ireland before joining the Motor Industry Research Association in Nuneaton, UK in 1983. Since 1986 he has operated as a senior consultant with de Montfort and has been responsible for the invention, design, and development of the Milner prismatic mirror now licensed by de Montfort to major first tier component suppliers in Europe, North America and Japan.

Richard E. Berg

In an extension of the G.M.I. Program he began his automotive career at the General Motors Tech Center in 1957. In 1963 he joined Inmont Corporation which through acquisitions and mergers became part of United Technologies Automotive. During a 33 year career at UTA he held positions in R&D, manufacturing, sales and marketing. In 1995 he took an early retirement from UTA and established his own company, Intertech Technology Marketing, and formed an alliance with de Montfort to market its prism mirror technology as well as its consultancy services. He also represents other client companies, marketing leading edge technology to auto manufacturers and suppliers.

980921

An Advanced Optic Rear Vision Device for Motor Vehicles

S. Li
Su Li Patent

S. FukSang
Su's Autoptic Limited

Copyright © 1988 Society of Automotive Engineers, Inc.

ABSTRACT

The outside rear view mirrors on motor vehicles are located outside a vehicle's solid body, normally attached to the side doors. Owing to their contour and positions, they present certain drawbacks which may cause dangers on the road, especially under unfavorable conditions when both the side door glasses and the surface of the outside mirrors are contaminated. An advanced optic rear view mirror unit has been designed and developed, with the aim to eliminate all the drawbacks of said outside rear view mirrors and hence to enhance the traffic safety. It shows that the said mirror unit eliminates the aerodynamic drags and sight blocks caused by the protrusion of conventional outside mirror bodies, places the rear vision inside a vehicle at a more logical physical position related to the driver, and provides an always clear view to the rear area regardless the outside weather conditions.

INTRODUCTION

The conventional rear vision system applied on motor vehicles today consists of one inside (interior) reflecting mirror and two outside (exterior) mirrors. The purpose of using such a rear vision system is to provide the driver of a motor vehicle a clear and reasonably unobstructed view to the rear, and hence to effectively reduce traffic accidents caused by the blindness and obstruction of vision to the concerned areas. Regulations regarding rear vision systems were made to guide their safety application on the road, by which one important aspect is to ensure that the system provides the driver a sufficient rear view along the side of a vehicle (1). To fit this requirement, two rear view mirrors are typically placed outside a vehicle body (hence the name "outside rear view mirrors"), with each of the mirrors having a body size which protrudes about 200mm to each side of the vehicle. As a consequence, outside mirrors directly or indirectly present certain drawbacks.

Although conventional outside rear view mirrors are optimally designed to reduce the aerodynamic resistance as much as possible, the large body of the two exterior mirrors are still one of the main problems for the automotive industry with regard to the maximum reduction of aerodynamic drag. This aerodynamic drag can result in reduced vehicle dynamic efficiency, extra gasoline consumption and environmental pollution.

The placement of the two outside rear view mirrors each forms a large angle relative to the driver's position and the straight-ahead front view. The driver needs to turn his head by the same large angle to check the rear view. In order to minimise the potential danger of this manoeuvre, the angle on the driver's side is rigidly restricted not to exceed 55° (2). All the current exterior mirrors are reaching the extremity. Since the human eyes each will provide a field of view of approximately 50°, a driver will lose the important front view considerably when he is looking at the rear view mirror, which would cause a temporary "blind spot" to the front area and may lengthen a driver's reaction time should a emergency occurs in that area.

Since the mirrors are placed outside a vehicle, the driver also needs to look at the mirror through the side door glass. Practically, there are many circumstances in which either the surface of the outside rear view mirrors, or side door glasses, or both are contaminated, for examples by water drops, moisture or road film etc., which makes rear vision difficult.

Among other drawbacks of the outside rear view mirrors are the dangers of collision with other road users; vulnerability to damages of the mirror construction themselves; and aerodynamic noise at high speed.

During recent years, certain improvements enabling the outside rear view mirrors to provide better performance were achieved. However, such improvements are unable to eliminate the drawback or drawbacks of the outside rear view mirrors. Other attempts, including electronic transmitting devices and other optic type mirrors are also under investigations.

To avoid the disadvantages caused by conventional outside rear view mirrors, prior attempts in the optic field have been trying to reduce the large protrusion of the outside mirror bodies while providing the rear vision within a vehicle with more proximity to the driver. These

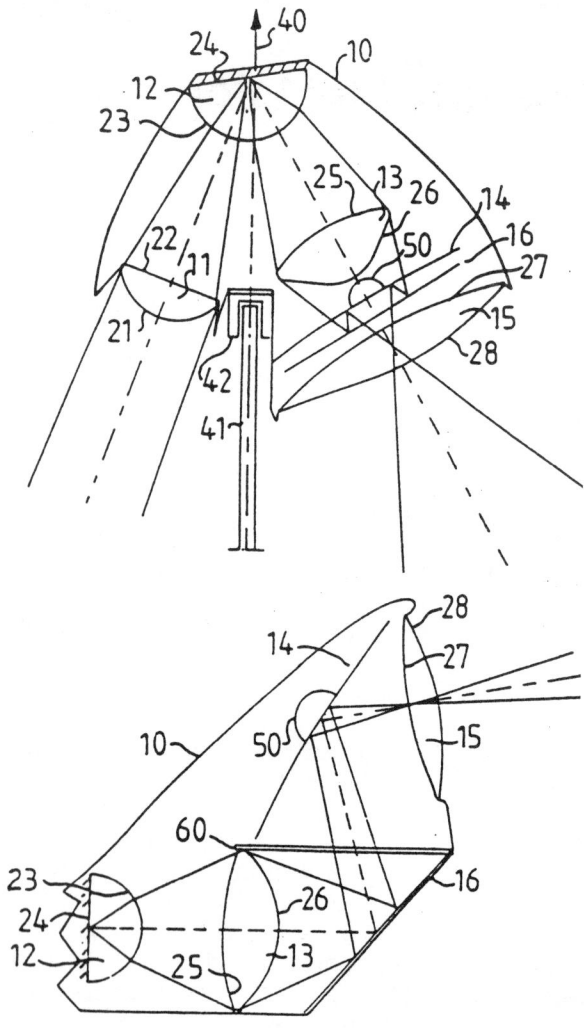

Fig. 1. Schematic sectional over-view and side-view of the advanced optic rear view mirror unit.

devices all concern multiple combination of optic element by means of refraction and reflection. Prior art patents (3, 4, 5) utilize prismatic refractors, to deflect the light from objects behind a vehicle towards a reflector, which is placed forwardly of an observer so as to be in his field of view. Other devices, disclosed by previous patents (6, 7, 8, 9) include plurality of lenses, or a combination of lenses and prismatic means as refractor, plus simple or prismatic reflector/reflectors to form a rear view unit. However, such devices suffer either from the optical defects which would destroy the quality of the image or complexity. Additionally, many of them require improper occupancy of space on the side door or within a vehicle which would have already been overstuffed by instruments. For these reasons, very few of them have achieved practical and commercial significance.

It is the aim of this paper to present an advanced optical rear view mirror unit, which would eliminate the drawbacks of the conventional outside rear view mirrors, and also eliminate the drawbacks of previous optic devices. It is another aim of the present paper to present such a optical mirror unit which would not require major reconstruction both on the side door and within a vehicle.

REPRESENTATION

The main characteristics of the advanced optic rear view mirror unit are hereby presented and discussed with the aid of relevant figures, photos and tables.

A PAIR OF SHARP "EYES"
-A BRIEF TECHNICAL PERSPECTIVE

A schematic sectional over-view and side-view showing the advanced optic rear view mirror unit in principle is displayed in Fig. 1.

The whole mirror unit was designed to imitate the function of the human eye with regard to the image formation and image proceeding. Its first objective lens 11 has a function exactly as the lens of a human eye, which forms an preliminary inverted image of the object. In the human eyes, the inverted image is focused on the retina tissue from where the light signals are transformed into bioelectric signal and further proceeded to the brain; the brain interpret the inverted image as an upright one. Similarly, the said image in the said optic mirror unit is focused on a optic receptive component 12 by which the light signals are collected and further transmitted to a third optic component 13. The optic component 13 is another *imaging* component which invert and revert the preliminary image by a single rotation and provides a secondary image which is upright and with desired left-right orientation. The upright image is being placed in or in the vicinity of the final optic receptive component 15, enabling it to serves as a screen monitor and presents the driver a realistic vision of the concerned area.

The second optic component 12 plays an important role in the optic mirror unit. Firstly, it acts both as a refractor as well as a reflector which receives and transmits the light as well as bends the light by one single optic component, which provides the mirror unit with a sufficient field of view and gives it an extremely compact package; Secondly, it applies a single reflection to the light from outside a vehicle into within, thus avoids the excessive occupancy of a vehicle's inside space; Thirdly, the light rays incident to its reflective surface form a narrowest optic pathway, which maximally utilizes the corresponding triangle construction on the side door; and lastly, its spherical surface constitutes exactly a half of a sphere, which enables the mirror unit to be easily adopted to different types of vehicles without introducing any additional optic aberrations.

Compared with other prior arts of similar type, another advantage of the mirror unit is its flexibility to achieve a sized image with different magnification without altering the image quality and the mirror package significantly. A simple formula characterizes the feature in this respect can be written as: **M = S2/S1 x S4/S3**, while **M** denotes image magnification (image size); **S1** denotes the distance between optic component 11 and 12; **S2** denotes the distance between optic component 12 and 13; **S3** denotes the distance between optic component 13 and 15; **S4** denote the distance between optic component 15 and the eyes of a driver. By changing the parameters **S1, S2,** and **S3** separately or simultaneously, sized images with different magnification are achieved. Similarly, it is also conceivable to maintain the image size while changing the mirror package, as an example, when it is desired to move the mirror body on the right side of a vehicle to the middle of the dash board and provide a symmetric rear vision corresponding to that of the left side (refer to Fig. 3a).

STREAMLINED CONCEPT
NO AERODYNAMIC DRAG

The advanced optic rear view mirror unit removes the huge bodies of conventional outside rear view mirrors; and provides the automotive engineers a totally new concept of modern vehicle design. The maximum sideway protrusion of the outside objective component in the mirror unit, at its broadest rim, is less than 1/3 of that of a conventional mirror. Furthermore, since the second optic component is located partly within and partly outside a vehicle body, the exterior part of the mirror unit can be built naturally to have an aerodynamically streamlined shape. A mirror prototype installed on a passenger car showing its outside contour and the difference of side protrusion between the advanced optic mirror unit and the conventional rear view mirror is presented in Fig. 2a-b.

A simulation test was carried out by one of the world's leading automobile manufacturers, using a 3-D sketches model. The model was tested for aerodynamic resistance in a wind tunnel on an actual vehicle. The results showed that there is virtually no difference between a 3-D model and a blank control in this respect (data not available), which means the installation of the advanced optic mirror unit totally eliminates the aerodynamic drags associated with the conventional rear view mirror.

TOTAL FREEDOM ON THE SIDE
ALL ROUND PERFORMANCE

The characteristic of total freedom on a vehicle's side brought about by the advanced mirror unit could be illustrated in several aspect.

1. The mirror unit provides a sufficient field of view with uniform image size, which eliminates the so called "blind spots" along the vehicle side. Evaluation under actual traffic conditions showed that when a driver notices a passing vehicle to the side, the tail of that vehicle is still visible in the mirror unit. In this case, the image is being retained to have a size demagnification corresponding to the ratio of a conventional mirror which still conforms to the requirement stated in the regulation (10).

Fig. 2a. Side view of an advanced optic rear view mirror unit on a passenger car showing its outside contour.

Fig. 2b. Front view of an advanced optic rear view mirror unit showing the difference of side protrusion compared with the conventional outside rear view mirror.

2. By removing the physical bodies of the conventional rear view mirror, the formerly visually blocked front area (area **f** in Fig. 3a) are also set free.

3. There could be one design in which the said optic mirror construction is rigidly fixed on the vehicle frame and not movable (Fig. 3b), which would enable the driver to still have a clear view to the rear, even when the side door opens.

The main body of the mirror unit within a vehicle can be placed at a corner made up with the "A" pillar frame on its top, the front door part on its outward side, and the edge of dash board on its inward side. Regional styling can be made accordingly to suit different types of vehicles and to form a neat lineament harmonious with the vehicle's entire inner construction.

The re-arrangement of rear vision from outside a vehicle into inside of it enables the driver to perceive information to the rear area in a more rational way. The angle at which a driver need to turn to see a image is

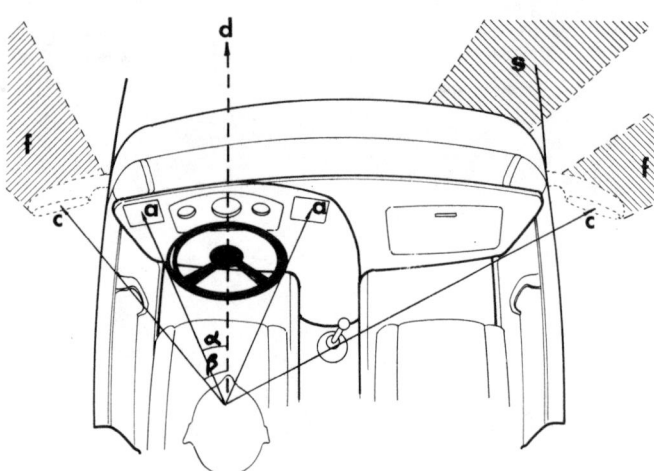

Fig. 3a. Schematic over-view showing the difference of head turning angle between the advanced optic rear view mirror unit and the conventional outside rear view mirror.

Fig. 3b. Schematic side-view showing a design in which the advanced optic rear view mirror unit is rigidly fixed onto a vehicle's body instead of on the side door.

being reduced from about 50° (angle β in Fig. 3a) in the conventional mirror to about 25° (angle α in Fig. 3a). Preliminary test showed that the average time which allows a driver to complete seeing a clear image and returning back to the front view is approximately 1.4 seconds for the advanced mirror unit at the driver's side mirror, while this value is doubled for the conventional mirror. The less time spent on the advanced mirror unit seems to be mainly owing to reduced head turning angle, more proximity between the mirror body and the eyes of a driver, and avoiding looking for the image through solid door glass. The mentioned time period was designated as "cruising time" and the distanced covered by this time period while driving as "cruising distance". Table.1. shows the difference between a conventional mirror and the advanced optic mirror with regard to cruising distance, braking distance, and the shortest stopping distance at different speeds. In actual traffic conditions, a driver's reaction time, which is normally one second, is also to be taken into consideration. This means that the stopping distance listed on the table will be even longer.

The human vision consist of two parts, the direct vision and the peripheral vision. Utilising peripheral vision can be a considerable advantage on vehicle applications when scanning the traffic environment, which brings objects along the road to the notice and identification by direct vision of the driver. This advantage, however, is handicapped in a motion when drivers tend to look at a conventional mirror and, at the same time, keep alert to the front view. In such cases, objects at the corner covered by the front wind screen frame on the opposite side (area **s** in Fig. 3a) will be out of the sight and will not be noticed, thus form a "blind spot" in that area. Contrarily, that area will be totally retained in the peripheral vision with the advanced mirror, which will give the driver a clue of that area and allow the driver to react promptly should an emergency occur.

Since the main body of the advanced mirror unit is closer to the human's inherent reading distance, the driver will also feel more assured and more confident about the image, especially in bad weather conditions.

ALWAYS CLEAR VIEW

In motor vehicle application, it appears that contamination will inevitably take place at the surface of optic component of a rear view system, especially if exposed to outside weather conditions. In this respect, it is important to maintain the image with adequate brightness and avoid any climate influence whenever possible. Since the conventional outside rear view mirrors are located outside the vehicle body, the mirror surfaces are easily subject to contamination of various kinds; Furthermore, side door glasses are also often contaminated at the same time. In such cases, the rear scene become very difficult to see.

Table.1. Comparisons of relevant distances

	Cruising Distance (m*)		Braking Distance (m)		Stopping Distance (m)	
	C*	A*	C	A	C	A
30km/h	25.2	12.6	5.4	5.4	30.6	8
50km/h	42	21	15	15	57	36
70km/h	58.8	29.4	24.5	24.5	83.3	53.9
90km/h	75.6	37.8	40.5	40.5	116.1	78.3
110km/h	92.4	46.2	72.6	72.6	165	118.8

*C=Conventional rear view mirror; *A=Advanced rear view mirror unit; *m=Meters

One distinct feature of the advanced optic rear view mirror unit is that it provides the driver rear vision which is less affected by outside weather conditions. This feature is closely related to its unique optic design and associated physical construction:

Firstly, the converging properties of the outside objective lens in the mirror unit enables it to act as the lens of the human eyes, which brings the light of interest into focus while leaving any contamination attached on its surface out of focus and hardly noticeable. This character is constantly maintained in the optic mirror unit which keeps the image clear and sharp. Furthermore, although the remaining optic components are inside the vehicle and ensured by a protective case, and are hence less affected by contamination compared to the objective component, they are designed to avoid such contamination as much as possible. One typical example is the second optic component, whose reflective interface serves as a plane on which the first image from the objective component is focused. It is conceivable that contamination on that surface will be magnified causing interruption to the main image. However, the mentioned surface is coated directly on the back side of the second optic component and is safeguarded by its substantial material in the front, ensuring that no contamination will reach that surface.

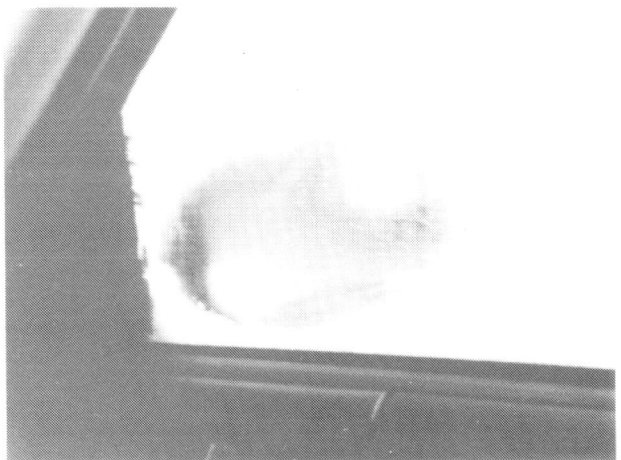

Fig. 4b. Vision comparison between a conventional outside rear view mirror (above) and an advanced optic rear view mirror (below) when the door glasses were misted up.

Fig. 4a. Vision comparison between a conventional outside rear view mirror (above) and an advanced optic rear view mirror (below) in a heavy snow fall.

Secondly, the characteristic of keeping a clear image by the advanced mirror unit is further enhanced with the addition of a sheltering cover to its exterior part. This sheltering cover seals the main body of the exterior part while leaves an aperture pointing to the rear with the objective lens well within the aperture canal. Considering the much smaller dimension of the objective lens compared to a conventional mirror, this sheltering cover will be able to prevent the objective lens from being contaminated by rain, snow etc. efficiently. The design and construction of the sheltering cover may also take advantages of the aerodynamic turbulence around its aperture, to further clean away the minor contamination on the surface of the outside objective lens.

Finally, compared to the conventional outside rear view mirrors, the advanced optic mirror unit avoids the drawback of having to look for the image through the solid door glass which is also easily subject to contamination along with conventional outside rear view mirrors.

Fig. 4c. Vision comparison between a conventional outside rear view mirror (above) and an advanced optic rear view mirror (below) in a rainy weather.

Comparisons between an advanced optic mirror unit and a conventional mirror are photographed under conditions of heavy snow fall, mist, rain and road film which are shown in Fig. 4a-b-c-d.

During the procedure of evaluations, it was observed that the advanced optic mirror unit provides a much clearer image to the rear under adverse weather conditions. It was also obvious that the more severe the outside environment, the more clear the difference between the advanced optic mirror and conventional mirrors with regard to the clarity of the image became. At times, water vapour can appear on the surface of inside optic components owing to the temperature difference within and outside. This phenomenon vanished rapidly through the utilization of the ventilation system which is normally equipped on a vehicle. The ventilation air, whether warm or cold, is blown into the protective case, via the ventilation hole on the corner of the dash board (Fig. 5) and the air holes at the bottom of the protective case (Fig. 6). The air turbulence is out-led through the holes on the side of the protective case and blown onto the side door glass (Fig.7).

During the winter when it is extremely cold, frost may occur on the window glasses, the surface of a conventional rear view mirror and the surface of the advanced rear view mirror unit. In this case, the frost take place only on the outside surface of the objective lens and appears to be less severe. A driver can still see a reasonably clear image in the advanced mirror unit while the rear vision is totally blocked to the conventional mirror. However, as the inside of the vehicle gets warmer, water vapour will occur on the surface of the inside optic components. This vapour can again be removed efficiently by the ventilation method mentioned above. The time needed to evaporate the surface of inside optic components was measured to be 2-5 minutes, well ahead of the heating procedure for cleaning up the surface of a conventional outside rear view mirror.

Theoretically, water vapour may even occur only on the outside surface of the objective lens. Although it may be dried by wind turbulence while driving, it may require additional means. Proposed solutions include air-dry method of blowing ventilation air directly on the surface, or electric heating of the objective lens.

Fig. 4d. Vision comparison between a conventional outside rear view mirror (above) and an advanced optic rear view mirror (below) at road film.

Fig. 5. The ventilation air is blown into the protective casing of the advanced optic rear view mirror unit, via the ventilation hole on the corner of the dash board.

Fig. 6. The ventilation air blown into the protective casing through the air holes on its bottom.

Fig. 7. The air turbulence is out-led through the air holes on the side of the protective casing and blown onto the side door glass.

DISCUSSIONS

The characteristics of the advanced optic rear view mirror unit, compared to a conventional outside rear view mirror, can be briefly summarised as following:

- Provides a streamlined concept of modern vehicle design;

- Eliminates the aerodynamic drag and noise;

- Eliminates the "blind spots" on the vehicle side;

- Removes the sight block caused by the conventional mirror bodies;

- Provides rear view inside a vehicle;

- Reduces the large head turning angle;

- Avoids seeing image through side door glass;

- Provides an always clear view.

A series of prototypes have been analyzed and evaluated since 1992, both in the optic laboratory with the assistant of highly advanced computer imaging programs and under the actual traffic condition with the aid of the world's leading automotive manufactures. It appears that the advanced optic rear view mirror unit presents a new concept of rear vision which adds safety factors, is more logical, and is environmental friendly.

The removal of the outside rear view mirror bodies eliminates the aerodynamic drags, which in turns means reduced fuel consumption and environmental pollution. Statistic analysis showed that at a speed of 90 Km/h, the aerodynamic resistance caused by two exterior mirrors is equal to carrying an average adult body weight of 60Kg. The extra fuel consumption owing to aerodynamic drag is about 1000 litres per average life of a passenger car during 20 years period. According to the available data, 0.003 kilograms of various harmful components are discharged upon consumption of one kilograms of fuel. The various harmful components converted by said extra fuel consumption per year only by the passenger cars in 1987 were as examples: Carbon Monoxide (CO) 21 million Kilograms, Carbon Dioxide (CO_2) 63 million Kilograms, Nitrogen Oxides (No_x) 420 million Kilograms, and unburned Hydrocarbons (HC) 2 million Kilograms.

The cost level regarding the advanced optic rear view mirror device is one of the most concerned topic both to automobile manufactures and rear view mirror suppliers. Since the multiple application of optic components, the cost of the advanced optic mirror unit is higher than that of the conventional outside rear view mirror. Table 2. represents some relevant information in this regard at the current time and rate.

Table.2. Optic components and cost levels

ITEMS	OPTIC COMPONENTS			
	No. 1	No. 2	No. 3	No. 4
Optic Material	BK7	BK7	BK7	BK7
Refractive Index (587.56 nm)	1.5168	1.5168	1.5168	1.5168
Density in g/cm^3	2.51	2.51	2.51	2.51
Diameter (mm)	80	80	80	126
Weight (g)	120	165	120	360
Cost level (USD)	4.1	5.6	4.1	12.3
UNIT COST (USD)	26.1			

BK7 is a borosilicate crown glass. It is one of the most widely used material by optics industry. There are several reasons for this: BK7 is relatively stable to chemicals; relatively hard; very low bubble content; and shows excellent transmitance down to 350nm.

There are a wide range of other optic materials which can be chosen for this application with consideration of the following facts:

- BK7 has a relatively low refractive index, while optic material with higher refractive indices are preferred in this respect;

- Since only two optic surfaces, the outward surface of component No. 1 and No. 4, are vulnerable to physical abuses, material hardness and chemical resistance are not critical for other inside components or inward surfaces. Furthermore, owing to the optic design, even if physical abuse occurs, such as finger prints, scratches and other contamination, they are hardly noticeable on the outward surface of component No. 1, and can easily be approached for treatment on the outward surface of component No. 4. If necessary, a specific coating can be applied uniquely to these two surfaces to increase their hardness.

- Optic components No.2 and No.4 do not contribute to image formation, but are mainly for light distribution; materials with lower quality, hence cheaper price, can be used without fearing the deterioration of the image. Furthermore, the weight of these two components, especially No.4, account for 68% of all optic material used. This might be a considerable advantage with regard to the reduction of unit cost.

To effectively reduced the unit cost and the total weight of the advanced mirror unit, a Fresnel lens has been tested as a candidate to replace the largest spectacle lens. Results showed that the quality of the primary image was unchanged. Since the Fresnel lens is located at a position which does not contribute to the image formation, and since the final image is overlapping with the substantial material of the Fresnel lens, the pitch grains on its inward interface were not obvious. The Fresnel lens also has a flat interface on one side, which was placed outwardly of the mirror unit facing to the driver. The secondary reflection on that outward interface appeared to be much less compared to the convex surface of the spectacle bi-convex lens. The tested Fresnel lens, having a focal length close to that of the spectacle lens, weighed just 22 grams compared to 360 grams of the spectacle lens. Another attraction of using Fresnel lens instead of glass lens is that the former costs only 1/3 of the latter at volume production.

The safety application of the advanced optic mirror concept within a vehicle is another concern from the automotive industries. Practically, the inside body of the mirror unit can be built to have a smoothly shaped contour itself and harmonious linkage with its adjacent structures. Safety materials which are used for the vehicle inside constructions can be also used for the casing of the advanced mirror. Since the configuration of the inside mirror body provides a relatively flat facade towards the driver, rather than a sharp corner at that area, the advanced mirror unit appears to be more attentive to the passive safety relative to the front collision or tumbling accidents. Furthermore, the last spectacle component can be built to be retreatable upon impact to further lessen the outcome of an accident.

Compared with another challenging alternative to the conventional rear view mirrors - the video camera, the advanced optic rear view mirror also possesses distinct advantages. Firstly, it does not need to be driven by electric power; secondly, the number of components involved is much less and more durable, and any default of individual component can be easily recognized and replaced; thirdly, although it presents a final image which hits on the spectacle component similar to that which hits on the screen monitor of a video camera, it avoids the loss of image fidelity by the latter owing to electronic signal conversion and representing procedure. This provides an image with better contrast and depth of field; This feature is important with regard to the judgment of the distance between one's own vehicle and the objects behind; and lastly, but not leastly, since its construction is much simpler, it is much lower in cost.

Further research regarding the advanced optic rear view mirror unit includes an anti-night glare program, which aims at prompt reduction of light intensity from the head lights of vehicles behind, by adding adequate chemical substrates in the optic raw material, or applying a specific optic coating on the surface of the optic component; These procedures are under way.

ACKNOWLEDGMENTS

The valuable assistance of providing relevant information by AB Svensk Bilprovning (The Swedish Motor Testing Center) and by Sveriges Trafikskolors Riksförbund-STR (The National Association of Swedish Driving Schools) are hereby gratefully acknowledged.

REFERENCES

[1]. Regulation No. 46., Uniform provisions concerning the approval of rear-view mirrors, and of motor vehicles with regard to the installation of rear-view mirrors. United Nations, Jan 1989 (hereinafter will be referred as Regulation No. 46)., Requirement, 16.5.3.1. and 16.5.3.2.

[2]. Regulation No. 46., Requirement, 16.3.4.

[3]. Milner, P. J., A Rear View System for A Vehicle. Pub. No. WO 9006866 A1, Jun 28, 1990.

[4]. Milner, P. J., A Rear View Mirror Unit, Pub. No. WO 9218353 A1, Oct 29, 1992.

[5]. Heber, K., Spiegaleinheit. Pub. No. DE 3146486 A1, Jun 1, 1993.

[6]. Schröder, W., Optisches Rückblicksystem. Jos. Schneider Optic Research GmbH & Co, Pub. No. DE 33 35 981 A1, April 18, 1985.

[7]. Alfred, U., Rückblickvorrichtung bei Kraftfahrzeugen. Pub. No. DE 2 014 696 A, Oct 21, 1971.

[8]. Odebrecht, W., Fahrzeug mit einem fahrerseitigen und einem beifahrerseitigen Außenspiegel. Mercedes-Benz Aktiengesellschaft, 70327 Stuttgart, Pub. No. DE 42 35 744 A1, May 5, 1994.

[9]. Maurin, J-F., Rétroviseur, notamment pour véhicules automobiles. Pub. No. FR 2240618 B1, Mar 7, 1975.

[10]. Regulation No. 46., Special Specifications, 7.2.3.1. and 7.2.3.2.

CONTACT

For more detailed technical information and discussion, please contact: Mr. Su Li, Armegatan 32-607, 171 71 Solna, Sweden. Tel: +46-8-735 53 12; Fax: +46-8-83 64 96; Mobile phone: +46-070-496 2250; e-mail address: suli@swipnet.se

980922

Added Feature Automotive Mirrors

Niall R. Lynam
Donnelly Corporation

Copyright © 1998 Society of Automotive Engineers, Inc.

ABSTRACT

Automotive rearview mirrors have numerous attributes that render them desirable hosts for a variety of added features beyond their principal function of providing a rearview field of vision. One attribute is location. The driver frequently looks at rearview mirrors as part of the normal driving task, and thus they are ideal locations for information display such as of directional information from a compass sensor and/or of temperature information from an exterior temperature sensor. Icons and indicia displaying status of, for example, passenger airbag enable/disable, are readily viewable by the driver when displayed at an interior rearview mirror or exterior sideview mirror. Rearview mirrors are desirable locations for automatic wiper activation rain sensors, automatic headlamp activation controllers, remote keyless entry receivers, garage door opener/home access transmitters, and antennae such as for global positioning satellite (GPS) systems. Mirrors are also excellent locations for lights such as map reading lights incorporated in interior mirror assemblies and ground illumination security lights located in exterior mirrors assemblies.

Another attribute is electrical service. Many interior and most exterior mirrors are electrically powered. For example, electrochromic mirrors that electrically dim in reflectivity when glaring conditions are detected are today commonplace. The circuitry to control the electrochromic dimming function and any other mirror-mounted electronic feature can be commonly housed in or on a rearview mirror assembly, and wholly or partially share components on a common circuit board.

A third attribute is flexibility. By hosting an added feature within a rearview mirror assembly, an automaker has wide latitude in option packaging. As car area networks (CAN) proliferate, the ease and convenience of incorporation of added features will be even further enhanced. Car area networks can provide a plug-and-play opportunity for incorporation of added feature functions whereby mirror mounted features such as, for example, a pyroelectric intrusion detector can output a signal to the CAN when a cabin intrusion is detected, that is received and reacted to by an alarm system elsewhere on the network.

This paper reviews added feature interior and exterior rearview mirrors, and outlines how they can enhance consumer safety, convenience and affordability.

INTRODUCTION

Automotive rearview mirrors have several salient attributes that make them an attractive choice for incorporation of added features. One is location. Looking at a rearview mirror is part and parcel of the driving task. Thus, information displayed at rearview mirrors is plainly and readily visible and interpretable to the driver. Furthermore, outside mirrors, protruding as they are from the main body of the vehicle, are plainly and readily visible to other road users, and thus are good locations for turn signals and brake lights. Lastly, the outside mirror housing is high mounted relative to the road surface, and so is a superior location for ground illumination lighting.

Incorporation of added features is aided whenever the mirror reflector is an electrochromic mirror[1] whose reflectivity is electrically variable depending on the glaring and ambient light conditions sensed by a rearward facing photosensor (detecting glare) and by a forward facing photosensor (detecting ambient light conditions), with both sensors usually incorporated into the interior rearview mirror assembly. Other electronic features such as headlamp activation controllers, remote keyless entry receivers, cabin intrusion detectors, compass sensors, rainsensor detectors, displays, garage door opener transceivers and their like can efficiently and economically access and share the circuitry already present in an automatic electrochromic mirror assembly.

With these advantages, it is not surprising that added feature use in rearview mirrors has grown rapidly over the last several years, a trend likely to continue.

LIGHTED INTERIOR MIRRORS

Light modules (shown in Figure 1) incorporated into the interior rearview mirror assembly provide glare-managed lighting that maximizes interior cabin visibility with minimum distraction to the driver. In excess of 1.5 million vehicles are produced annually with map lights incorporated into the interior mirror assembly. Because they emit below the driver's line of sight, mirror-mounted lights provide excellent illumination of the driver and passenger lap areas without projecting glaring light to the driver's eyes. Also, because the lights are mounted in the bottom of the mirror housing (or, in some European models, are attached to the mirror mount), the rearward facing glare sensor found in automatic electrochromic mirrors is largely unaffected by operation of the mirror-mounted lights. Operation of lights mounted in the header or dome area of the vehicle generally causes a spurious dimming of an automatic electrochromic mirror; an undesirable event avoided in the lighted electrochromic mirror shown in Figure 2.

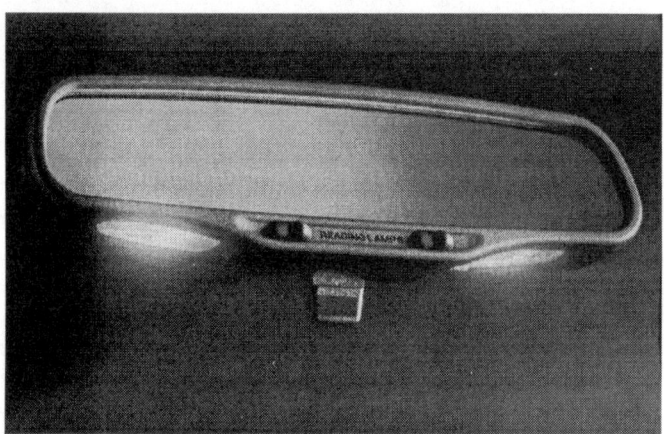

Figure 1. Interior Mirror Incorporating Map Light.

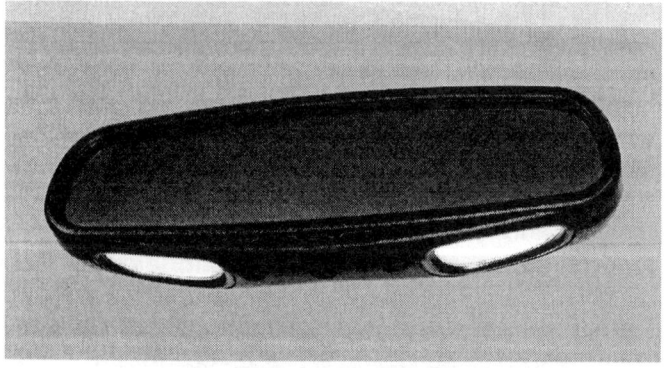

Figure 2. Lighted Electrochromic Mirror.

A recent development, available today on the Chevrolet Corvette, is addition of a high intensity light emitting diode to an interior mirror to provide low level, directed illumination of the center floor console (including the transmission selector) such as is illustrated in Figures 3A and B.

Whenever the ignition is on, the LED light is powered. Being a low current, low level illuminator, its operation is invisible to the driver by day, but at night the LED casts low level lighting onto the center floor console, enabling the driver determine gear selection, find coins for toll booths, etc. Automakers like this feature because it obviates providing dedicated lighting at the floor console.

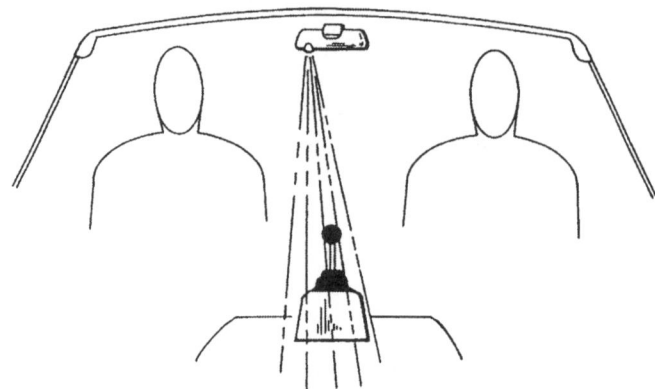

Figure 3A. High Intensity LED in Interior Mirror Housing.

Figure 3B. Interior Mirror Incorporating Low Level, Directed Illumination of a Center Floor Console.

LIGHTED EXTERIOR MIRRORS

The Lincoln Mark VIII vehicle was the first vehicle to provide security lighting at the vehicle entrances by utilizing ground illumination lights located in the outside mirror housings (Figure 4).

Personal security is important to everyone, and is particularly important to females approaching a parked vehicle in a dark and isolated lot. Newspapers have reported instances where malcreants have hidden at or under the vehicle awaiting an opportunity to tackle the driver as he or she enters the vehicle. In some instances, the driver was disabled by having ankle tendons slashed.

Assaults in or around the vicinity of a parked vehicle have increased significantly in recent years. Most frightening has been the rise in carjackings (see Figure 5). Many occur at night. Parking lots are favored areas for carjackers, and handguns are frequently used.

Figure 4. Mirror-housed Security Lights Providing Vehicle-side Ground Illumination.

Automotive Security - Annual Statistics

- 1.5 Million Stolen Vehicles
- $8 Billion Cost
- A Car is Stolen Every 30 Seconds
- 30% of Incidents are at Night Close to Home
- 66% of the Criminals are Under the Age of 21; 88% are Male
- 25,000 Carjackings
- Parking Lots Most Favored by Carjackers, Followed by City Street, Residential Driveways, Car Dealerships and Gas Stations
- Most Carjackings Occur Between 8:00 P.M. and 11:00 P.M.
- When There are Weapons Used, 90% Involve a Handgun

Figure 5 Annual Statistics on Automotive Security.

When approaching at night a vehicle equipped with lighted exterior mirrors, the situation is different. When the driver remotely unlocks the vehicle doors utilizing his/her key fob, floodlights packaged into each of the outside mirror housings illuminate and flood the side and underside of the vehicle with light. The driver can determine whether it is secure to approach, and on entering the vehicle, can avoid puddles and debris at the entrance doorway.

Locating a security light within an outside mirror housing is a challenge, both for packaging and for optical design. The interior cavity of a mirror housing can be crowded. The mirror reflector is typically attached to an electrical actuator, and clearance must be maintained sufficient to allow freedom of adjustment of the reflector field of view to suit each driver's individual need. Increasing the housing size or shape is often not an option for styling and aerodynamic reasons.

As illustrated in Figure 6, the security light utilized on the Mark VIII is a module that is insertable and removable into the mirror housing. The lamp assembly is fully serviceable, is moisture impervious to withstand car washes, rain, road splash and the like, and is fully integrated into the mirror housing so as not to interfere with the external styling of the assembly, and so as not to interfere with aerodynamic performance.

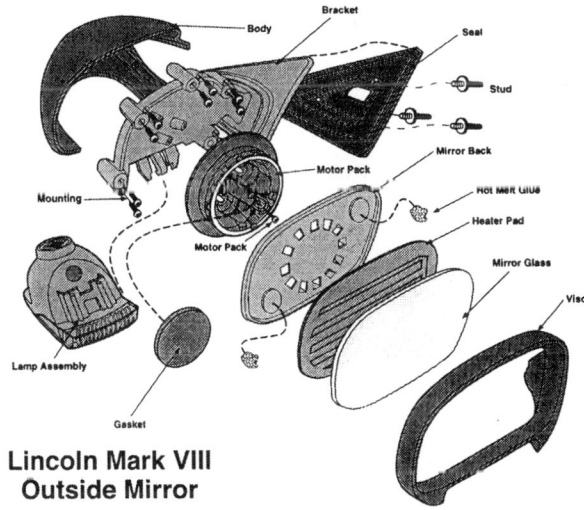

Lincoln Mark VIII Outside Mirror

Figure 6. Construction of Lincoln Mark VIIII Lighted Outside Mirror.

When mounted in the vehicle, a lockout is provided to obviate actuation of the security light during normal driving. This is an important feature, as it would be both undesirable and potentially unsafe if the security lights were activated while normally driving down the highway.

Use of security lighting around the perimeter of a vehicle will grow in consumer popularity, and use of security lights in outside mirror housings will likewise correspondingly proliferate.

To test this, a telephone survey was conducted by Market Facts Incorporated, a market research company which has conducted weekly interviews of this nature for over 20 years. Of 1000 people polled in the U.S. market, 53% expressed an interest in security lighting when approaching a parked vehicle. This is a high initial level of interest given that the respondents had no details on cost, configuration or complexity. Minivan owners were most interested. Given their popularity with females, such a result is not surprising.

One further application of exterior lighting is as a parking aid (see Figure 7). Curbside parking can often be challenging and today many luxury vehicles provide a park tilt feature where the outside mirror reflector tilts to face slightly downward when reverse gear is selected. Thus, when reversing curbside into a tight parking space, the driver can see how close to the curb the vehicle is parked by reflection in the sideview mirror.

Figures 7. Illuminated Park Tilt.

This works fine by day when there is high ambient light to see by, but at night the feature is difficult, even impossible, to use as the front and rear vehicle lighting casts scant light curbside. However, with a floodlight mounted in the outside mirror housing, parking can be as easy by night as by day.

When reverse gear is selected, the lockout preventing actuation of the exterior mirror lights is overridden, the floodlight in each housing illuminates in tandem with the mirror reflector tilting downward, and the driver at night can now see the reflection of the illuminated curb in the tilted reflector of the sideview mirror.

TURN SIGNALS IN EXTERIOR MIRROR ASSEMBLIES

Drivers are familiar with occasions when they signal to change lanes and then initiate a lane change unaware that there is an adjacent vehicle in the side lane overtaking in a blind spot. Even though the driver has signaled a lane change, the adjacent driver is unaware of this intent to change lanes as neither the front nor the rear turn indicator light is visible to the adjacent driver.

This is a potential safety hazard that can be obviated by mirror-mounted signal indicators. A Signal Mirror™, developed by Muth Corporation of Sheboygan, Wisconsin, is available on Ford Bronco and Expedition vehicles (see Figure 8). An array of high intensity light emitting diodes is placed behind a dichroic mirror reflector. The dichroic reflector is designed such that it has a bandpass of high light transmission to the wavelengths of light output by the LEDs therebehind, but is low transmitting at all other wavelengths. Thus, the Muth mirror acts akin to a one-way mirror with the presence of the array of LEDs behind the mirror reflector being undetected by the driver until actuated when the turn signal is selected.

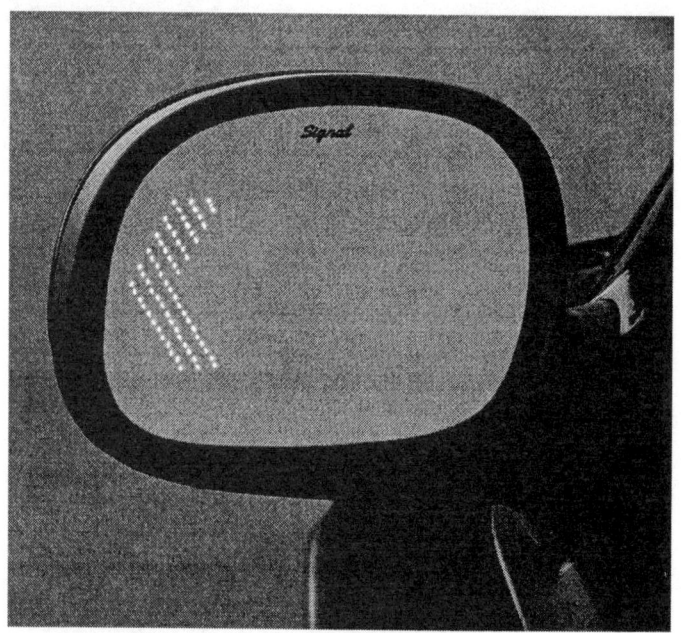

Figure 8. Ford Expedition Signal Mirror.

Figure 9. Turn Signal Mounted in Outside Mirror Housing.

An alternate design for a mirror-mounted turn signal is shown in Figure 9. A linear array of six high intensity LEDs are provided in the lower, rearward facing portion of the mirror housing. The LEDs are orientated at an angle of about 25° to 30° relative to the longitudinal centerline of the vehicle, so that their light output is principally directed to be highly visible to vehicles approaching in adjacent sidelanes. Plastic louvers are used to separate the LEDs one-from-another, and to shield their light output from the line of sight of the host driver. These louvers help protect the driver from distraction or glare when the turn signal indicators are actuated at night to signal a lane change.

This design of mirror-mounted turn signal is equipped on the Lincoln Mark VIII. Being fixedly mounted to the mirror housing rather than mounted behind the movable mirror reflector, turn signal indicator modules of this design can be provided economically, with the automaker retaining full flexibility in the choice of mirror reflector (electrochromic or standard chrome).

INFORMATION MIRRORS

Currently, compass direction is the most widely provided mirror-mounted information display, with in excess of 600,000 vehicles annually so equipped. A variety of display options (see Figures 10A, B, and C) are available, including locating behind the mirror element with the display being viewed through a window created by removing the mirror reflector in the local area; locating within a pod that attaches to the mirror mount at its point of attachment to the mirror mounting button on the windshield; and locating within the bezel below the mirror reflector. Location of the information display within the mirror bezel has several advantages, particularly as shown in Figure 11, where multiple displays of compass direction and of exterior temperature are desired.

Placing the display in the bezel rather than behind the mirror element eliminates removal of mirror reflector surface and consequent local loss of field of view. Thus, safety is enhanced by providing an unobstructed rearward field-of-view. Consumer satisfaction is also enhanced by enabling simultaneous display of compass direction and temperature together, rather than requiring the driver to toggle between one and the other (as is more typical when a display is mounted behind the mirror reflector, where desire to avoid excessive loss of mirror reflector typically leads to display of only one information item at a time).

Figure 10A. Pod Mounted Display.

Figure 10B. Information Displayed Through Window Created in Mirror Reflector.

Figure 10C. Information Displayed Below Mirror Reflector.

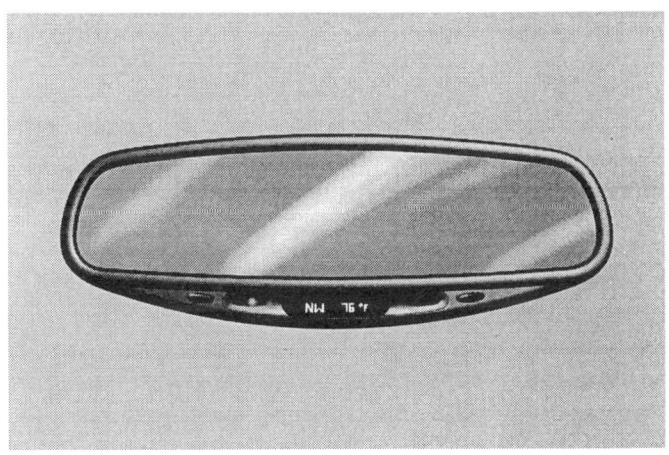

Figure 11. Simultaneous Display of Compass and External Temperature Information.

Location of an information display within a pod (Figure 12) has several advantages of its own. The information display pod can readily attach to a variety of interior mirror assemblies, including prismatic mirrors. Thus, the automakers and consumer retain full freedom of choice in terms of the option to select. Display pods are also advantageous for installations in the aftermarket.

Information that are candidates for mirror-mounted display include compass direction, external and internal temperature, altitude and incline (of particular interest in sports utility vehicles), turn signals, pager display, tire pressure status, trip computer, fuel/oil level, hazard warnings, status indicators (such the icon displaying the status of the passenger-side airbag activation shown in Figure 13) and their like. Location of an information display either in the interior or in the exterior mirror is particularly advantageous when it is critical to catch the driver's attention. Information that may potentially pass unnoticed by the driver when displayed in an already cluttered instrument panel will more certainly be noticed by the driver as he/she constantly and repetitively looks at the rearview mirror during normal driving.

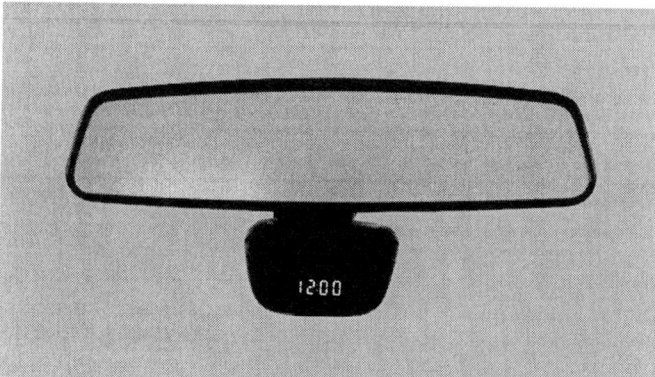

Figure 12. Prismatic Interior Mirror Equipped with Pod-mounted Clock Display.

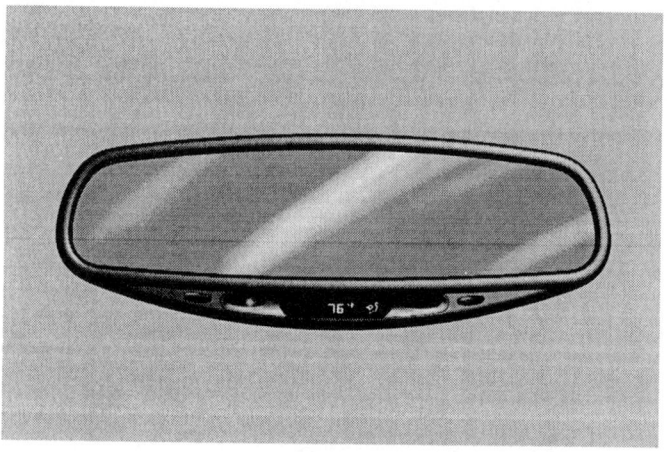

Figure 13. Simultaneous Display of External Temperature and Status of Passenger-side Airbag Activation.

RAINSENSOR MIRRORS

Rainsensors[3] detect the presence of moisture on the exterior of the windshield, and automatically activate the wipers to remove. Over the last several years, they have grown in popularity, with current usage on vehicles exceeding 500,000 units annually. Rainsensors currently on vehicles are windshield contacting sensors whereby the rainsensor module is either bonded to the inner windshield surface by an optical adhesive or has an optical polymer surface that is pressed intimately against the glass surface mechanically[4] (and is thus removable for service). Rainsensors are typically mounted to the area of the windshield behind the interior mirror housing, thus providing a good location to detect moisture but also one that is unobtrusive to the driver's forward field of view.

A rainsensor located within the support arm of an interior mirror assembly is equipped on the Volkswagen Golf. This is a compact design that renders the presence of the rainsensor largely undetected by the average consumer. As illustrated in Figure 14, the mirror mounting button to which the interior mirror attaches is an annulus with a solid outer ring and with a hole at its center. The rainsensor unit mounts within the cavity of the mirror support arm, and the act of attaching the mirror assembly to the windshield button causes the rainsensor module within the mirror support arm be pressed to the windshield glass surface at the center, hollow portion of the donut-like button mount. The rainsensor thus views the outer surface of the windshield via the hole at the center of the mirror mounting button.

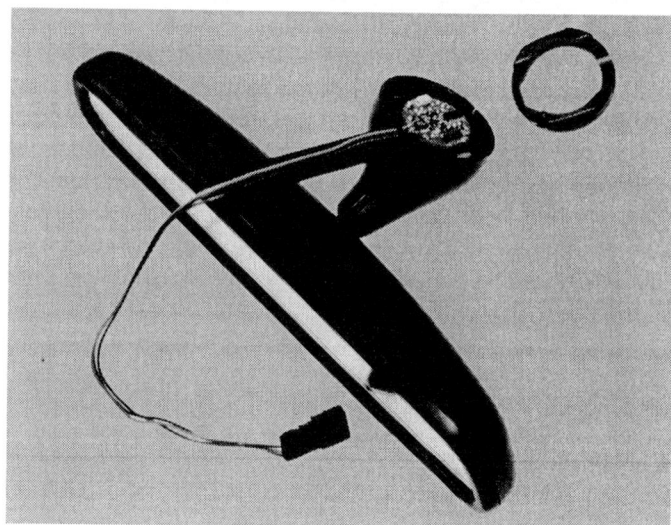

Figure 14. Compact Rainsensor Mounted in Mirror Support Arm.

Non-contacting rainsensors[5] are currently under development where the rainsensor module is mounted within the interior mirror assembly with its detection surface stood off the windshield inner glass surface. There are many advantages to not having direct contact between the rainsensor and the glass surface of the windshield. Windshields are frequently replaced due to damage from road debris such as chips and stones. When a contacting rainsensor is bonded to the glass, replacement during service can be costly and complex. This is not the case with non-contacting rainsensors. Also, the non-contacting rainsensor has opportunity to detect moisture not only on the outer windshield surface, but also on the inner surface as well. Thus, a defroster blower can be automatically activated to remove any condensation or frost build-up, a significant benefit for those driving in humid/frigid conditions.

AUTOMATIC HEADLAMP ACTIVATION MIRRORS

Automatic headlamp activators, often referred to as Twilight Sentinels, are commonplace in vehicles. These automatically turn headlamps on and off at dusk and dawn. Frequently, a skyward facing photosensor is mounted in a module attached to the interior mirror assembly at its point of attachment to the mirror mounting button adhered to the vehicle windshield. An automatic headlamp activation added feature electrochromic mirror of this type that uses a dedicated skyward facing, mirror-button mounted photosensor is available on MY98 Ford Windstar and Ford Explorer vehicles.

An alternate design that utilizes the forward and rearward facing photosensors already on-board in an automatic electrochromic mirror assembly to both control automatic dimming of the electrochromic mirror element and to automatically activate the headlamp at dusk is standard on MY98 Jaguar automobiles (Figure 15). Since the electrochromic automotive mirror circuitry already utilizes two photosensors, this design is economical, aesthetically appealing (attaching a third photosensor facing skyward to the mirror-button mount is plainly visible through the windshield), and is adaptable to both mirror-button mounted interior mirrors and to header mounted interior mirrors (where attachment of a third skyward facing sensor may be problematic).

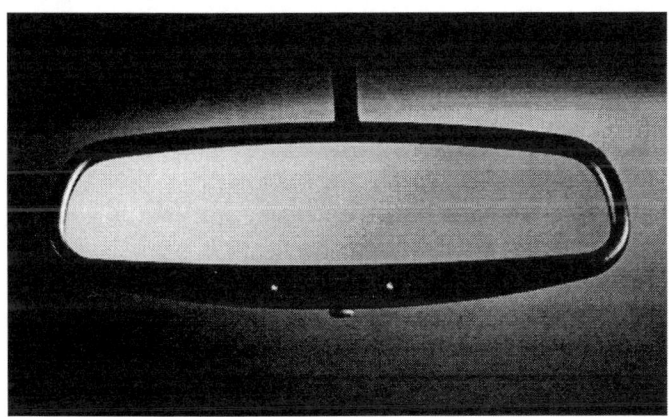

Figure 15. Automatic Headlamp Activation Electrochromic Mirror.

HOME ACCESS MIRRORS

The interior mirror assembly is a convenient location for a myriad of added features that are manually actuated, drivers being long used to reaching to flip the manual actuator toggle on prismatic mirrors. Thus, a mirror located switch is readily accessed by drivers (a significant advantage for the switches on lighted mirrors versus lights placed elsewhere). Several automakers have placed, or are contemplating placing, circuitry within the interior mirror that allows the driver open/close a garage door, security gate, and the like. Two very different systems are currently used or proposed in vehicles. Prince Corporation of Holland, Michigan has developed a universal garage door opener system. Prince's HOMELINK™ system[6] is initially "trained" by the vehicle owner using the garage door opener hand transmitter originally provided by the manufacturer of the garage door opener (GDO) mechanism installed at the owner's home. Once the HOMELINK™ unit is mounted in the vehicle and has learned the owner's garage door opener code, the driver of the vehicle simply actuates the vehicle mounted HOMELINK™ switch to open the garage door from the vehicle when approaching.

TRW Automotive Electronics Group of Cleveland, Ohio has proposed a very different universal home access system. TRW's KWIKLINK™ operates in a totally different way to that of the HOMELINK™ approach. In the KWIKLINK™ system, the wall mounted switch used to manually open/close the garage door from the driver's home is replaced with a KWIKLINK™ wall unit. The KWIKLINK™ wall unit includes circuitry that communicates with a vehicle mounted transmitter that allows the driver, in essence, to open the garage door from the vehicle by transmission to the wall switch, bypassing the remote receiver mechanism originally installed in the garage opener mechanism.

The KWIKLINK™ system has several advantages. Most of the existing installed base of GDO remote control units operate on transmissions that are vulnerable to electronic eavesdropping whereby criminals can electronically listen and record the homeowner's entry to the home when the owner actuates his/her GDO remote control. The criminal can later return when the owner is away, and playback the recorded signal to open the garage and potentially gain entry to the home. Potential eavesdropping has long been a security problem with remote keyless entry systems to vehicles, a problem overcome by the introduction of rolling code technology whereby a new transmission code is selected each time the remote unit is used. For enhanced security, the KWIKLINK™ universal home access system also operates on a rolling code, ensuring that an eavesdropper cannot benefit from decoding a particular transmission as the code will automatically change rendering repeat transmission by the potential burglar of the decoded transmission futile.

Another advantage for the KWIKLINK™ system is that the vehicle mounted system is relatively simple (it fulfills no "learning" function as with the HOMELINK™ system), it is compact, economical and uses low current. The circuitry will operate off a lithium button-type battery and it is planned to add a garage door opener actuation button to the vehicle owner's key fob (used for remote keyless entry to the vehicle) so that a homeowner can open/close the garage door when outside the vehicle.

Figure 16 shows an automatic electrochromic interior mirror that incorporates a universal home access unit as an added feature. The driver can select from three different, bezel-mounted switches to gain access to, for example, a home residence, a summer cottage and a security gate.

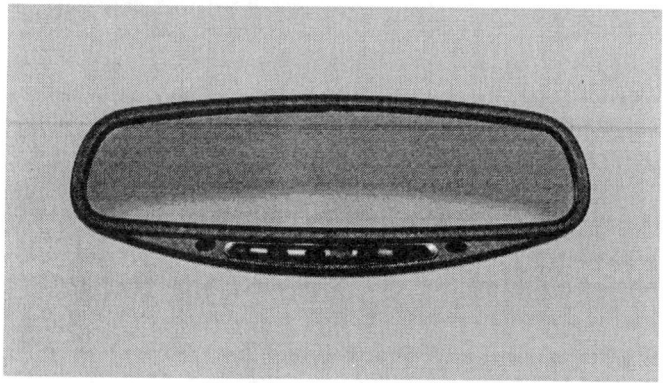

Figure 16. Interior Electrochromic Mirror Incorporating Universal Home Access Unit.

REMOTE KEYLESS ENTRY MIRRORS

Since the early '90s, remote keyless entry receivers (both infrared receivers and radio frequency receivers) have been incorporated into interior rearview mirror assemblies. Being centrally and high mounted within the interior vehicle cabin, and with wide-angle unobstructed reception of transmitted signals, the interior mirror is a desirable location for a remote keyless entry (RKE) receiver. This is particularly so for infrared operating RKE units where placement of the IR detector in a bulb protruding from the lower portion of the interior mirror housing (see Figure 17) greatly enhances range and width of signal reception.

Figure 17. Mirror-mounted IR Receiver for RKE Transmitter.

Recently, vehicles have been equipped with an automatic proximity detector that detects the approach of the vehicle owner and unlocks the doors automatically without the owner operating any button. An antenna in the vehicle transmits to a receiver carried on the driver's key ring. Once a link is established as the driver approaches the vehicle, the system verifies the identity of the driver, and the doors automatically unlock.

The outside mirror housing is a good location for an antenna, given that the housing is non-metallic and that the mirror housing protrudes away from the vehicle body in the direction of approach and at the height where the key ring is either handbag or pocket carried by the approaching driver. When combined with a security light in the mirror housing, the driver can securely approach the vehicle at night and enter without taking out the ignition key.

INTRUSION DETECTOR MIRRORS

Vehicle theft is an ongoing problem. Recent advances in electronic tagging of ignition keys and sensing of vehicle location using satellite tracking technology has made it more difficult and less fruitful to steal a vehicle. However a need continues to prevent theft of articles left in the interior cabin, and to protect against entry by potential carjackers and similar intruders.

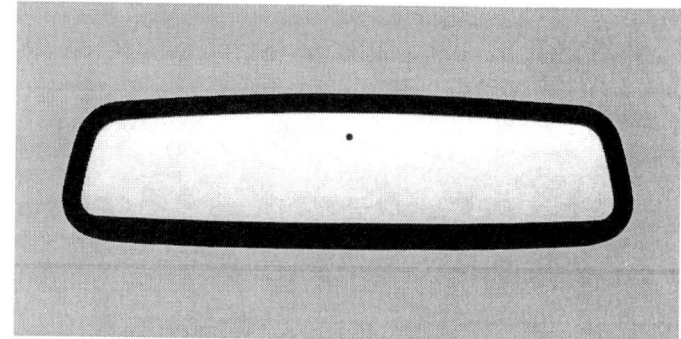

Figure 18A. Camera Vision Intrusion Detector Mirror

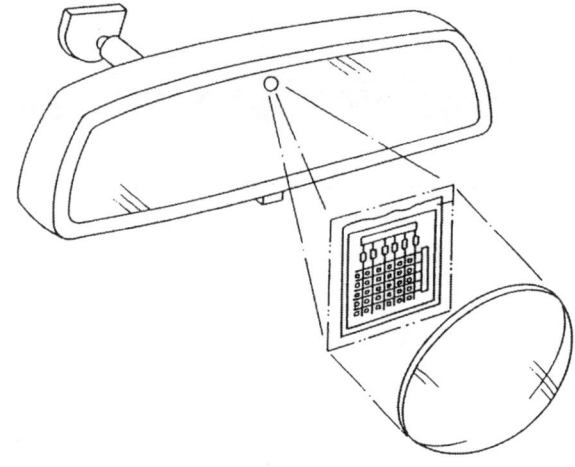

Figure 18B. Video Camera Mounted within Interior Mirror Assembly.

A variety of detectors have been proposed or are in use to protect against cabin intrusion. These include ultrasonic and radar based detectors. Such systems tend to be expensive and potentially subject to false alarms.

Location of a detector within or at an interior mirror assembly is a convenient and effective means to provide protection again cabin intrusion. A system utilizing a video microchip camera device located within the interior mirror housing (Figure 18) has been developed[7]. With the interior cabin illuminated by day with daylight and by night with an infrared floodlight, the camera-captured photoimage is analyzed and any intrusion detected triggers a security response..

An inexpensive mirror-mounted pyroelectric intrusion detection system is shown in Figure 19. This pod mounted system uses a pyrodetector that reacts to any change in cabin temperature such as would be caused by an intruder entering or partially entering the vehicle. This unit has high reliability and operates on a current of less than 0.25 milliamps, thus providing continuous intrusion protection even when the vehicle is parked for weeks without itself draining the vehicle battery.

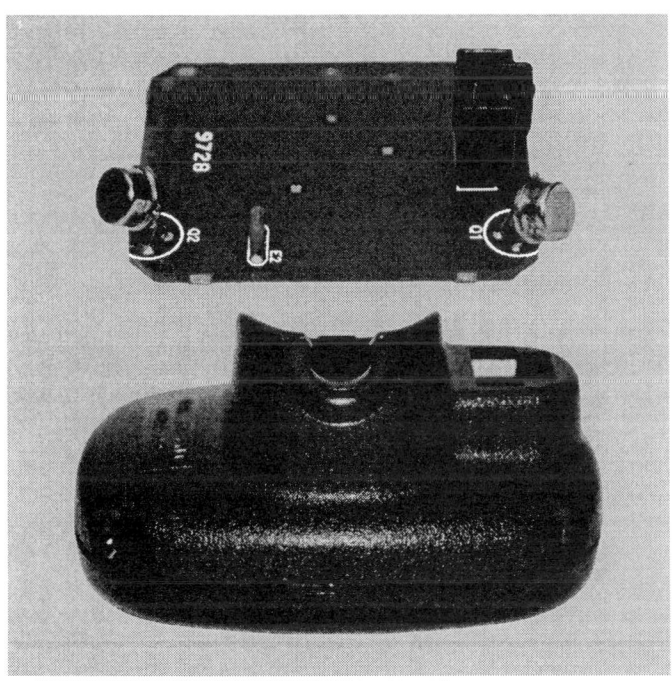

Figure 19. Pyroelectric Intrusion Detector.

GPS ANTENNAE IN MIRRORS

Navigational and vehicle security aids that track vehicle location utilizing geographic positioning satellite (GPS) systems are growing in use. To establish geographic location, the vehicle receives signals from multiple orbiting satellites via a vehicle mounted GPS antenna. Preferably, such an antenna is located with an unobstructed skyward line of sight, and for optimum reception, the antenna should be housed away from body sheet metal.

Vehicle mirrors, and particularly exterior mirrors, are desirable locations for GPS antennae (Figure 20). Mirror housing assemblies offer opportunity for unobstructed skyward view with good reception. This is particularly so for exterior mirrors which protrude clear of the vehicle and which offer an opportunity to house the GPS antenna unobtrusively and with minimum wiring length to connect with the navigational display unit mounted in the vehicle dashboard. A GPS antenna mounted at an interior mirror is equipped on Renault vehicles beginning production 1998.

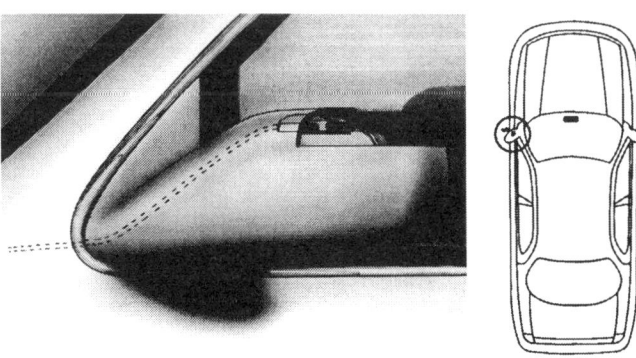

Figure 20. GPS Antenna Mounted in Outside Mirror Housing.

CAR-AREA-NETWORKS

The concept of local area networks is well established for interconnecting computers and their accessories. The vehicular equivalent is a car-area-network (CAN) where nodes controlling the features and accessories in various parts of the vehicle are interconnected on a network. Various CAN protocols are in use or are contemplated such as Motorola's msCAN protocol and Siemens's FULL-CAN controller.

A CAN node located within the interior mirror assembly will be standard on a European luxury vehicle in 1998. The mirror mounted node will control electrical function in the upper half of the vehicle such as control of the sunroof, interior lighting and the CHMSL rear brake light. The interior mirror assembly (which will be provided in both a prismatic version and an electrochromic version) includes the automatic glare detection circuitry (if an electrochromic unit), a remote keyless entry receiver, a security system that uses an ultrasonic intrusion detector, an information display on seat belt use, as well as serving as the electronic controller for sunroof operation and interior lighting in the upper portion of the vehicle. When launched in Summer 1998, it will be the most technologically advanced interior rearview mirror in production.

There are many advantages to the CAN concept for the automaker, for the accessory supplier, and for the consumer. Addition of features such as a rainsensor or intrusion detector to the interior mirror conventionally requires agreement and acceptance by diverse, often distinct, suppliers. For example, introduction of a mirror-mounted rainsensor requires that the mirror supplier, the windshield wiper controller supplier, and diverse groups at the automaker coalesce and coordinate together, which sometimes can be challenging. With a CAN network, implementation of new features is greatly facilitated. For example, when a mirror-mounted intrusion detector senses a breach of cabin security, its local node outputs a signal to the network. This signal, in turn, is captured by the node to which the car alarm system is connected, and is acted on accordingly. This allows the automaker add features as plug-and-play accessories, and provides wide flexibility to the automaker in terms of choice of sub-system supplier. The supplier, in turn, can focus on its area of best expertise.

For the consumer, the plug-and-play opportunities of a CAN environment offers not only broader access to diverse features, but also more convenience in location of the switches required to operate various functions. Conventionally, the need to hard wire switches to the units they operate has meant that they be located close together. With a CAN, the switch need only connect to its local node, and it can control remote units via signals passed along the network and processed at local nodes.

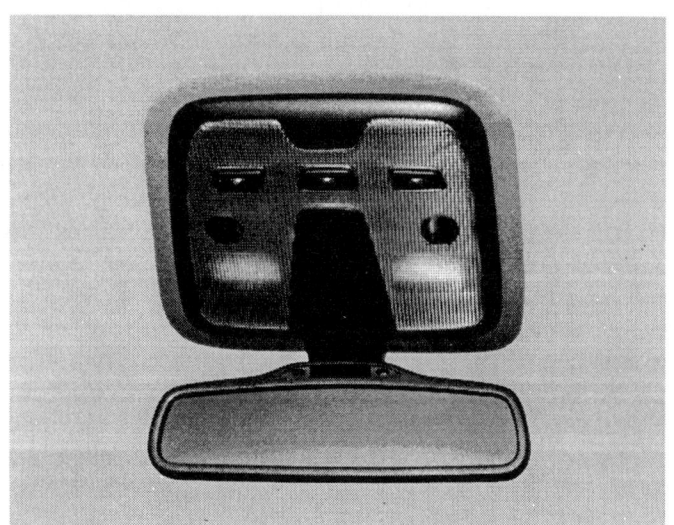

Figure 21. Interior Mirror/Overhead Console System Incorporating CAN Node.

There is no direct wire connection between the map light switches adjacent the map lights in the overhead console shown in Figure 21. When an occupant depresses the map light switch, a signal is sent from the overhead console down to the CAN node in the mirror. Here it is interpreted, and a command to turn on a map light is sent back to the overhead console. However, since the node in the mirror is microprocessor controlled, the turn off of the map light can be controlled to a fade-out over several seconds, an illustrative example of how use of a CAN economically enhances added features in the vehicle.

CONCLUSION

An ever increasing proportion of interior mirrors are electronic (sales of automatic interior electrochromic mirrors in 1997 exceeded three million units). Given their location in the vehicle, and given that addition of further electronic features is more economical where circuitry is already on-board, use of added feature interior mirrors will continue to rise. Furthermore, many outside mirrors are already electrically serviced for power actuation and defrosting. Added outside mirror features such as ground illuminator security lighting will grow in popularity. Once a consumer is equipped with such a feature, it is likely that it will be looked for in future vehicles that driver buys.

Use of added feature mirrors will continue to grow, with an ever increasing variety for the automaker and consumer to choose from.

The interior mirror assembly can host seat occupancy detectors; video cameras for internal cabin surveillance and/or video telephone function; messaging displays that relay paging, traffic status or hazard warning information to the driver; links to the World Wide Internet Web via a modem/cellular phone/alphanumerical display; a digital recorder for recording and/or playing back messages including e-mail messages; loudspeakers such as for a vehicle audio system or for a cellular phone.

The exterior mirror assembly can house blind-spot detectors that detect the presence of approaching vehicles in adjacent sidelanes; and a transducer that receives and/or transmits information to a component of an intelligent highway system or an automatic toll booth system.

REFERENCES

1. "Electrochromic Automotive Day/Night Mirrors" Niall R. Lynam, SAE Technical Paper #870636 (1987).

2. United States Patent 5,361,190.

3. United States Patent 4,859,867; 4,916,374; 4,973,844.

4. United States Patent 4,871,917.

5. International Patent Publication NO: WO 94/27262.

6. United States Patent 5,442,340.

7. European Patent EP 0 683 738 B1.

School Bus Visibility: Driver's Field of View and Performance of Mirror Systems on a Conventional Long-nosed School Bus

Paul Lemay and Alex Vincent
Transport Canada

Copyright © 1988 Society of Automotive Engineers, Inc.

ABSTRACT

This report presents the results of the driver's field of view and the performance of six crossview and two sideview mirror systems on a conventional long-nosed school bus. It also contains an evaluation of the image quality of the crossview mirrors in terms of the angular length and width of their reflected images. The measurements of the field of view and the evaluation of image quality were done at two driver eye locations, one representative of the cyclopian view of a 95th percentile adult male and the other one representative of the cyclopian view of a 5th percentile adult female. Measurements were taken considering that there were no head movements. For the purposes of the study, the term "blind spot" was defined as meaning any area that could not be seen directly by the driver. The performance of the mirrors was judged in terms of their capacity to provide a complete and clear view of the blind spots.

The results demonstrated that none of the crossview mirrors on the conventional long-nosed bus performed adequately in that they did not eliminate the blind spots and provide good quality images to the front and sides of the bus. Furthermore, no crossview mirror reflected all the cylinders along the rear axle, and where they were viewed, the image quality was not always acceptable. The Double Nickel sideview mirror system, which is composed of one pair of flat and one pair of convex mirrors, had a narrower field of view than the mirrors installed by the bus manufacturer. Although the image quality of the sideview mirror systems was not formally evaluated, the Double Nickel had better image quality.

It should be noted that the results presented in this publication were extracted from a comprehensive school bus visibility study that was conducted by Transport Canada on four school bus configurations. In addition to the conventional long-nosed bus, the configurations included the flat-nosed bus, the short-nosed bus, and the minibus.

INTRODUCTION

According to a study conducted by Transport Canada, the school bus is one of the safest means of transportation.[1] Statistically, per passenger-kilometre of travel, school bus occupants are sixteen times less likely to be injured in a collision than the occupants of other motor vehicles. Despite Canada's high level of school bus safety, a number of accidents have occurred in recent years in which children were struck by the school bus they were boarding or disembarking. A lack of visibility to the front and sides of the bus appeared to be the major contributing factor in these accidents.

As part of its initiative to improve the Federal standard for school bus mirror systems, Transport Canada conducted a study of school bus visibility problems. The study charted the driver's field of view and evaluated the performance of six crossview and two sideview mirror systems. Specifically, the study measured the field of view (FOV) at two driver eye locations for four configurations of school bus. This publication only presents the results for the conventional long-nosed bus. The FOV included both the direct view from the driver's eye position and the indirect views provided by the mirror systems. In order to ensure that objects on the reflecting surface were recognizable to the driver, the study also assessed the image quality of the crossview mirrors.

STUDY CRITERIA

The choice of model for the long-nosed bus was based on the criterion that it had to have the poorest

---

[1] Paul A. Gutoskie, *An Analysis of Canadian School Bus Accident Records: 1982-83 - 1984-85*, August 1987, Transport Canada TP 8740E.

direct visibility for the driver. The following model was used:

Conventional long-nosed bus: 1992 Corbeil built on a Navistar International chassis
Capacity: 72 passengers;

In all, the fields of view of eight mirror systems were charted: six crossview mirrors and two sideview mirror systems. A crossview mirror is a convex mirror installed on one or both front corners of a bus to provide a seated driver with a view to the front and side. Crossview mirrors are used to detect pedestrians when the vehicle is stopped. They do not provide information by which to judge the position of objects in relation to the bus and, consequently, are not used during driving. The study evaluated the following six crossview mirrors: the quadrispheric, a mirror called Banana, the elliptical, a convex mirror 8 inches (20.3 cm) in diameter (Φ) with a 23-inch (58.4-cm) radius of curvature (RC), a mirror 10 inches (25.4 cm) in diameter with a 31-inch (78.7-cm) radius of curvature, and a mirror 8 inches (20.3 cm) in diameter with an 87-inch (221-cm) radius of curvature.[2]

The fields of view of two sideview mirror systems were also charted, one of which was the system installed by the bus manufacturer. For the conventional long-nosed bus, this system consisted simply of a pair of West Coast mirrors, which are rectangular mirrors of one unit of magnification.

The other sideview mirror system studied was the Double Nickel™, which consisted of one pair of mirrors of one unit of magnification and one pair of convex mirrors with a long radius of curvature installed beneath. This combination permits the detection of both nearby objects and vehicles approaching from the rear.

The measurements for each mirror were taken at two driver eye locations that were chosen to approximate the cyclopian view of a tall male driver and a short female driver with no head movements. The exact eye positions were established using three sources: Society of Automotive Engineers (SAE) Recommended Practice J941, *Motor Vehicle Drivers' Eye Locations*, June 1992, for class B vehicles; data on driver eye heights compiled as part of a private study; and the eye positions of two Hybrid II anthropomorphic test devices representing a 95th percentile adult male and a 5th percentile adult female, designed in accordance with the requirements of section 100 of the *Motor Vehicle Safety Regulations*. Both eye positions were established in relation to the seating reference point in order to compensate for variations in design between the different bus configurations.[3]

For the purposes of this study, a distinction was made between the direct field of view of the unaided eye and the indirect field of view provided by the mirrors. The direct field of view was defined as the area of the ground visible from the driver's eye location, and the indirect FOV as the area of the ground that was reflected on the mirror's surface, as seen from the driver's eye position. Conversely, a blind spot was defined as any area on the ground around the bus that could not be seen directly from the driver's position. In order to assess whether a small child lying on the ground would be visible in a blind spot, the dimensions of a six-year-old Canadian girl of the fifth percentile were estimated using anthropometric data.[4] The standing height of this theoretical child was estimated to be 104.5 cm (41.1 in.) and her diameter 30.48 cm (12 in.). The height of blind spots was measured, and their dimensions were compared with those of the theoretical child.

The image quality was evaluated using the reflected images of cylinders that were 30.5 cm (1 ft.) in diameter and either 30.5 cm (1 ft.) or 91.4 cm (3 ft.) in height.[5] These cylinders, intended to represent children, were positioned in visually critical areas around the bus and arranged as shown in figure 1. This configuration closely resembles that established by the U.S. in its Federal Motor Vehicle Safety Standard No. 111 (FMVSS), which governs rearview mirrors.[6] Three one-foot high cylinders were placed in a row 30.5 cm (1 ft.) in front of the bumper, three at a distance of 2.13 m (7 ft.), and three more 4 m (13 ft.) away. (A direct view to the front of the bus does not necessarily begin at the furthest row of cylinders.) In addition to the 9 cylinders in front of the bus, one was also placed 91.4 cm (3 ft.) away from the centre of each front axle, and three more were placed on each side of the rear axle at distances of 61 cm (2 ft.),

---

[2] In order to simplify reference to the latter three mirrors, only the imperial measures will be used; these mirrors are manufactured and sold in accordance with imperial measures.

[3] The term "seating reference point" is defined in the *Motor Vehicle Safety Regulations*.
[4] The anthromorphic data for estimating the dimensions of the theoretical six-year-old Canadian girl of the 5th percentile used in this study were taken from *Anthropometry Report: Height, Weight and Body Dimensions*, a report from Nutrition Canada prepared by Dr. A. Demirjian, University of Montreal, 1980, p. 24-27.
[5] For the sake of brevity, future references to the height of the cylinders will be in imperial units only.
[6] *U.S. Federal Register,* Vol. 57, No. 232, Wednesday, December 2, 1992, p. 57000-57020.

1.85 m (6 ft.), and 3.7 m (12 ft.).[7] A first set of measurements was made with the one-foot cylinders in all the positions, and a second set was made along the rear axle using the three-foot cylinders.

The image quality was assessed in terms of the angular width and length of a cylinder's image, as reflected on the mirror's surface and measured at the driver's eye position. The chosen limits were 3 and 9 minutes of arc for the shortest angular width and length, respectively, which are the same as those of FMVSS 111, and which were intended to ensure a minimum image size. Image quality was evaluated on a pass/fail basis, the quality being inadequate when the visual angle criteria were not met in either crossview mirror. The image quality of the convex mirrors installed below the flat mirrors of the sideview mirror systems was not evaluated.

TEST PROCEDURE

The bus was stabilized on a surface 24.38 m (80 ft.) by 9.14 m (30 ft.) that was marked with a one-foot grid pattern. A coordinate of origin was determined that corresponded to a point on the ground located at the intersection of two planes perpendicular to the ground, one passing through the vehicle's centerline and the other passing through the centerline of the front axle. Using a three-dimensional manikin[8], the seating reference point was established, and the eye positions were determined in relation to the seating reference point. The driver's seat was replaced by a camera equipped with a 50-mm lens mounted on a tripod and positioned at each eye location in turn. The mirrors were adjusted to provide the widest possible field of view using a procedure that was based on those recommended by the different manufacturers. In order to optimize the performance of the mirrors, the crossview and sideview mirrors were adjusted so that the blind spots were minimized and their views overlapped. Where a mounting bracket prevented a mirror from being properly

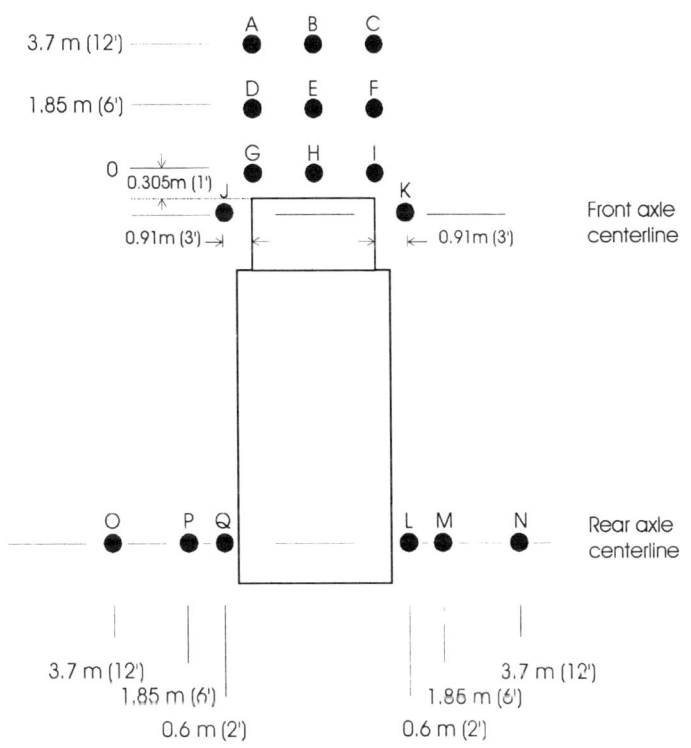

Figure 1: Top view of the cylinders in relation to the bus

---

[7] FMVSS 111 does not require the three cylinders to the left of the rear axle.

[8] For further details on the characteristics of the manikin used, see Society of Automotive Engineers (SAE)

adjusted, it was installed on a tripod equipped with a flexible bracket.

The fields of view and the blind spots were charted by moving markers 0.6 cm (0.25 in.) high along the floor's surface until they were visible through the 50-mm camera lens at the driver's eye position.[9] The limits of the direct view were drawn for the areas visible through the windshield, the entrance door, and the side windows up to 90° on either side of the bus's longitudinal axis. In addition, the obstructions created by the mirrors were charted. The limits of the indirect views reflected in the mirrors were also drawn, for which a 250-mm telephoto lens was used in order to enlarge the mirror's surface without changing the distance between the lens and the mirror. The height of the areas that were not visible, either directly to the driver or indirectly on a mirror's surface, was also recorded.

The image quality was evaluated by photographing the reflection of the cylinders on the mirror's surface using the 250-mm telephoto lens. A graduated ruler mounted above the mirror, perpendicular to the camera's line of sight and tangent to the mirror's geometric centre, was used as a reference for measuring the angular width and length. Each photograph was scanned and analyzed using Adobe Photoshop 2.5 software. Based on the dimensions of the image and the distance between the eye position and the mirror's surface, the visual angles were calculated using the following equations:

$$X \geq D \times 0.000873$$

$$Y \geq D \times 0.002618$$

Where,

X, Y = the shortest angular width and length of the image reflected on the mirror's surface,

D = the distance between the eye position and the geometric center of the mirror,

0.000873 = the tangent of 3 minutes of arc, and

0.002618 = the tangent of 9 minutes of arc.

Especially on the periphery, where aberration is greatest, the crossview mirrors deformed the reflected image to the point that parallel lines on the floor grid converged. The limit of visibility was taken to be the line along which two adjacent parallel lines on the floor grid converged in the reflected image. A driver is unlikely to perceive the movement of an object reflected in the mirror beyond this limit or when the image falls too close to the edge of the mirror.

RESULTS

A diagram outlining the field of view of each mirror is presented, for each eye location, in Appendix B to this paper. Each diagram shows: a top view of the outline of the bus; all the areas that were visible, both directly and indirectly, through the camera's lenses; any overlap that occurred between the left and right mirror views or between the direct and indirect views; and most important, the blind spots, some of which were created by the mirrors. Appendix A provides a key for interpreting the diagrams. Tables are provided in the body of this report that show which cylinders were visible in each mirror and the image quality results, by eye position.

For the conventional long-nosed bus, none of the crossview mirrors provided both a complete view and good image quality for the area from the entrance door forward to the point of direct visibility. The quadrisphere mirror provided the widest FOV to the front, with only a narrow four-foot-long blind spot on the side opposite the driver. The height of this blind spot was 15.2 cm (6 in.), for a 95th percentile male, and 17.8 cm (7 in.), for a 5th percentile female, which would allow a child lying on the ground to be detected. Although the front field of view provided by the quadrisphere mirror was adequate, the mirror did not meet the deformation criteria for the images of cylinders A, B, and C located 3.96 m (13 ft.) in front of the bus, which is a significant failure. Since the area of direct visibility started between 3.05 m (10 ft.) and 6.40 m (21 ft.) from the front bumper, for the male eye position, and between 4.27 m (14 ft.) and 9.30 m (30.5 ft.), for the female, it would not be possible for a bus driver to clearly distinguish a child who was beyond the third row of cylinders. None of the cylinders located along the rear axle was visible on the left for a 95th percentile male, and although the cylinders along the right were visible, only the image of the 3-foot cylinder L met the deformation criteria. For a 5th percentile female, all the cylinders along the rear axle were visible, but only the images of the 3-foot cylinders P and Q were clear.

The Banana mirror provided the best image quality of all the crossview mirrors on the conventional bus; however, for a 95th percentile male, there was a triangular blind spot in the front approximately 1.83 m

---

Recommended Practice J826, *Devices for use in defining and measuring vehicle seating accommodation*, June 1991.
[9] The dimensions of the markers were: 10.2 x 7.6 x 0.6 cm (4 x 3 x 0.25 in.).

(6 ft.) long and 38.10 cm (15 in.) high, and for a 5th percentile female, a blind spot 2.54 m (8.4 ft.) long and 45.72 cm (18 in.) high. A small child lying on the ground in this blind spot would not be visible. In addition, at the male eye position, the Banana mirror did not reflect any of the cylinders along the left rear axle, while along the right rear axle, 4 cylinders were not reflected and the other 2 failed to meet the criteria. For the female eye position, only the 3-foot cylinders Q and L had acceptable image quality, with four cylinders not reflected.

The elliptical mirror offered the widest FOV for both the front and side views at both eye locations, but for a 5th percentile female, there was a blind spot in the front 66 cm (26 in.) high. There was also blind spot 13 cm (5 in.) high for a 95th percentile male. Because of its short radius of curvature, the elliptical mirror had the highest image deformation. The shorter a mirror's radius of curvature, the closer to the reflecting surface an object must be so that its image will not fall on the periphery, where deformation is greatest. The images of the cylinders in the third row in front and those along the rear axle did not meet the 3 and 9 minutes of arc criteria.

The 8-in. Φ (23-in. RC) convex mirror had large blind spots in front at both eye positions and poor image quality for the cylinders in the third row. In addition, for a 95th percentile male, cylinders E and F in the second row did not meet the deformation criteria, and at both eye positions, both the 1-foot and 3-foot cylinders P, Q, and L along the rear axle could not be viewed clearly. Because of their long radii of curvature, the 10-in. Φ (31-in. RC) and the 8-in. Φ (87-in. RC) convex mirrors had excellent image quality, but limited fields of view, both to the front and sides. The 10-in. Φ (31-in. RC) convex mirror did not reflect cylinders A and B in the front for a 95th percentile male, and it reflected none of the cylinders in the farthest row for a 5th percentile female. At the male eye position, the 10-in. Φ (31-in. RC) mirror provided acceptable image quality for only the 3-foot cylinder L along the rear axle, and at the female eye position, for only the 3-foot cylinders P and Q. For a 95th percentile male, the 8-in. Φ (87-in. RC) reflected only cylinders G, H, I, and J in front and none to the rear, and for a 5th percentile female, it reflected only cylinders H and I.

As already mentioned, the sideview mirrors installed on the conventional long-nosed bus by the manufacturer consisted of one pair of flat mirrors. Because the field of view provided by these mirrors commenced beyond the limits of the diagram, no depiction of their FOV has been included at the appendices. The Double Nickel sideview mirror system provided a complete view of both sides to the rear, with the exception of a blind spot one foot wide along the driver's side that was created by the stop arm.

Overall, the quadrisphere mirror provided the widest field of view to the front of the conventional long-nosed bus, at both eye positions, followed by the Banana and the elliptical mirrors. However, all three left sizeable blind spots on one or both sides. The 8-in. Φ (23-in. RC), 10-in. Φ (31-in. RC), and 8-in. Φ (87-in. RC) mirrors had large blind spots in front and to the sides in which it would not be possible to detect a child. With regard to image quality, the Banana mirror was superior to the quadrisphere and elliptical mirrors. The deformation of the 8-in. Φ (23-in. RC) mirror was about the same as that of the quadrisphere and elliptical mirrors for the front cylinders, except that two additional cylinders failed to meet the criteria at the 95th male eye position. The 10-in. Φ (31-in. RC) provided a clear image of most, but not all, the reflected cylinders, while the 8-in. Φ (87-in. RC) mirror met the criteria for all the cylinders it reflected. With few exceptions, the crossview mirrors did not provide good quality images for the cylinders along the rear axle.

CONCLUSION

This study of the performance of school bus mirrors revealed that, on a conventional long-nosed bus, none of the mirror systems both eliminated all the blind spots and met the deformation criteria.

The study also found that no crossview mirror reflected all the cylinders along the rear axle, and that where the cylinders were viewed, the image quality did not always pass the deformation test. The Double Nickel mirror system had a narrower field of view than the mirrors installed by the bus manufacturer, but the Double Nickel system had better image quality, which was apparent even though a formal evaluation was not done.

In systematically evaluating the performance of six commonly used crossview mirrors, this study demonstrated that, while most provided an adequate view to the front of the bus, their image quality was often poor for the areas along the sides, and the view to the rear did not usually extend as far as the rear axle. It was concluded that the crossview mirrors cannot be relied upon to detect children near the back wheels of a bus and that the convex sideview mirrors must be used. An additional finding was that, while safety standards usually stipulate a long radius of curvature for convex mirrors, which minimizes image deformation, the reflected image may be too small to be recognizable by a driver. It may be necessary to set a minimum image size for convex sideview mirrors in order to ensure the detection of children near the rear wheels.

The results of this study underscore the importance of adjusting the mirrors to provide a continuous view along the side of the bus rearward to the horizon. To eliminate blind spots between the mirror systems, the views of the crossview mirrors must overlap with those of the convex sideview mirrors, and the views of the convex sideview mirrors must overlap with those of the flat mirrors.

Table 1: Summary of Visibility and Image Quality Results
Conventional Long-Nosed Bus
95th Percentile Male Eye Position

CYLINDER ⇒ MIRROR TYPE ⇓	A	B	C	D	E	F	G	H	I	J	K
QUADRISPHERE	X	X	X								
BANANA			X								
ELLIPTICAL	X	X	X								
8" φ (23" RC) CONVEX	X	X	X		X	X					
10" φ (31" RC) CONVEX	NV	NV	X								
8" φ (87" RC) CONVEX	NV	NV	NV	NV	NV	NV					NV

CYLINDER ⇒ MIRROR TYPE ⇓	O (1')	P (1')	Q (1')	O (3')	P (3')	Q (3')	L (1')	M (1')	N (1')	L (3')	M (3')	N (3')
QUADRISPHERE	NV	NV	NV	NV	NV	NV	X	X	X		X	X
BANANA	NV	NV	NV	NV	NV	NV	X	NV	NV	X	NV	NV
ELLIPTICAL	X	X	X	X	X	X	X	X	X	X	X	X
8" φ (23" RC) CONVEX	NV	X	X	NV	X	X	X	NV	NV	X	NV	NV
10" φ (31" RC) CONVEX	NV	NV	X	NV	NV	X	X	NV	NV		NV	NV
8" φ (87" RC) CONVEX	NV	NV	NV	NV	NV	NV	NV	NV	NV	NV	NV	NV

X = The image did not meet the criteria of 3 and 9 minutes of arc.

NV = The cylinder was not visible in the mirror.

Table 2: Summary of Visibility and Image Quality Results
Conventional Long-Nosed Bus
5th Percentile Female Eye Position

CYLINDER TYPES ⇒ MIRROR ⇓	A	B	C	D	E	F	G	H	I	J	K
QUADRISPHERE	X	X	X								
BANANA											
ELLIPTICAL	X	X	X								
8" φ (23" RC) CONVEX	X	X	X								
10" φ (31" RC) CONVEX	NV	NV	NV								
8" φ (87" RC) CONVEX	NV	NV	NV	NV	NV	NV	NV			NV	NV

CYLINDER TYPES ⇒ MIRROR ⇓	O (1')	P (1')	Q (1')	O (3')	P (3')	Q (3')	L (1')	M (1')	N (1')	L (3')	M (3')	N (3')
QUADRISPHERE	X	X	X	X			X	X	X	X	X	X
BANANA	NV	X	X	NV	X		X	X	NV		X	NV
ELLIPTICAL	X	X	X	X	X	X	X	X	X	X	X	X
8" φ (23" RC) CONVEX	NV	X	X	NV	X	X	X	NV	NV	X	NV	NV
10" φ (31" RC) CONVEX	NV	X	X	NV			X	NV	NV	X	NV	NV
8" φ (87" RC) CONVEX	NV	NV	NV	NV	NV	NV	NV	NV	NV	NV	NV	NV

X = The image did not meet the criteria of 3 and 9 minutes of arc.

NV = The cylinder was not visible in the mirror.

APPENDIX A

KEY FOR INTERPRETING THE FIELD-OF-VIEW DIAGRAMS

Interpretation Key for the Diagrams of the Fields of View

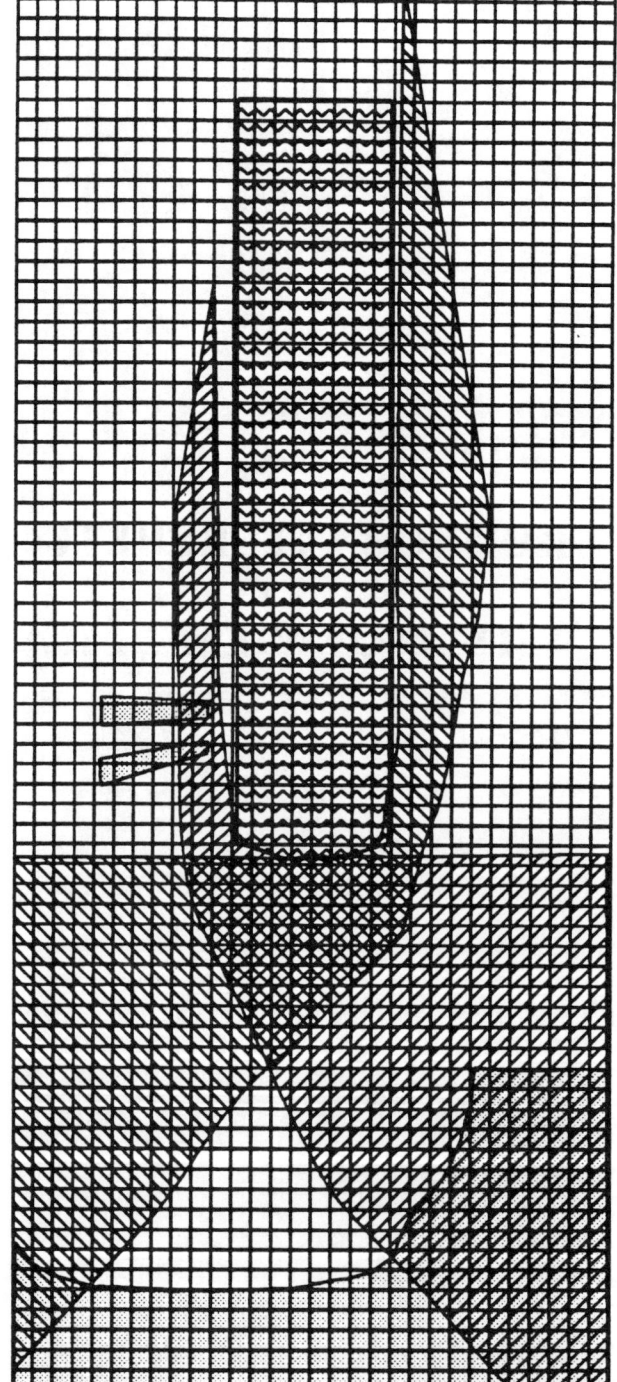

One-foot square grid.

Empty grids indicate the presence of a blind spot on the ground. There may be visibility above the ground.

Diagonal crosshatching represents an indirect view, in this case that provided by the crossview mirrors.

The views provided by the left and right crossview mirrors overlap in front of the bus.

Dotted squares show the direct view from the driver's eye location.

In conventional buses, the driver has two narrow direct views of the ground through the passenger side windows, but no direct view of the ground through the driver's side window, except at a distance beyond that depicted in the diagram. Other bus configurations have a direct view of the ground through the driver's side window.

For bus configurations other than the conventional, the direct view is sometimes obstructed by design features, such as the A-pillar.

The direct view overlaps with the indirect view provided by the crossview mirror.

A blind spot is created by the crossview mirror.

APPENDIX B

SCHOOL BUS VISIBILITY STUDY:

FIELDS OF VIEW

for the

CONVENTIONAL LONG-NOSED BUS

SCHOOL BUS VISIBILITY STUDY

BUS TYPE

Conventional long-nosed 1992 Corbeil bus built on a Navistar International chassis

MIRROR TYPE

Crossview: Quadrisphere

EYE LOCATION

95th percentile adult male

COMMENTS

- Provided the widest field of view to the front;
- Poor image quality at 4.0 m (13 ft.) in front to the point of direct visibility;
- Blind spot in the front 15.2 cm (6 in.) high;
- None of the cylinders along the left rear axle was reflected;
- Only 3-ft. L along the right rear axle met the criteria; all were reflected.

B-1

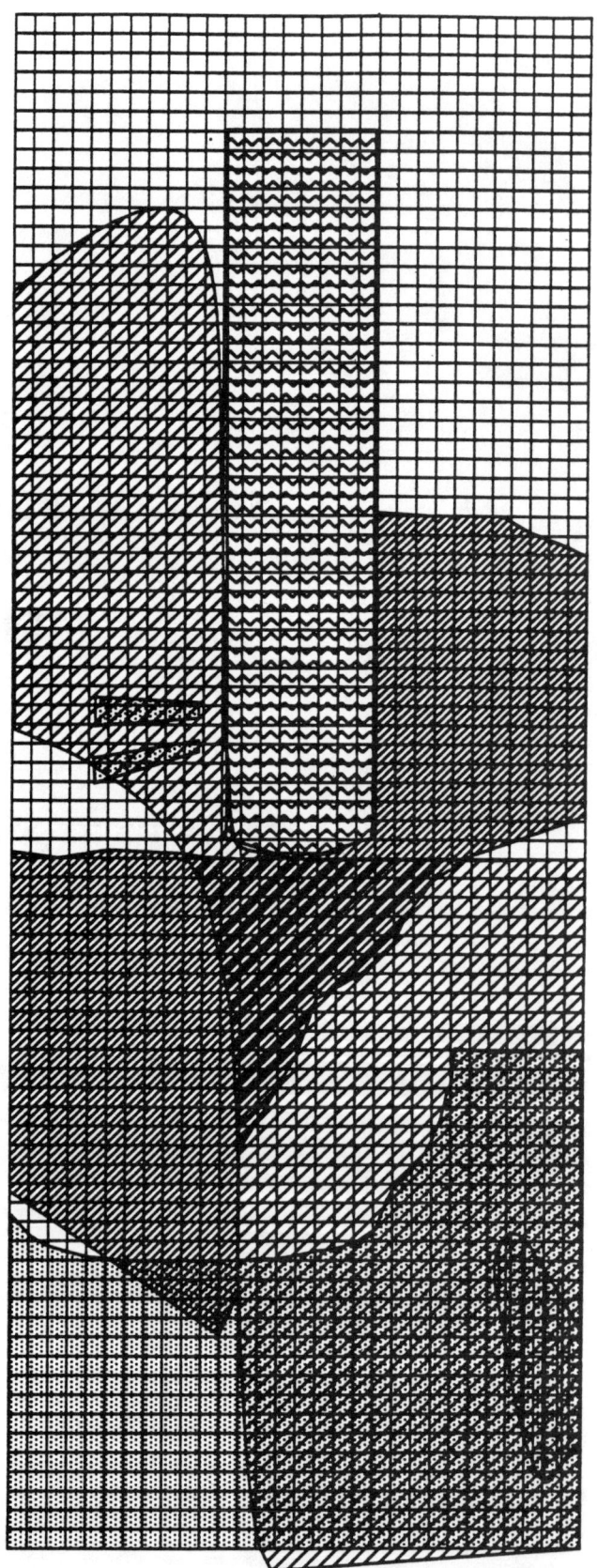

SCHOOL BUS VISIBILITY STUDY

BUS TYPE

Conventional long-nosed 1992 Corbeil bus built on a Navistar International chassis

MIRROR TYPE

Crossview: Banana

EYE LOCATION

95th percentile adult male

COMMENTS

- Provided the best image quality;
- Triangular blind spot in the front 1.83 m (6 ft.) long and 38.1 cm (15 in.) high;
- 1-ft. and 3-ft. L did not meet the deformation criteria;
- The other cylinders along the rear axle were not reflected.

B-2

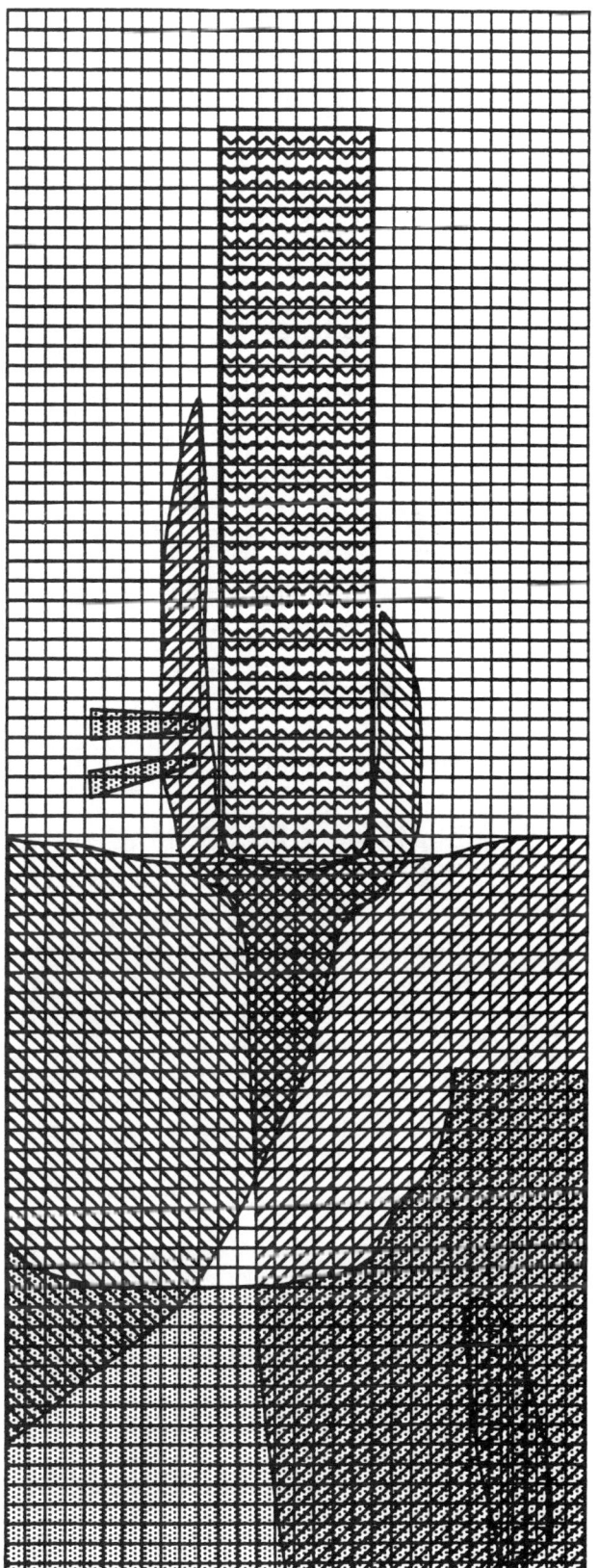

SCHOOL BUS VISIBILITY STUDY

BUS TYPE

Conventional long-nosed 1992 Corbeil bus built on a Navistar International chassis

MIRROR TYPE

Crossview: Elliptical

EYE LOCATION

95th percentile adult male

COMMENTS

- Provided the widest field of view to the front and sides;
- Blind spot 12.7 cm (5 in.) high in the front;
- Highest image deformation because of the mirror's short radius of curvature;
- None of the cylinders along the rear axle met the criteria.

B-3

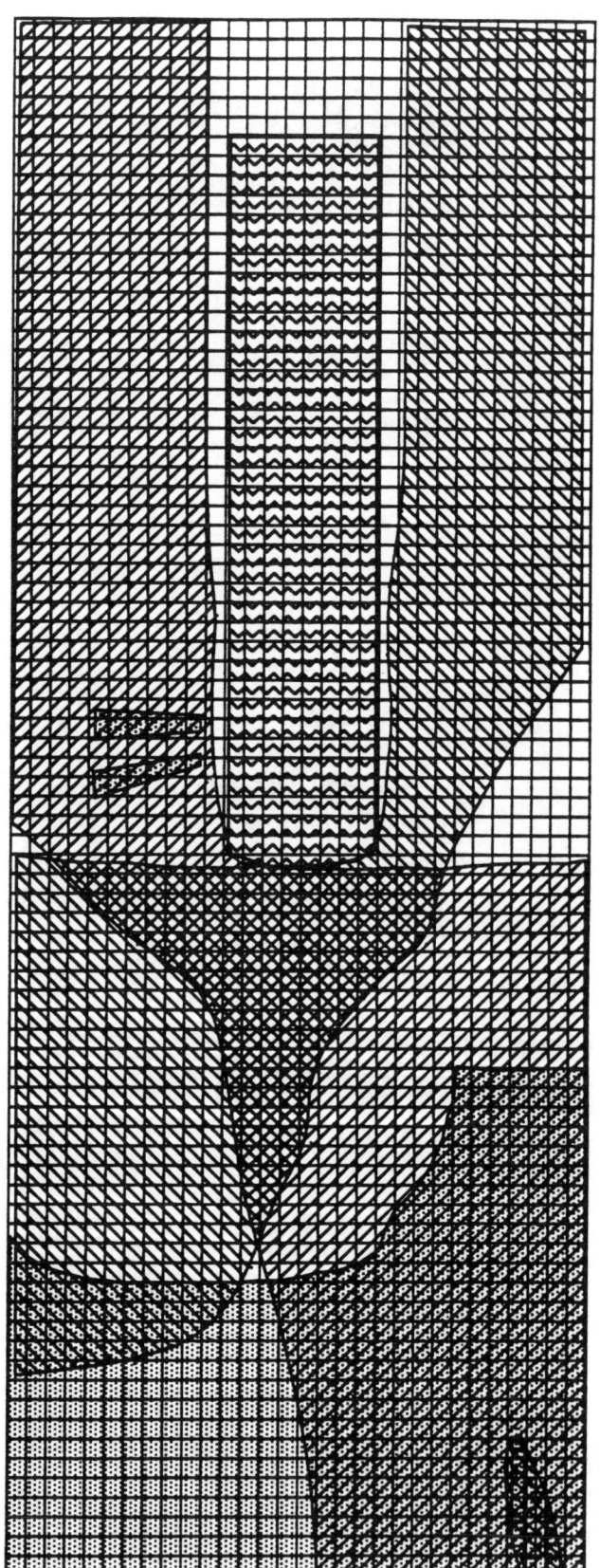

SCHOOL BUS VISIBILITY STUDY

BUS TYPE

Conventional long-nosed 1992 Corbeil bus built on a Navistar International chassis

MIRROR TYPE

Crossview: Convex mirror 8-in. Φ (23-in. RC)

EYE LOCATION

95th percentile adult male

COMMENTS

- Large blind spot in the front;
- Considerable image deformation of the 2nd and 3rd row cylinders;
- 1-ft. and 3-ft. O, M, and N along the rear axle were not reflected; the other 6 did not meet the deformation criteria.

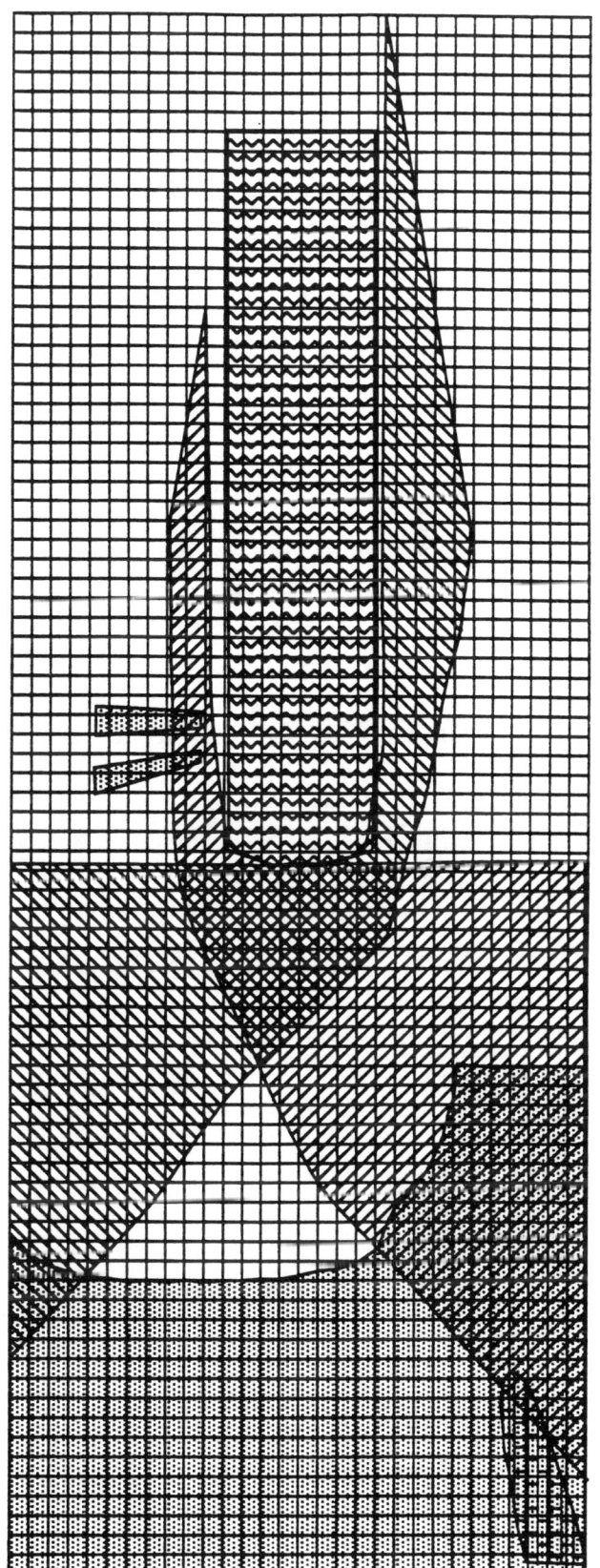

B-4

SCHOOL BUS VISIBILITY STUDY

BUS TYPE

Conventional long-nosed 1992 Corbeil bus built on a Navistar International chassis

MIRROR TYPE

Crossview: Convex mirror 10-in. Φ (31-in. RC)

EYE LOCATION

95th percentile adult male

COMMENTS

- Limited field of view with a large blind spot in front;
- Excellent image quality;
- 3-ft. L along the rear axle met the criteria; 8 cylinders were not reflected.

B-5

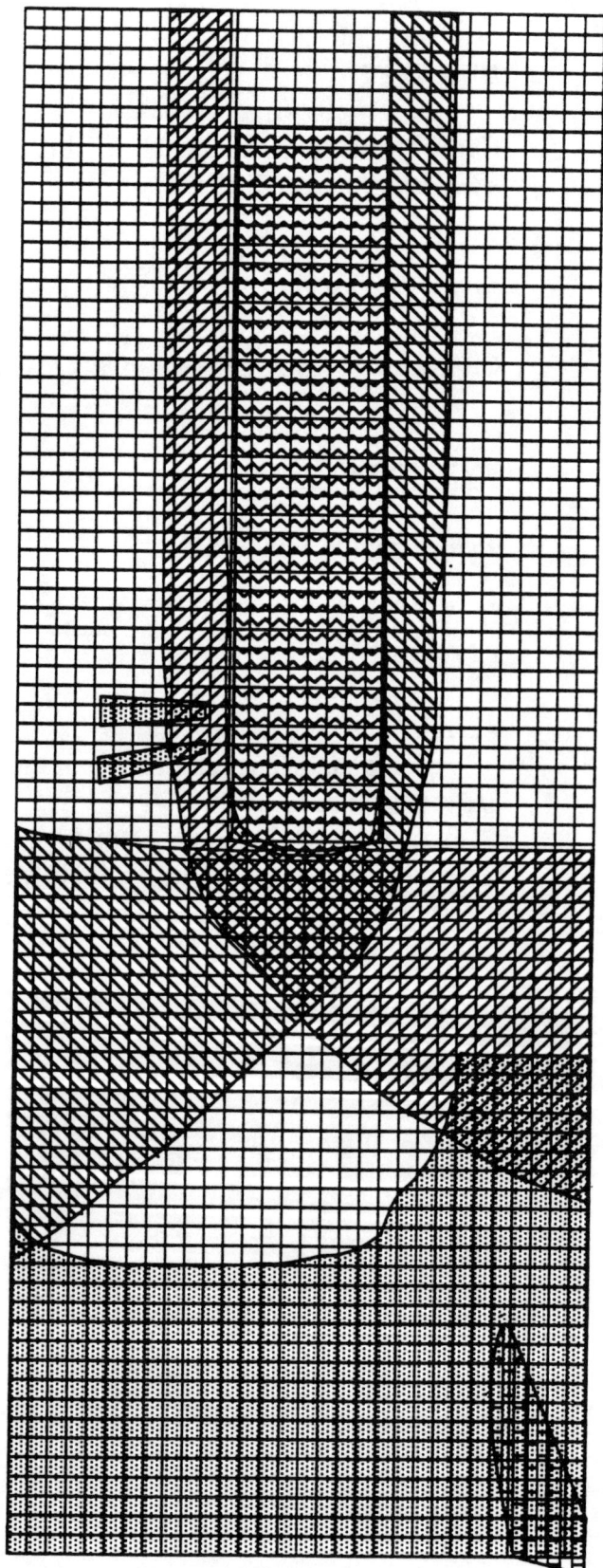

SCHOOL BUS VISIBILITY STUDY

BUS TYPE

Conventional long-nosed 1992 Corbeil bus built on a Navistar International chassis

MIRROR TYPE

Crossview: Convex mirror 8-in. Φ (87-in. RC)

EYE LOCATION

95th percentile adult male

COMMENTS

- Very limited field of view; only G, H, I, and J were reflected;
- Excellent image quality.

B-6

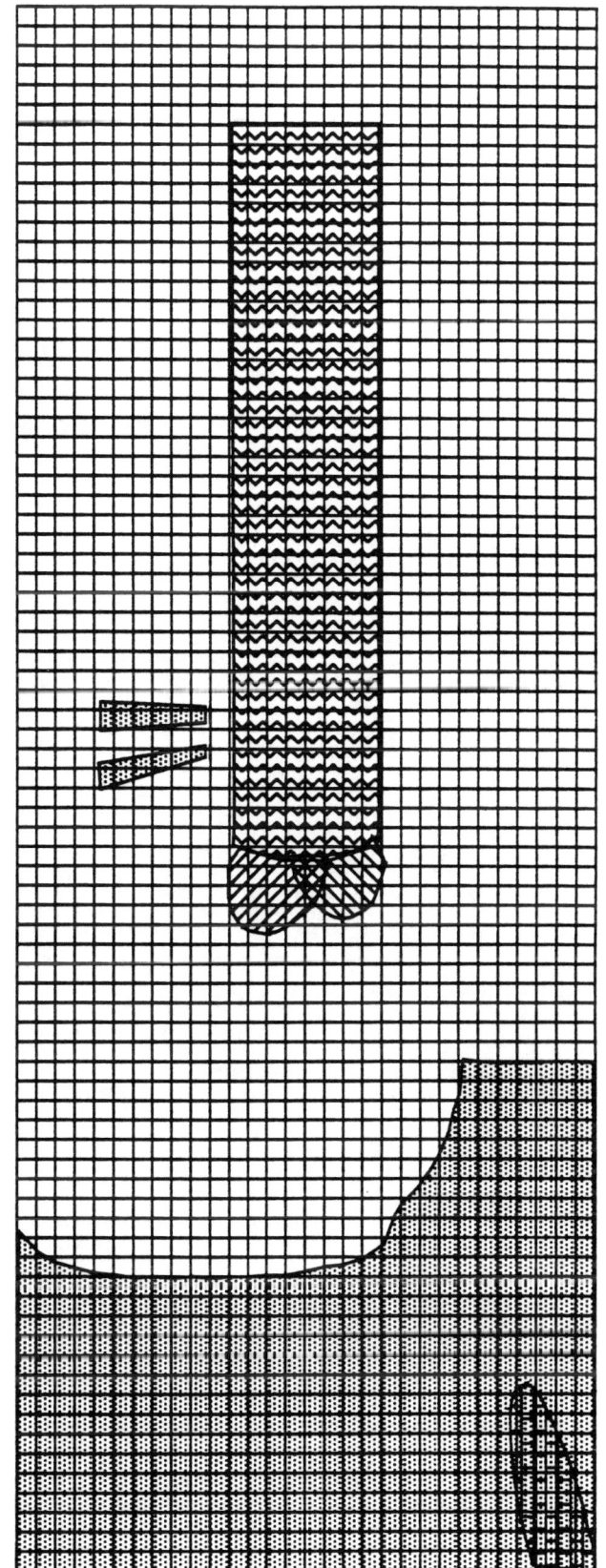

SCHOOL BUS VISIBILITY STUDY

BUS TYPE

Conventional long-nosed 1992 Corbeil bus built on a Navistar International chassis

MIRROR TYPE

Double Nickel sideview mirror system:
- one pair of flat mirrors
- one pair of convex mirrors with a long radius of curvature installed beneath

EYE LOCATION

95th percentile adult male

COMMENTS

- View of the sides starting at about 3.0 m (10 ft.) rearward of the front bumper;
- Small blind spot on the left side of no significance caused by the stop arm;
- The long radius of curvature provides good image quality, which was not evaluated;
- The view provided by the flat mirrors fell outside the limits of the diagram.

B-7

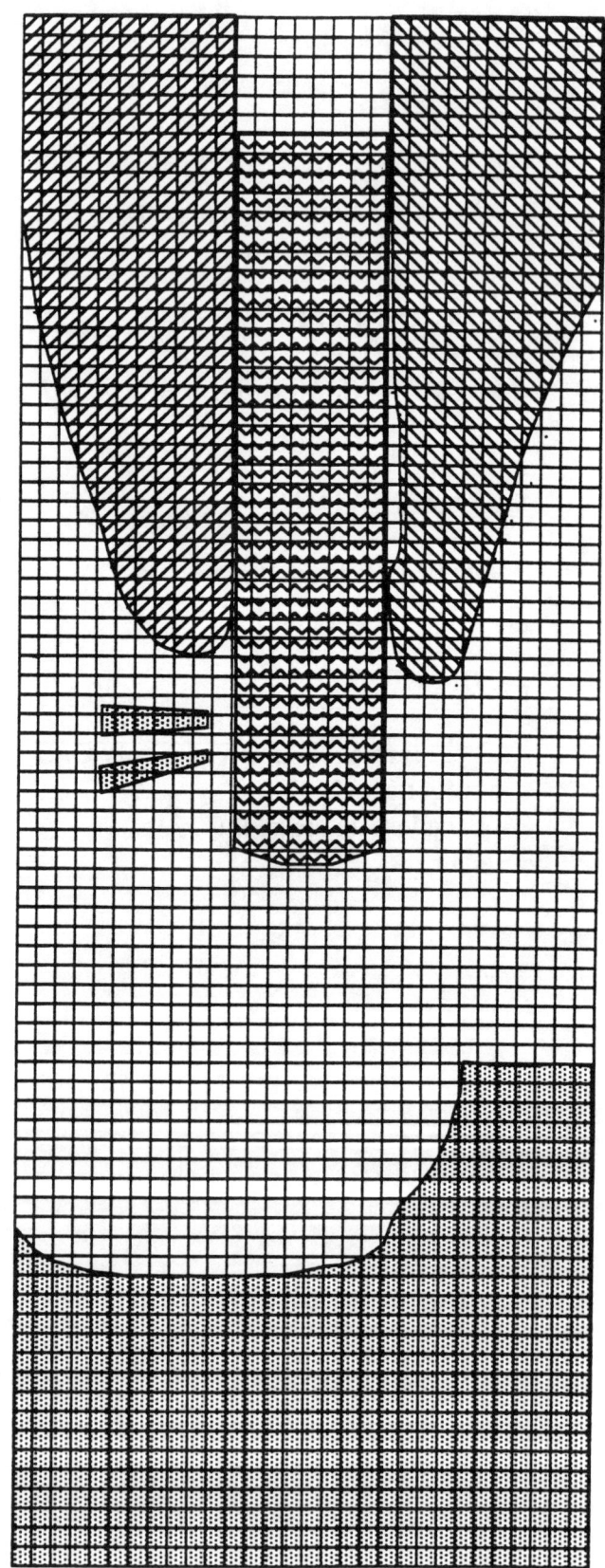

SCHOOL BUS VISIBILITY STUDY

BUS TYPE

Conventional long-nosed 1992 Corbeil bus built on a Navistar International chassis

MIRROR TYPE

Crossview: Quadrisphere

EYE LOCATION

5th percentile adult female

COMMENTS

- Provided the widest field of view to the front;
- Poor image quality at 4.0 m (13 ft.) in front to the point of direct visibility;
- Blind spot 17.8 cm (7 in.) high;
- Only 3-ft. P and Q met the criteria;
- The other cylinders along the rear axle were reflected.

B-8

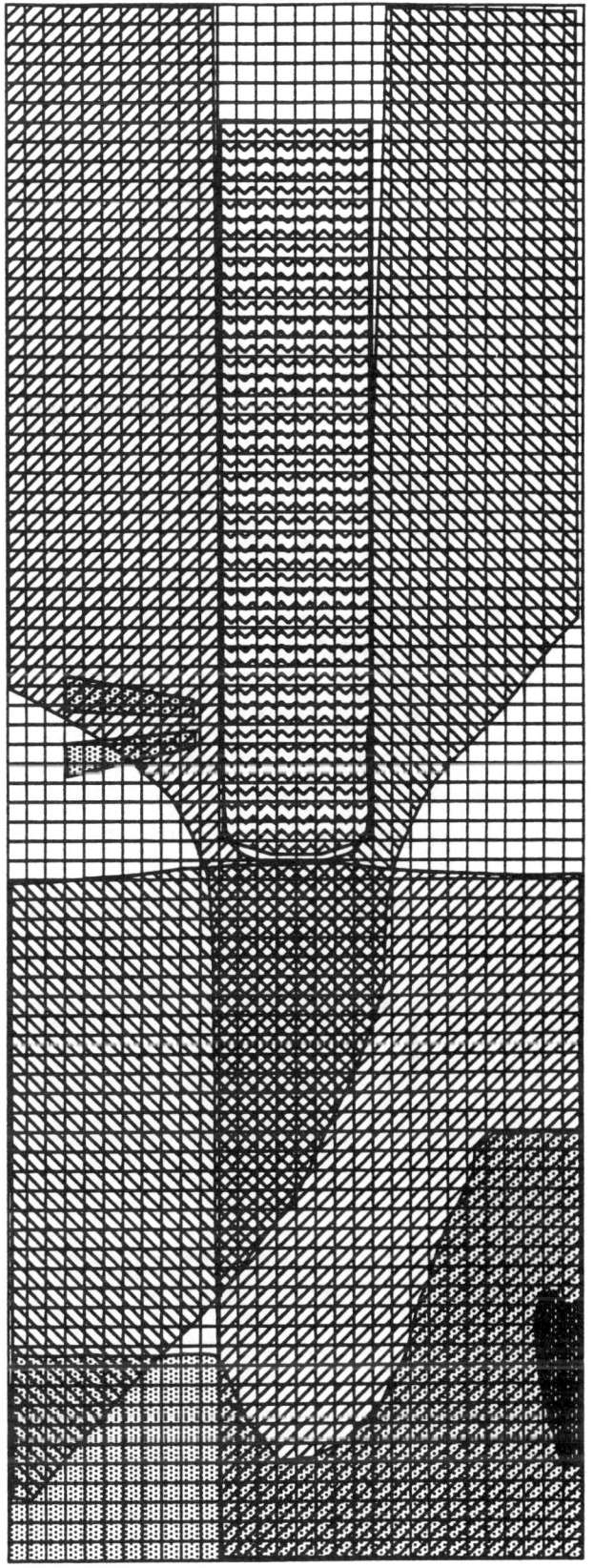

SCHOOL BUS VISIBILITY STUDY

BUS TYPE

Conventional long-nosed 1992 Corbeil bus built on a Navistar International chassis

MIRROR TYPE

Crossview: Banana

EYE LOCATION

5th percentile adult female

COMMENTS

- Provided the best image quality;
- Triangular blind spot in the front 2.5 cm (8.4 ft.) long and 45.7 cm (18 in.) high;
- Only 3-ft. Q and L along the rear axle met the criteria;
- 1-ft. and 3-ft. O and N were not reflected.

B-9

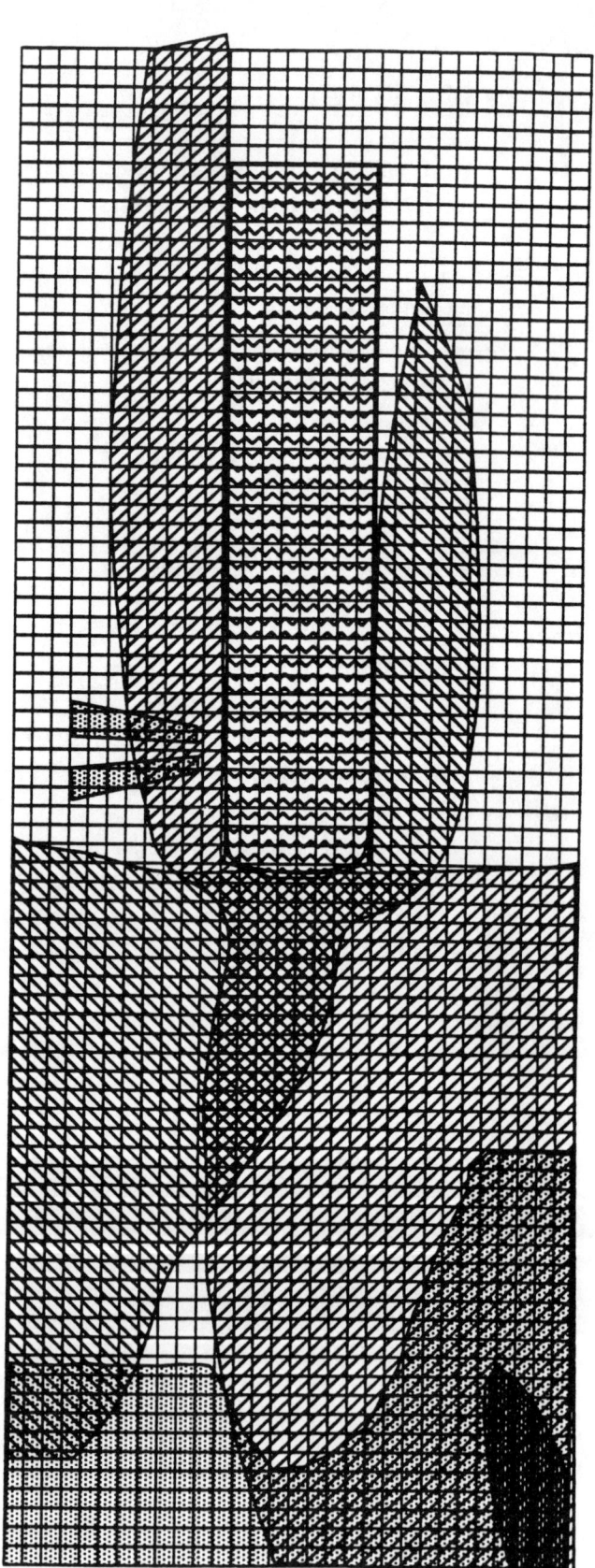

SCHOOL BUS VISIBILITY STUDY

BUS TYPE

Conventional long-nosed 1992 Corbeil bus built on a Navistar International chassis

MIRROR TYPE

Crossview: Elliptical

EYE LOCATION

5th percentile adult female

COMMENTS

- Provided the widest field of view to the front and sides;
- Blind spot 66 cm (26 in.) high in the front;
- Highest image deformation because of the mirror's short radius of curvature;
- None of the cylinders along the rear axle met the criteria.

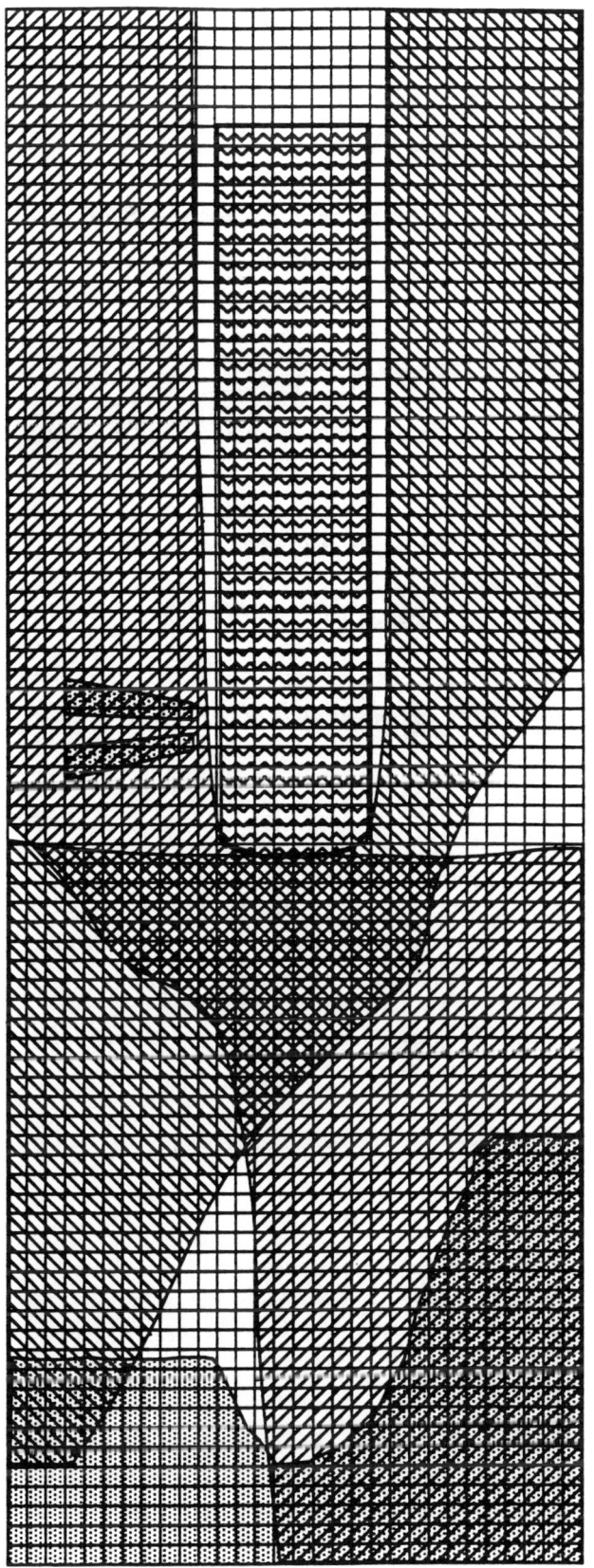

B-10

SCHOOL BUS VISIBILITY STUDY

BUS TYPE

Conventional long-nosed 1992 Corbeil bus built on a Navistar International chassis

MIRROR TYPE

Crossview: Convex mirror 8-in. Φ (23-in. RC)

EYE LOCATION

5th percentile adult female

COMMENTS:

- Large blind spot in the front;
- Considerable image deformation of the 3rd row cylinders;
- 1-ft. and 3-ft. O, M, and N along the rear axle were not reflected; the other 6 did not meet the deformation criteria.

B-11

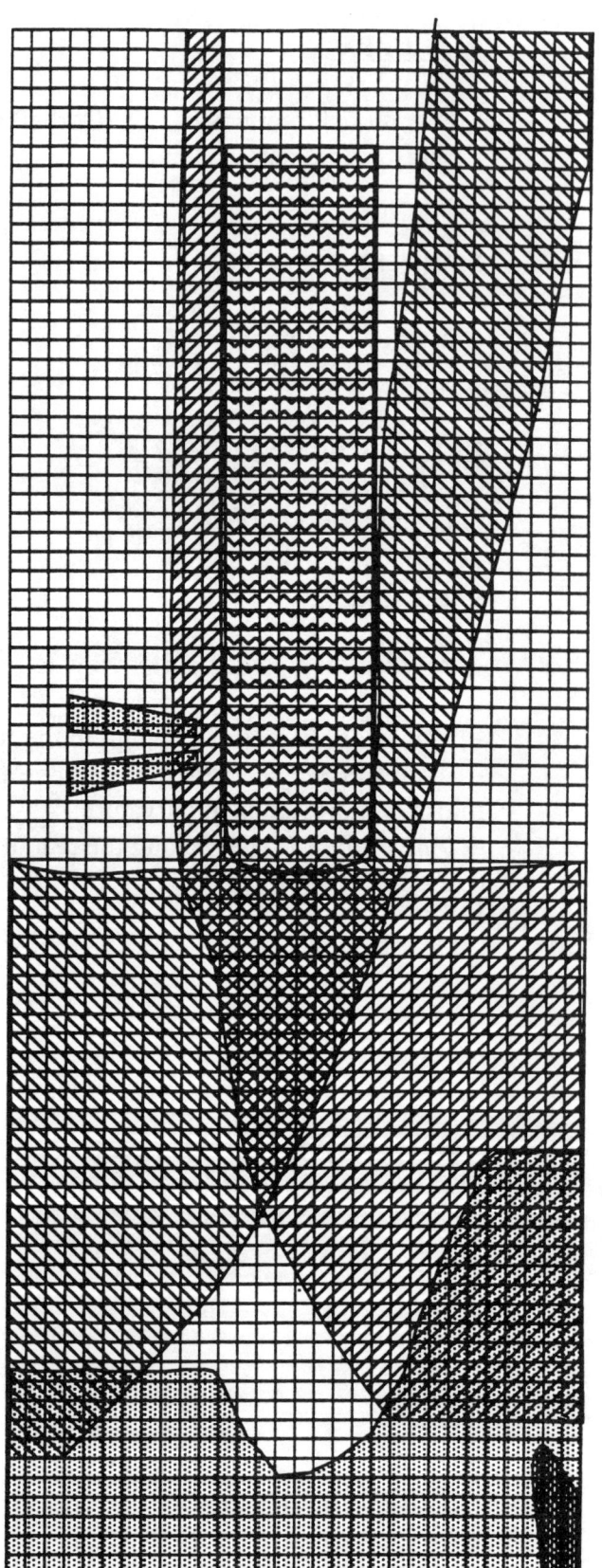

SCHOOL BUS VISIBILITY STUDY

BUS TYPE

Conventional long-nosed 1992 Corbeil bus built on a Navistar International chassis

MIRROR TYPE

Crossview: Convex mirror 10-in. Φ (31-in. RC)

EYE LOCATION

5th percentile adult female

COMMENTS

- Limited field of view with a large blind spot in front;
- Excellent image quality;
- 3-ft. P and Q along the rear axle met the criteria;
- 1-ft. and 3-ft. O, M, and N were not reflected.

B-12

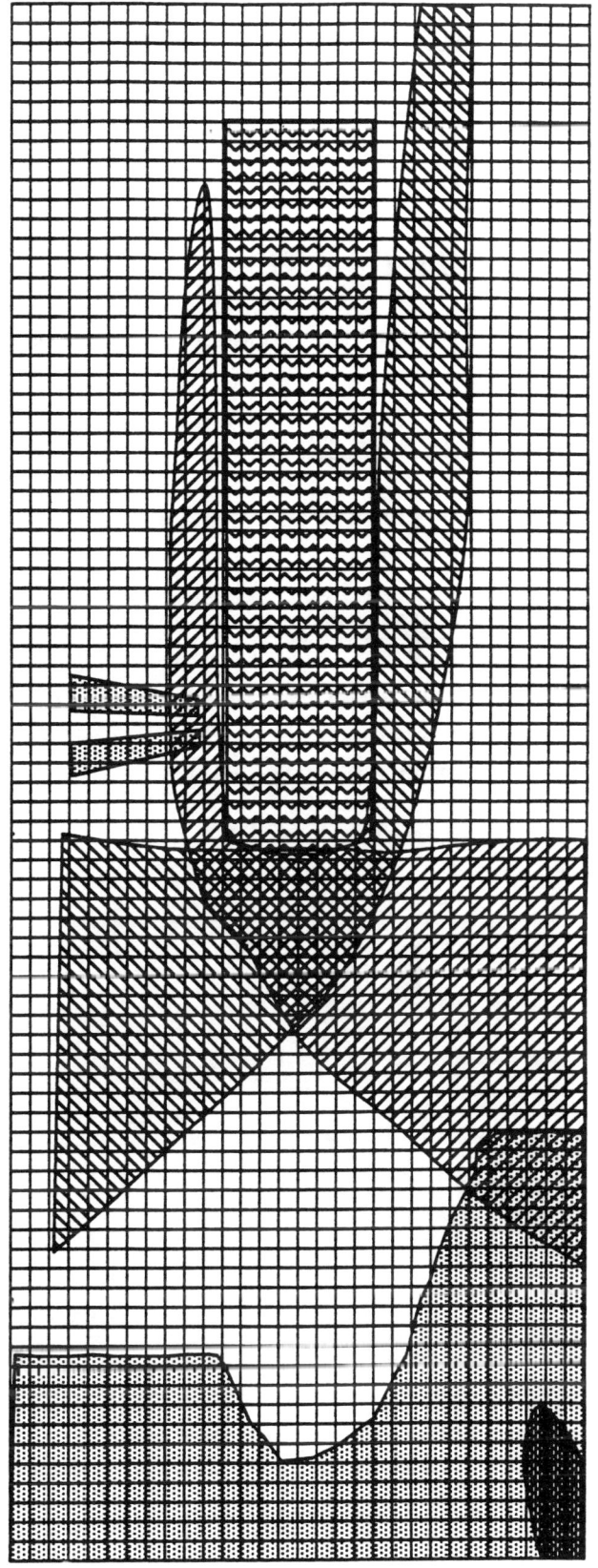

SCHOOL BUS VISIBILITY STUDY

BUS TYPE

Conventional long-nosed 1992 Corbeil bus built on a Navistar International chassis

MIRROR TYPE

Crossview: Convex mirror 8-in. Φ (87-in. RC)

EYE LOCATION

5th percentile adult female

COMMENTS

- Very limited field of view; only H and I were reflected;
- Excellent image quality.

B-13

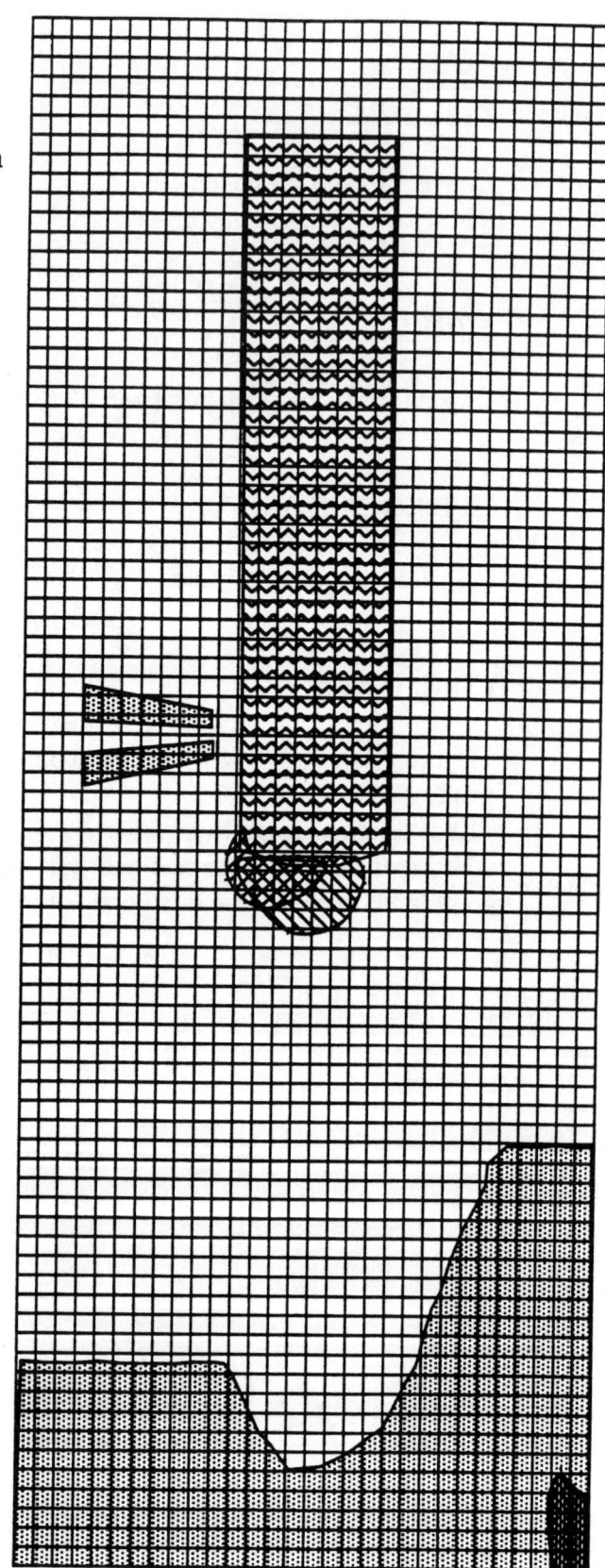

SCHOOL BUS VISIBILITY STUDY

BUS TYPE

Conventional long-nosed 1992 Corbeil bus built on a Navistar International chassis

MIRROR TYPE

Double Nickel sideview mirror system:
- one pair of flat mirrors
- one pair of convex mirrors with a long radius of curvature installed beneath

EYE LOCATION

5th percentile adult female

COMMENTS

- View of the sides starting at about 3.7 m (12 ft.) rearward of the front bumper;
- Small blind spot on the left side of no significance caused by the stop arm;
- The long radius of curvature provides good image quality, which was not evaluated;
- The view provided by the flat mirrors fell outside the limits of the diagram.

B-14

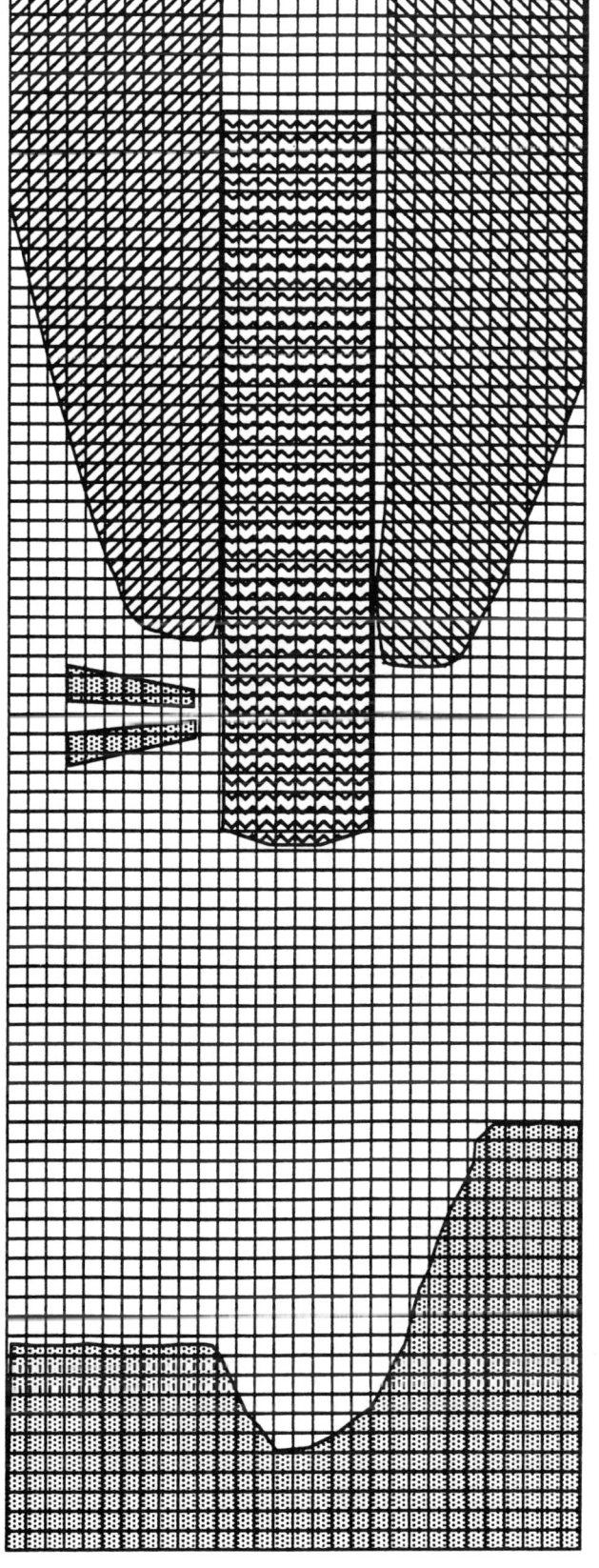